Genetic Variation

Edited by Rafael Trindade Maia and Magnólia de Araújo Campos

Published in London, United Kingdom

IntechOpen

Supporting open minds since 2005

Genetic Variation
http://dx.doi.org/10.5772/intechopen.87548
Edited by Rafael Trindade Maia and Magnólia de Araújo Campos

Contributors
Mahmood A. Al Azzawi, Moustafa Sakr, Gilberto Manzo-Sánchez, Marco Tulio Buenrostro-Nava, Carlos Luis Leopardi-Verde, Mario Orozco-Santos, Mauricio Guzman Quesada, Safa A Faraj, Naeem Al- Abedy, Faisal Saeed Awan, Bushra Sadia, Kiran Hameed, Madiha Habib, Javaria Altaf, Shabbir Hussain, Barahima Abbas, Rajeswar Nath Srivastava, Amar Chandra Sharma, Sudeepti R. Srivastava, Saloni Raj, Lavini Raj, Fozia Saleem, Farwa Ashraf, Abdullah Bin Umar, Mariam Nasir, Muhammad Nadeem, Samra Hameed, Jelena M. Milenkovic, Dijana R. Stojanovic, Jelena G. Najdanović, Juliano Lino Ferreira, Leila Ferreira, Eveline Teixeira Caixeta, Tesfahun Alemu Setotaw, Fernanda Fátima Caniato, Gustavo César Sant'Ana, Favour Osazuwa, Paulo Vitor Marques Simas, Sanghoo Lee, Jimyeong Park, Jinwoo Ahn, Hyeonkyun Na, Youngkee Lee, Yejin Kim, Gayeon Hong, Kyoung-Ryul Lee, Arain Saima Mir Saima Mir, Sajjad Muhammad, Sial Mahboob Ali, Meer Maria, María Eugenia Barrandeguy, María Victoria Garcia, Ali Ishaq, Ayesha Noor, Fatima Noor, Rahmawaty Samad, Fatima Jalal, Faiza Jabeen, Tahira Younis, Afshan Syed Abbas, Sara Mumtaz, Rafael Trindade Trindade Maia, Magnólia De Araújo de Araújo Campos, Clarice Weis Arns

First published in London, United Kingdom, 2021 by IntechOpen
IntechOpen is the global imprint of INTECHOPEN LIMITED, registered in England and Wales, registration number: 11086078, 5 Princes Gate Court, London, SW7 2QJ, United Kingdom
Printed in Croatia

British Library Cataloguing-in-Publication Data
A catalogue record for this book is available from the British Library

Additional hard and PDF copies can be obtained from orders@intechopen.com

Genetic Variation
Edited by Rafael Trindade Maia and Magnólia de Araújo Campos
p. cm.
Print ISBN 978-1-83881-096-2
Online ISBN 978-1-83881-097-9
eBook (PDF) ISBN 978-1-83881-101-3

For EU product safety concerns:
IN TECH d.o.o., Prolaz Marije Krucifikse Kozulić 3, 51000 Rijeka, Croatia, info@intechopen.com or visit our website at intechopen.com.

Meet the editors

Dr. Rafael Trindade Maia studied biological sciences at the Federal Rural University of Pernambuco, Brazil (2005). He received a master´s degree in Genetics, Conservation, and Evolutionary Biology from the National Institute of Amazonian Research, Brazil, in 2008, and a Ph.D. in Animal Biology from the Federal University of Pernambuco, Brazil, in 2013. He is currently an adjunct professor at the Center for the Sustainable Development for Semiarid (CDSA) at Federal University of Campina Grande (UFCG), Brazil. He has experience with population genetics, bioinformatics, molecular docking, and modeling and molecular dynamics of proteins. He works in science and biology education. Dr. Maia also leads the research groups Computational and Theoretical Biology (CTB) and Education in Sciences and Biology (ESB).

Dr. Magnólia de Araújo Campos is a biologist with a master's degree in Agronomy/Plant Breeding from the Federal University of Pelotas, Brazil, and a Ph.D. in Biological Sciences/Molecular Biology from the University of Brasília (2002). She has experience in genomic sciences as a postdoctoral researcher at the Federal University of Lavras/Agronomic Institute (IAC), Brazil. Dr. Campos has been a professor at the Federal University of Campina Grande (UFCG), Brazil, since 2008. She mainly works on the following topics: genomics, bioinformatics, tissue culture and plant cells, genetic transformation of plants, gene expression during plant-microbe interactions, and expression of heterologous proteins in bacteria.

Contents

Preface

This book addresses the topic of genetic diversity, which is a measure of biodiversity that encompasses genetic variation within species. This book was edited by geneticist researchers and provides academics (professors, graduates, students, scientists) with up-to-date and quality information on the subject.

The work consists of revised chapters and is divided into three sections: Section 1 "About Genetic Variation" (containing five chapters); Section 2 "Molecular Markers in the Detection of Genetic Polymorphism" (containing three chapters); and Section 3 "Genetic Diversity and Health" (containing eight chapters).

This collection of scientific papers was chosen and analyzed to offer readers a broad and integrated view of the importance of genetic diversity in the evolution and adaptation of living beings, as well as practical applications of the information needed to analyze this diversity in different organisms.

Chapter 1, "Introductory Chapter: Genetic Variation - The Source of Biological Diversity", is written by the editors and serves as an introduction to the topics addressed in the book. It presents the main mechanisms of genetic diversity, such as the different types of mutations. The chapter situates the reader on the theme of the book. Chapter 2 systematically addresses biotechnological tools to induce mutations to generate genetic variability in plants of economic interest, with examples and concepts of cutting-edge techniques. Chapter 3 takes a formidable approach to the genetic aspects involved in the evolution and improvement of the sago palm. In Chapter 4, the authors analyze the sensitivity and effectiveness of heterozygosity and allelic richness in estimating the genetic diversity of populations. Chapter 5 is the result of a systematic review of the genetic diversity of the coffee tree (*Coffea arabica*), one of the most important cultivars in the world.

In Chapter 6, the reader will come across an essay on the relevance of the production of mutations in the genetic improvement of cultivars. Chapter 7 analyzes the potential of morphological and molecular markers in the evaluation of the genetic variation of strawberries. Chapter 8 provides an overview of the genetic diversity of one of the most important pathogens in bananas.

In an analysis of the Mediterranean region, the authors of Chapter 9 explore how important the role of genetic diversity is in adapting to the local environment. Chapter 10 presents a fascinating approach to the genetic and molecular aspects of co-evolution between susceptible populations and the new coronavirus (SARS-CoV-2). Chapter 11 provides the first report of a new strain of norovirus in Nigeria. Chapter 12 discusses the genetic aspects of insulin resistance and metabolic syndrome. In Chapter 13, the authors review the genetic bases associated with arthritis. Chapter 14 is an essay on polymorphism in genes related to the hormone Hepcidin. In Chapter 15, the authors use a brief survey to examine the genetic and environmental factors of periodic and episodic attacks of articular inflammation in humans. Finally, Chapter 16 highlights the importance of metagenomics in understanding and coping with pandemic viruses.

In these chapters, the reader will find numerous interesting pieces of information related to the topic of genetic diversity, from unprecedented data to classic concepts and feasible hypotheses.

Genetic Variation is an interdisciplinary and integrated work that will contribute to the knowledge of academics from different areas of biological sciences.

Rafael Trindade Maia
Center of Sustainable Development for Semiarid,
Sumé-Paraiba State, Brazil

Magnólia de Araújo Campos
Federal University of Campina Grande,
Brazil

Section 1

About Genetic Variation

Chapter 1

Introductory Chapter: Genetic Variation - The Source of Biological Diversity

Rafael Trindade Maia and Magnólia de Araújo Campos

1. Introduction

Genetic diversity is usually defined as the number of genetic characteristics (alleles and genotypes) in a species [1]. In this context, analyzing the genetic diversity in populations is essential to understand evolutionary and adaptative process for most species [2]. The genetic diversity is also very useful to implement conservation strategies and crop management. This is also the source of disease resistance of natural populations, as it is the font of drug resistance by many pathogens.

Genes are DNA fragments that encodes some biological information, usually coding a protein or a RNA. Genes can be represented as a sequence of nucleotides that can be expressed in a living organism. Most genes have small nucleotide sequence differences among individuals. These differences are called genetic polymorphism [3]. Some of these polymorphisms may affects how proteins works and how the proteins interacts with subtracts and other proteins. The different gene forms caused by genetic polymorphisms are called alleles.

The genetic diversity has three different sources: mutation, recombination and immigration of genes. Mutation is the driving force of genetic variation and evolution. There are three types of DNA mutations: base substitutions (also called point mutations), deletions and insertions (**Figure 1**) [4].

2. Base substitutions (point mutations)

The point mutations are also subdivided in three groups: (1) Silent mutations: when the nucleotide substitution does not change the aminoacid in the polypeptide sequence; (2) Missense mutations: when occurs aminoacid change, that can be classified in conservative (when the change results in an amino acid from the same physical–chemical group) and non-conservative (when the change results in a different physicochemical aminoacid group); (3) Non-sense mutations: when the nucleotide modification results in a stop codon (**Figure 2**).

Single mutations are very important in population genetics and evolution. They are the mainly source of DNA polymorphic sites, which provides information for many inferences and analysis such as nucleotide and haplotype diversity, allelic diversity, genetic distance, heterozygosity, etc. These parameters are very important to elucidate evolutionary process in populations across time and space. Genes with high levels of polymorphism can be applied to genetic population studies, while genes with moderate and low polymorphic levels can be used for phylogeographic and phylogenetic inferences [5].

Figure 1.
Illustrative scheme of DNA mutations types. Source: Google Images.

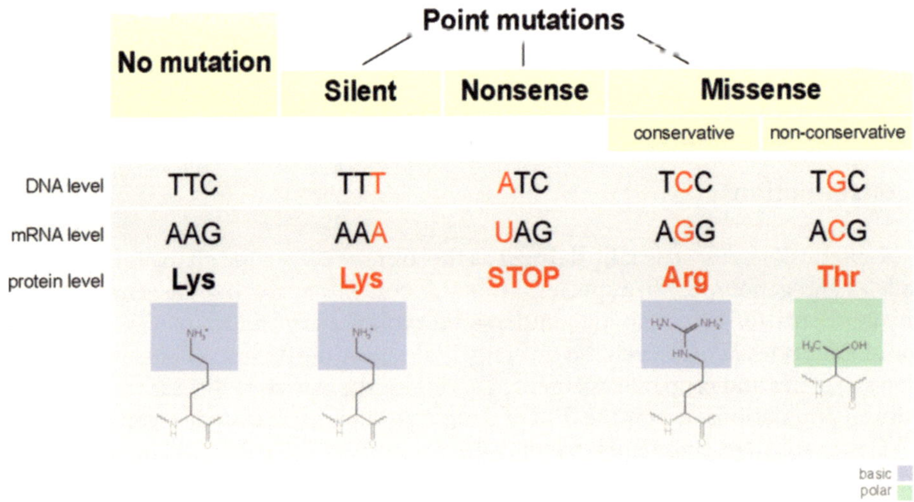

Figure 2.
Types of point mutations (nucleotide substitution) in DNA molecule. Source: Google Images.

Single mutations are also important for health: many missense mutations can be deleterious and resulting in a disease or metabolic disorder. Another thing to be considered is that single mutations can also provide adaptative vantages such as pathogen resistance, xenobiotic tolerance and fitness improvement [6]. So, detecting point mutations in the organisms can be very useful to implement many strategies such as biodiversity conservation, crop management and infectious disease monitoring.

3. Recombination

In eukaryotes, genetic recombination is the aleatory change of genetic material resulting from the meiosis process, also called crossing-over or permutation. This type of recombination consists in break and rejoining homologous regions of pared chromosomes between the Prophase I and Metaphase I from meiosis division (**Figure 3**). Many combinations can be performed among gene exchange between two individuals [7]. Although prokaryotic species does not have chromosomes conjugation is performed by these beings by one of these three process: (a) Conjugation: when the DNA is transferred by tube after cells contact; (b) transduction: the DNA is inserted accidentally from one bacterium to another by a virus; and (c) transformation: when the bacterium receives exogenous DNA from the environment [8].

Recombination is very important because it makes new combinations of the existent alleles. Many effects of the DNA rearrangement can be good for species and populations, once it can improve adaptation. However, some recombination events can be unfavorable if it breaks apart important and beneficial alleles in the sisters chromatids [9]. The recombination rate is positively correlated with

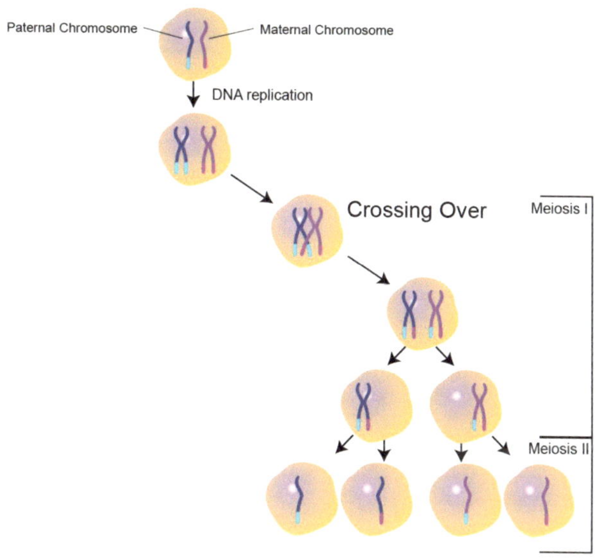

Figure 3.
Crossing-over scheme in homologous chromosomes. Source: Google Images.

nucleotide diversity, which increases genetic variation and resulting in purifying deleterious mutations.

4. Deletions and insertions

Nucleotide deletions and insertions are types of mutation that changes the number of DNA base in genome. Deletions changes the base number by removing pieces of DNA and insertions alters the base number by adding pieces of DNA [10]. Often, these kind of mutations results in a gene that encodes a protein that does not function properly (**Figure 4**). When the deletions/insertions change the gene's reading frame they are called *frameshit* mutations.

Insertions and deletions can be particularly hazardous when occurs in an exon region, which is the coding segment of a gene. Due to multiple new aminoacid after translation, the protein function may be affected [11]. Knowing the rate of insertion–deletion mutations are crucial to understanding evolutionary process, such as natural selection, especially in coding regions due to the protein disruption that is usually caused.

5. Gene immigration

Gene immigration, or gene flow, is the transfer of genetic material from one population to another by migration of individuals or gametes. This can alter genetic diversity by changing allelic frequencies in populations [12]. Gene flow is essential to prevent population diverging. When gene flow is interrupted by physical (geographical) barriers, allopatric speciation tends to occur. Population gene flow can be measured by the formula:

$$Nm = \frac{\left(\frac{1}{Fst}\right) - 1}{4}$$

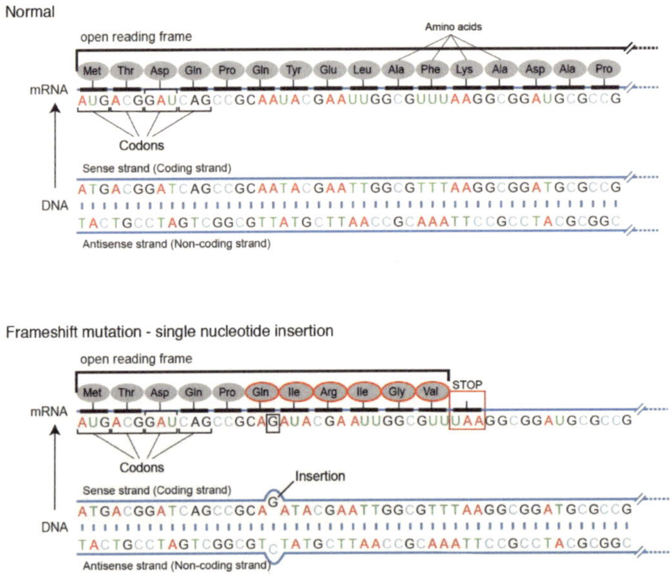

Figure 4.
Illustrative example of a DNA insertion modifying all the subsequent codons. Source: Google Images.

Where *Nm* refers to the number of migrants per generation; *Fst* is the degree of genetic differentiation.

When *Fst* is 1, there is a strong differentiation among populations. Gene flow is also very important to reduce genetic drift effects. Due to this particularity, the gene flow is extremely important for conservation genetics.

6. Final considerations

Genetic diversity is a very important feature of living organisms. It serves for population adapting to environment, once that how higher is the allelic variation, it is more likely that individuals display adaptative characteristics that suits to the environment. So, genetic diversity is essential for species survival.

Molecular markers, amplification and DNA sequencing technologies are improving an incredible advance in access genetic variation. The number of sequences and genomes deposited in the databases grows exponentially, generating an enormous amount of information to be studied and analyzed. The knowledge resulting from this information has innumerous applications and has been promoting a huge revolution in several areas of the biological, agrarian and health sciences. The way we understand, analyze and deal with biodiversity has been intensely modified and deepened by the advancement of genetics. In this context, knowledge about the genetic diversity of organisms will bring solutions to various problems and issues involving living beings.

Author details

Rafael Trindade Maia[1*] and Magnólia de Araújo Campos[2]

1 Federal University of Campina Grande, Center for Sustainable Development of Semiarid, Sumé, Paraiba State, Brazil

2 Federal University of Campina Grande, Center of Education and Health, Cuité, Paraiba State, Brazil

*Address all correspondence to: rafael.rafatrin@gmail.com

References

[1] Ellegren H, Galtier N. Determinants of genetic diversity. Nat Rev Genet 2016. 17, 422-433. https://doi.org/10.1038/nrg.2016.58

[2] Wright, A F. Genetic Variation: Polymorphisms and Mutations. In eLS, (Ed.). 2005. https://doi.org/10.1038/npg.els.0005005.

[3] Singh R S, Kulathinal R J. Polymorphism. Brenner's Encyclopedia of Genetics. 2013. 398-399. doi:10.1016/b978-0-12-374984-0.01189-x.

[4] Griffiths A J F, Miller J H, Suzuki D T, et al. An Introduction to Genetic Analysis. 7th edition. New York: W. H. Freeman; 2000. Sources of variation. Available from: https://www.ncbi.nlm.nih.gov/books/NBK22012/.

[5] Cutter A D. Integrating phylogenetics, phylogeography and population genetics through genomes and evolutionary theory. Molecular Phylogenetics and Evolution. 2013. Volume 69, Issue 3. https://doi.org/10.1016/j.ympev.2013.06.006

[6] Markert J A, Champlin D M, Gutjahr-Gobell R. et al. Population genetic diversity and fitness in multiple environments. BMC Evol Biol 2010. 10, 205. https://doi.org/10.1186/1471-2148-10-205.

[7] Bogdanov Y F, Grishaeva T M. Meiotic Recombination. The Metabolic Pathways from DNA Double-Strand Breaks to Crossing Over and Chiasmata. Russ J Genet. 2020.56, 159-176. https://doi.org/10.1134/S1022795420020039

[8] Bobay, L M. CoreSimul: a forward-in-time simulator of genome evolution for prokaryotes modeling homologous recombination. BMC Bioinformatics. 2020. 21, 264. https://doi.org/10.1186/s12859-020-03619-x

[9] Tumini E., Aguilera A. 2021. The Sister-Chromatid Exchange Assay in Human Cells. In: Aguilera A., Carreira A. (eds) Homologous Recombination. Methods in Molecular Biology, vol 2153. Humana, New York, NY. https://doi.org/10.1007/978-1-0716-0644-5_26.

[10] Iengar P. An analysis of substitution, deletion and insertion mutations in cancer genes, Nucleic Acids Research. 2012. Volume 40, Issue 14, 1, Pages 6401-6413, https://doi.org/10.1093/nar/gks290

[11] Nesta A V, Tafur D, Beck C R. Hotspots of Human Mutation. Trends in Genetics. 2020. https://doi.org/10.1016/j.tig.2020.10.003.

[12] Slatkin M. Annual Review of Ecology and Systematics 1985. Vol. 16, pp. 393-430

Chapter 2

Genetic Variability through Induced Mutation

Faisal Saeed Awan, Bushra Sadia, Javaria Altaf, Madiha Habib, Kiran Hameed and Shabbir Hussain

Abstract

The success of plant breeding is based on the accessibility of genetic variation, information about desired traits with well-organized approach that make it likely to develop existing genetic resources. Food security demands to break the yield barrier through increasing new cultivars which can adapt to wide range of environment. It is especially important to observe the character association for yield along with its components before recognizing novel technique to break the yield barrier. There are numerous methods for improved exploiting of the inherent genetic makeup of crops with heritable variations. It is recommended that recognized parental resources can also be induced to mutate for unmasking novel alleles of genes that organize the traits suitable for the crop varieties of the 21st century world. Chemical mutagens have extensively been applied to make genetic changes in crop plants for breeding investigation as well as genetic studies. Ethyl methane sulphonate (EMS) is the most frequently applied as chemical agents in plants. EMS normally induces GC → AT transitions in the genome causing mutated protein that performed different functions rather than normal. It is exposed that the utilization of EMS is an efficient approach for developing novel gene pool.

Keywords: ethyl methane sulphonate, mutation, genetic variability, singe nucleotide polymorphism

1. Introduction

Plant breeding involved in rapid introduction of genetic variability in plants, to divulge them with desirable characteristics, through genetic mutation. Plant evolution and genetic breeding depends on the genetic variation that, not all come from spontaneous mutation rather it comes by genetic recombination within population and their interactions with environmental factors [1]. Conventional breeding depends on prevailed environmental genetic variations in wild and cultivated plants that require a large-scale backcrossing to incorporate it and stabilize it while, new mutation breeding strategy is less time consuming and easy that enhances the selection of desirable mutants [2]. Mutation breeding is an advancement of plant breeding where the induction of physical and chemical mutagens cause genetic variation. These variations are transferred to next generation through recombinant hybridization in meiosis [3]. Selection of breeding individuals only probable when there is a significant genetic variability exists [4]. Mutation breeding depends upon the transfer and stabilization of heritable characters that cause the variability [5].

Spontaneous and Induced mutations are the primary source of all variations exists in an organisms that may be plants or animals [6]. Genetic variability endorses the differences among the same species and its existence in population is essential for its survival with changing environment. Induced mutation with specific mutagens generates the individuals with desirable characters that can be further exploit in breeding to generate new verities in plants [7, 8]. Mutation was first recognized in the late nineteenth century by Hugo de Varies, when he was working on the 'rediscovery' of Mendel's laws of inheritance [6]. Chemical mutagens are less harmful and easily available for work. In plants, widely used chemical mutagen is EMS that is very effective in causing point mutation in genome [9]. Mutation approaches produces huge and minute effect on all types of phenotypic traits [6]. Induced mutation is helpful in growing novel cultivars of plants as seedless grapes and edible bananas [3, 10] also it bring out the novel color variants of tuber and root crops [11]. Mutagenesis also apply to improve dwarfness, early growth, resistance to biotic and abiotic stresses, and yield improvement as well as quality enhancements in plants [12–15].

Novel breeding techniques based on biological mutagenic agent are widely introduce in plants for targeted variation is also known as targeted genome editing [16]. These genome editing techniques cause the specific and precise genome mutations. It introduce targeted mutation by either insertion, deletion that disturbs the function of gene. CRISPER/Cas9 is a novel technique that introduce desired targeted mutation permanently inside the genome [17].

2. Brief description of induced mutation

Mutation breeding carried out through three types of mutation as induced mutation, site-directed mutation and insertion mutation [1]. Induced mutation is a tool of generating variability artificially [18]. Mutagenesis is sudden and heritable changes in genetic sequence that stimulated by some mutagens like physical, chemical as well as biological agents [19]. Mutagenesis became well-known in 1950s when various crop species were largely induced through irradiation to enhance trait divergence [20]. Natural mutants are typical type of spontaneous mutations that generate modern phenotypes without human beings interference such as seed dispersal, thin seed coat, seed dormancy and reduced seed length. Heritable mutants are appropriate for human utilization for example loss of bitterness in various types of nuts, almonds, watermelons, potatoes, lima beans, egg plants as well as cabbages. Dwarfing genes exploited to increase grain yield in 1960s. It was completed with the introgression of natural mutant alleles into rice as well as wheat genome. The main disadvantage of this mutation is the loss of numerous wild features in crops [21].

Physical and chemical mutations are collectively called as induced mutation. Induced mutations have a record of 83 years as the first reported [22, 23] in plants. In 1927, Muller illustrated that X-ray induction could enhance the mutation rate in a *Drosophila* up to 15,000% [22] then Stadler examined a powerful phenotypic divergence in barley and maize by induction of X-rays and radium [23]. Induced mutations in plants originated directly from X-rays, radioactivity as well as radioactive elements through Roentgen (1895), Becquerel (1896) as well as Marie with Pierre Curie (1898) respectively. The Nobel Prize was awarded to Roentgen, Becquerel, Marie and Pierre Curie for successful mutation induction [24]. Nitrogen mustard is composed of poisonous mustard gas that applied in World Wars I and II. It is a chemical mutant that exposed to introduce mutations in cells [25]. Agents that cause artificial mutation are called mutagens that includes physical mutagens (X-ray, Gamma rays, Neutrons, Alpha/beta particles) and chemical mutagens (Alkylating agents, Azide, Hydroxylamine, Antibiotics, Nitrous acid) [1].

Replace the entirety of this text with the main body of your chapter. The body is where the author explains experiments, presents and interprets data of one's research. Authors are free to decide how the main body will be structured. However, you are required to have at least one heading. Please ensure that either British or American English is used consistently in your chapter.

2.1 Physical mutagens

Radiation is described as energy transfer in the sort of particles and waves [26]. These radiations are types of the electromagnetic (EM) spectrum that generates ions so it is also called as ionizing radiation. Ionizing radiations are the most frequently used physical mutagens [27]. Approximately, seventy percent of mutant varieties were generated by action of ionizing radiations in past eighty years [27]. These radiations consist of cosmic, gamma (γ) as well as X-rays [28]. Practical mutations by cosmic radiation have been reported in rice, cotton, wheat, tomato, sesame and pepper [29] as well as in maize [30]. The most universally applied physical mutagens are gamma and X-rays [27]. X-rays were the primary mutagens that applied to stimulate mutations [26]. However, gamma rays have gained popularity when these rays were accessible by the in several developing countries [31]. Gamma rays are less harmful produce point mutations with minute deletions while, fast neutron produces chromosome losses, translocations with huge deletions [32]. Additional physical agents are subatomic particles known as alpha (α), beta (β) particles, neutrons as well as protons. These particles are ionizing agents [26]. Ultraviolet (UV) rays are non-ionizing. These rays have potential of tissue penetrability for mutagenesis. Recently, plant materials have been thrown out into space for analysis of mutagenesis. Nevertheless, information about genetics of space induction is so far insufficient [1].

2.2 Chemical mutagens

Researchers search for another source for producing mutations due to the high chromosomal irregularity from ionizing emission. Consequently, a group of chemical induction has been exposed [33]. There are some chemical mutagens namely alkylating agents, base analogues as well as intercalating agents [25]. Alkylating agents were the primary group of chemical mutagens to be exposed by Auerbach and Robson [34] when they discovered the mutagenic result in mustard gas throughout World War II. Chemical mutagens consist of nitrogen mustards, sulfur mustards, ethyl methane sulphontes, ethyleneimines, epoxides, alkyl methane sulphonates, ethyleneimides, alky lnitrosoamines and alkyl nitrosoureas [35]. Chemical mutagens are more applicable for introduction of in-vitro mutation as compared to radiation approaches [36]. Chemical mutagens introduces single base pair (SNPs) change as compared to translocations and deletions as occurred in physical induction that induce more damage with harshly decrease viability. They are simple to apply rather than physical agents [32]. However, undesirable changes are usually high in chemical induction as compared to physical induction [26]. However, these mutagens are usually carcinogenic. Mustard gas, ethyl methane sulfonate (EMS), methyl methane sulfonate (MMS) as well as nitrosoguanidine are powerfully carcinogenic that should be used carefully [37].

2.2.1 Introduction and mode of action of ethyl methane sulfonate (EMS)

Ethyl methane sulfonate is the most frequently applied in plants among chemical mutagens [37] due to its efficiency and accessibility [38]. It has capability to generate the high and stable nucleotide substitution in diverse genomes of organisms [39, 40].

This chemical generates a huge quantity of point mutations in relatively little mutant population. This chemical is enough to develop the genome mutations [40]. EMS has major role in forward genetic for screening of various organisms. It is also applied in model animal and plant for mutagenesis named *Drosophila melanogaster* as well as *Arabidopsis thaliana* respectively. EMS is extraordinarily reliable due to similar levels of induction have been attained in model organisms for example base replacement are analogous for *Arabidopsis* seeds immersed in EMS [41, 42] as well as EMS-fed males *Drosophila* [43]. EMS causes suitable levels of lethality as well as sterility [40]. Genome size does not show to be a significant issue in EMS mutagenesis. Nevertheless, EMS toxicity may differ from species to species [44]. It is also applied in high throughput selection such as TILLING populations [37] in plants

Ethyl methane sulfonate forms an abnormal base of O^{-6}-ethyl guanine due to alkylation of guanine bases. During DNA duplication, it located a thymine residue above a cytosine residue result in an accidental point mutation. Approximately 70–99% alterations in EMS-treated populations are due to GC → AT base pair conversion [37, 40].

2.2.2 Dose of mutagen

LD_{50} is the percentage of test material that are killed by a specific dosage of chemical or radiation mutagen in which half test material will be die. Fixation of LD_{50} is important before the start of an experiment in induced mutation. These doses vary according to fluctuation in treatments duration, quantity, pH as well as solvent used. Mutagen dose can be caused low or high mutation frequency as a result of ignoring the importance of LD_{50} [45–47]. Doses lower than LD_{50} favor plant's recovery after treatment, while the use of high doses increases the probability to induce mutation either in positive or in negative direction. The efficiency of mutation is determined by concluding the accurate doses of mutagens if the dose is random it creates higher number of harmful mutations in each plant [48]. The mutation quality of practical mutation is not absolutely correlated to dose rate. High mutagen doses did not produced the excellent results of yield [26]. Seeds of *Oryza sativa* L. spp. *Indica* cv. MR219 were mutagenized to different doses of EMS from 0.25–2%. Seeds were incubated for ten to twenty hours for establishment of kill curve as well as sensitivity of the tested genotypes [49].

2.3 Mutagenesis with biological agents

Insertion mutageneis with biological agents involved in insertion or deletion of some sequence in genome. It may cause random mutation at genome as transfer DNA, retro-transposon and transposon. Also the insertion mutations can be site specific or targeted that cause genomic variation at specific site included the novel genome editing techniques.

Transfer DNA (T-DNA) insertion mutation helps in identification of gene function in plant genome. This insertion mutagenesis cause loss or gain of gene function that can be observed by phenotypic response [50]. It can also identified through whole genome sequencing or using the *Agrobacterium tumefaciens* machinery that insert the T-DNA at specific flanking sites.

2.3.1 Mutation by genome editing tools

Plant breeding relies on incorporation of genetic variation for desired traits. The innovative strategies are exploited from many years to reduce the off-target random mutations caused by physical and chemical mutagens [16]. These technologies includes: site-directed nucleases, RNA-dependent methylation, oligonucleotide directed mutagenesis, agro infiltration, cisgenec/intragenic and reverse genetics. Site-directed nucleases including Zinc finger nucleases (ZFN) [51], transcription activator-like effector nucleases (TALEN) [52] and Clustered regularly inter-spaced short palindromic repeats (CRISPER) and CRISPER-associated nuclease 9 (CRISPER/Cas9) [53] system had revolutionized the mutation breeding strategy by introducing targeted genome editing. TALEN and CRISPER/Cas techniques precisely and permanently incorporate the desired DNA into the genome and hence cause genetic variation [54].

CRISPR system is a well settle defense system that generate acquired immune response for resistance to bacteria, fungi and phages [55]. It was originated from bacteria and consist of repetitive DNA genetic codes as well as proto-spacer DNA (defensive genetic codes formed during exposure to pathogens) [56, 57]. Cas9 is an endonuclease that generate double stranded break in DNA through its two active domains [58]. Cas9 creates the break at targeted site by utilizing the guided RNA sequence. The CRISPER/Cas system recognized the specific site at genome through guide RNA (gRNA) [59] and at targeted site Cas9 creates a double stranded break (DSB). These break are repaired by DNA repairing system that ultimately cause mutation by either non-homologous end joining (NHEJ) or homologous recombination system [HR] [60, 61]. Specific base-pair change occur during DNA repair system as Cytosine to thymine (C/T) [62] and Adenine to guanine (A/G) [63] that observed in several crop plants as canola, rice, tomato, wheat and corn [57]. CRISPR/Cas9 system was used in rice to generate semi-dwarf mutants in rice from T2 to T4 generation. Stable indels passed through generation producing homozygous mutant [64]. In plants, CRISPR/Cas system generates induced mutation through gene knockouts, insertion or generating single nucleotide polymorphism (SNP) in plants [56, 57]. Some latest gene editing mutations using CRISPR machinery enlisted in **Table 1**.

Gene targeted	Vector: promoter	Transformation method: promoter	Plant variety	Mutation nature	Reference
BnaMAX1s	Gateway 100 vectors; BGK01 vector: 35S promoter	Agrobacterium GV3101	Rapeseed RS862	Knockout mutation	[65]
GmFT2a; G,FT5a	pTF101.1: 35S promoter	Agrobacterium strain EHA101	Soybean jack variety (*Glycine max*)	Knockout by CRISPR	[66]
OsRR22	pYLCRISPR/ Vas9Pubi-H; Cas9-OsRR22-gRNA: OsU6 promoter	Agrobacterium EHA105: OsU6	Rice japonica WPB106	Knockout mutation	[67]
MaGA20ox2	pYLCRISPR/ Cas9P$_{ubi}$-H vector: U3 promoter	Agrobacterium strain EHA105	Banana cultivar Gro Michel) (Musa acuminate)	Mutation as insertion and deletion	[68]
Exon of *SD1* gene	pBIN-sgR-Cas9-OsU3 vector: 35S PROMOTER	Agrobacterium strain LBA4404, EHA105	Rice variety 9815B, JIAODA138, HUAIDAO1055	On target and off target mutations	[64]
BnSFAR4; BnSFAR5	pCas9-TPC: pMP90RK	Agrobacterium GV3101	Rapeseed RS306	Knockout mutation	[69]

Gene targeted	Vector: promoter	Transformation method: promoter	Plant variety	Mutation nature	Reference
63 immunity associated genes	P201N-Cas9: U6 promoter	Agrobacterium ID1249 strain	Tomato RG-PtoR or RG-prf3	Short Indels	[70]
VvMLO3; VvMLO4	pYLCRIPSR/Cas9-N vector: *AtU3b* and *AtU6–1* promoter	Agrobacterium strain GV3101	Grape wine PN40024 (*Vitis vinifera*)	Short Indels	[71]
HvITPK1	pYLsgRNA-OsU6 pYLCRISPR/ Cas9Pubi-H: U6 promoter	Agrobacterium strain AGL1	Barley model cultivar Golden Promise	Insertion mutation	[72]
Clpsk1	pRGEBB320cas9-gRNA-Clpk1: AtU6 promoter	Agrobacterium strain EHA105	Watermelon Sumi 1	Lnockout	[73]
GhCLA and GhPEBP	pRGEB32-GhU6.7: cotton U6 promoter	Agrobacterium strain GV3101	Cotton Jin668 (*Gossypium hirsutum*)	Point mutation	[74]

Table 1.
Induced mutation in plants using targeted genome editing method CRISPR/Cas9.

3. Identification of mutagenic site through molecular marker

Markers have been used for cultivar recognition as the first light of forward genetics. Markers fall into three major categories as visually measurement of traits, gene product as well as DNA test known as morphological markers, biochemical markers as well as molecular markers respectively [75]. Classical breeding can be fast-tracked using molecular marker approaches for identification of mutagenic regions and to access the variations inside genome [76]. These markers lies inside the genetic region or nearby it. Newest genotyping approaches like genotyping by sequencing (GBS) made it easier to identify even a single base pair change as single nucleotide polymorphism (SNP).

3.1 Morphological markers

Most of the induced mutants have been released as cultivars by selection through morphological markers. These markers based upon agronomic traits such as maturity, height, early flowering, fruit appearance, seed quality as well as resistance to diseases that can be monitored easily as a result of their epiphytotic character. Huge morphological, physiological as well as ecological differences has been existed in cultivated rice genomes. It is a general approach employed to determine genotypic relationship [77].

Detection of morphological traits is performed by statistically method. Multivariate methods have statistical approaches that widely used in telling the intrinsic variation among various crop genotypes. Multivariate analysis has been reported for study of genetic diversity in numerous crops such as barley [78], sorghum [79], wheat [80], peanut [81] and rice [82].

3.2 Genetic markers

Mutant phenotypes were usually recognized depend upon their morphological characteristic. But morphological markers are not steady due to less heritability along with complex genetic nature [83]. Then novel method have been developed depend upon genetic information of DNA. This procedure is a faster and more reliable as compared to other methods. Molecular markers depend upon genetic variant in the genome [84]. DNA-based markers have basically replaced previous biochemical markers as importance of DNA analysis has been reported in plants [85]. DNA markers are largely scattered across the entire genome due to larger in quantity [75]. Genetic markers perform main role for reorganization of heritable traits in plant breeding as well as genetics [86].

3.2.1 Types of genetic marker

Numerous types of molecular markers have been utilized. Molecular markers include restriction fragment length polymorphism (RFLP) [87], random amplified polymorphic DNA (RAPD) [88], amplified fragment length polymorphism (AFLP) [89], inter-simple sequence repeat (ISSRs) [90], microsatellite or simple sequence repeats (SSRs) [91] and single nucleotide polymorphisms (SNPs) [92] are currently accessible to evaluate the diversity and variability at the DNA level.

3.2.1.1 SSRs as a sequence based marker

These markers are group of tedious DNA chain typically two to six base pairs. It is a form of VNTRs (Variable Number Tandem Repeats) [93]. These markers are well-known as STRs (short tandem repeats). The rate of different STRs length is feature of microsatellite loci in rice [75]. They consist of dinucleotide; trinucleotide as well as tetranucleotide repeats for genetic analysis. Dinucleotides are the key form present in most vertebrates. Trinucleotide repeat are rich in plants [94]. Microsatellite markers are believed to be suitable over different array of markers due to following reasons. These are scattered all over the genome of extremely conserved region. These markers have various qualities of simplicity, high polymorphism, rapidity as well as stability. These markers have been model for examination of germplasm, genetic diversity [95], heterosis, purity test, gene mapping, fingerprints assembly, phylogenetic comparison as well as marker aided selection [75, 95]. A random collection of SSRs assist in estimation of rice genetic diversity and rice cultivar classification without mistakes [96]. Particularly SSRs markers have been extensively employed in rice genetic analysis for high allelic detection [97]. Microsatellite exposed unreliable level of genetic relationship among the domesticated as well as wild collection of rice [98].

3.2.1.2 SNPs as genetic markers from high-throughput sequencing

SNPs (single nucleotide polymorphisms) signify a strong group of genetic markers [99] among different categories of molecular markers due to following reason [100]. These markers detect single-base pair location depend upon sequence variation in genomes [101]. SNPs markers offer a huge marker density in genomes [102]. SNPs markers have achieved significant importance in plant genetic analysis due to their brilliant genetic qualities, genetic diversity, evolutionary interaction [103], high throughput genetic mapping [104], population substructure [105], genome-wide linkage disequilibrium [106] as well as association mapping [107]. Availability of high quality reference genome sequence made it easier to scan out mutation by re-sequencing the species genotypes through next generation sequencing (NGS) approaches and to identify the variation in targeted genotype through mapping techniques as genome-wide association mapping [108].

Classically, a quite large sequencing attempt is faithful to recognize polymorphic location in a genome among a set of various breeding lines [109]. A precise multiplexed SNPs genotyping analysis is necessary to utilize the huge SNPs source for high-throughput genetic test in rice [110]. It will become routine to re-sequence the plants genome with current SNPs platforms as the price of genome sequencing keep on to reduce [111].

SNPs genotyping have been applied in many organism including rice [112–114], *Arabidopsis* [115], maize [116], soybean [117] and wheat [118]. The high class order of the rice genome has offered genome-wide SNPs source [119]. Polymorphic loci (5.41 million) were detected between the two main domesticated rice subspecies (*Indica* and *Japonica*) by SNPs genotyping [120].

A complete map of rice genome builds 6,119,311 SNPs variants for 1529 genome orders. SNPs (213,188) were located in *Indica* and *Japonica* rice. Asian and African rice were established 9595 SNPs [121]. Three thousand rice genomes project [122] submit for rice clustering of aus/boro genotypes. Only 208 accessions are categorized as aus/boro depends upon SNPs markers of 200,000. It is also exposed from further study that aus group was genetically related by 376,000 SNPs markers [123].

4. Conclusions

The considerable amount of phenotypic variability can be identified by employing highly sophisticated Molecular approaches like SSR and SNPs within the mutant populations. These genetic changes indicated that EMS might be helpful for the development of desired genetic changes in crop plants. It was also recommended that current SSRs and SNPs markers could be suited in further analysis for estimation of genetic diversity of rice mutants.

Acknowledgements

The authors acknowledge Higher Education Commission (HEC), Islamabad, Pakistan for providing funds for research activities at Genomics lab, CABB. The authors are thankful to Plant Breeding Division (NIAB) for providing assistance in analysis of EMS mutagenesis.

Conflict of interest

"The authors declare no conflict of interest."

Author details

Faisal Saeed Awan[1*], Bushra Sadia[1], Javaria Altaf[2], Madiha Habib[1], Kiran Hameed[1] and Shabbir Hussain[1]

1 Centre of Agricultural Biochemistry and Biotechnology, University of Agriculture Faisalabad, Pakistan

2 Department of Zoology, Government College University Faisalabad, Pakistan

*Address all correspondence to: faisal.saeed@uaf.edu.pk

References

[1] Oladosu Y, Rafii MY, Abdullah N, Hussin G, Ramli A, Rahim HA, Miah G, Usman M. Principle and application of plant mutagenesis in crop improvement: a review. Biotechnology & Biotechnological Equipment. 2016; 30(1):1-6. DOI.org/10.1080/13102818.2015.1087333

[2] Rani R, Yadav P, Barbadikar KM, Baliyan N, Malhotra EV, Singh BK, Kumar A, Singh D. CRISPR/Cas9: a promising way to exploit genetic variation in plants. Biotechnology letters. 2016;38(12):1991-2006.DOI 10.1007/s10529-016-2195-z

[3] Shu QY, Forster BP, Nakagawa H. Nakagawa H, editors. Plant mutation breeding and biotechnology: CABI; 2012. DOI: 10.1079/9781780640853.0301

[4] Ragvendra T, Suresh BG, Mishra VK, Ashutosh K, Ashok K. Genetic variability and character association in direct seeded upland rice (*Oryza sativa*). Environment and Ecology. 2011; 29(4A):2132-5. orcid.org/0000-0002-1812-3311

[5] Sumanth V, Suresh BG, Ram BJ, Srujana G. Estimation of genetic variability, heritability and genetic advance for grain yield components in rice (*Oryza sativa* L.). Journal of Pharmacognosy and Phytochemistry. 2017;**6**(4):1437-1439

[6] Kharkwa MC. A brief history of plant mutagenesis Plant Mutation Breeding and Biotechnology ed Q Y Shu, B P Forster and H Nakagawa (Austria: Food and Agriculture Organization of the United Nations).2012; 21-30.

[7] El-Degwy IS. Mutation induced genetic variability in rice (*Oryza sativa* L.). International Journal of Agriculture and Crop Sciences. 2013;5(23):2789-2794

[8] Dewi AK, Dwimahyani I. Application of induced mutation technique to improve genetic variability of Indonesian traditional rice varieties. E&ES. 2020; 482(1): DOI 012016.10.1088/1755-1315/482/1/012016

[9] Kostov K, Batchvarova R, Slavov S. Application of chemical mutagenesis to increase the resistance of tomato to *Orobanche ramosa* L. Bulgarian Journal of Agricultural Science. 2007;**13**(5):505-513

[10] Pathirana R. Plant mutation breeding in agriculture. Plant sciences reviews. 2011;**6**(032):107-126. DOI: 10.1079/PAVSNNR20116032

[11] Kondo E, Nakayama M, Kameari N, Tanikawa N, Morita Y, Akita Y, et al. Red-purple flower due to delphinidin 3, 5-diglucoside, a novel pigment for cyclamen spp., generated by ion-beam irradiation. Plant biotechnology. 2009;**26**(5):565-569. DOI: 10.5511/plantbiotechnology.26.565

[12] Parry MA, Madgwick PJ, Bayon C, Tearall K, Hernandez-Lopez A, Baudo M, Rakszegi M, Hamada W, Al-Yassin A, Ouabbou H, Labhilili M. Mutation discovery for crop improvement. Journal of Experimental Botany. 2009; 60(10): DOI 2817-25. 10.1093/jxb/erp189

[13] Ali HM, Shah SA. Evaluation and selection of rapeseed (*Brassica napus* L.) mutant lines for yield performance using augmented design. J Anim Plant Sci. 2013; 23:1125-30.

[14] Lee YH, Park W, Kim KS, Jang YS, Lee JE, Cha YL, Moon YH, Song YS, Lee K. EMS-induced mutation of an endoplasmic reticulum oleate desaturase gene (FAD2-2) results in elevated oleic acid content in rapeseed (*Brassica napus* L.). Euphytica. 2018; 214(2):28. DOI 10.1007/s10681-017-2106-y

[15] Channaoui S, Labhilili M, Mouhib M, Mazouz H, El Fechtali M, Nabloussi A. Development and evaluation of diverse promising rapeseed (*Brassica napus* L.) mutants using physical and chemical mutagens. OCL. 2019; 26:35.DOI 10.1051/ocl/2019031

[16] Holme IB, Gregersen PL, Brinch-Pedersen H. Induced genetic variation in crop plants by random or targeted mutagenesis: Convergence and differences. Frontiers in Plant Science. 2019;**10**. DOI: 10.3389/fpls.2019.01468

[17] Jain M. Function genomics of abiotic stress tolerance in plants: A CRISPR approach. Frontiers in plant science. 2015;**6**:375. DOI: 10.3389/fpls.2015.00375

[18] Wei FJ, Droc G, Guiderdoni E, Yue-ie CH. International consortium of rice mutagenesis: Resources and beyond. Rice. 2013;**6**(1):39. DOI: 10.1186/1939-8433-6-39

[19] Roychowdhury R, Tah J. Mutagenesis—A potential approach for crop improvement. InCrop Improvement 2013 (pp. 149-187). Springer, Boston, MA.DOI 10.1007/978-1-4614-7028-1_4

[20] Leitao JM. Chemical mutagenesis. p. 135-158. In: Shu, Q. Y., B. P. Forster and H. Nakagawa (Eds.). Plant mutation breeding and biotechnology. Wallingford: CABI.2012. DOI 10.1079/9781780640853.0135

[21] Lönnig WE. Mutation breeding, evolution, and the law of recurrent variation. Recent Res. Devel. Genet. Breeding. 2005;**2**:45-70

[22] Muller HJ. Artificial transmutation of the gene. Science. 1927;**66**(1699):84-87. DOI: 10.1126/science.66.1699.84

[23] Stadler LJ. Some genetic effects of X-rays in plants. Journal of heredity. 1930;**21**. DOI: 10.1093/oxfordjournals.jhered.a103249

[24] All Nobel Prizes in Physics. 2012. http://nobelprize.org/nobel_prizes/physics/

[25] Auerbach C. Genetical effects of radiation and chemicals. Experientia. 1957; 13(6):217-24.DOI 10.1007/BF02157426

[26] Acquaah G. Principles of Plant Genetics and Breeding. Wiley-Blackwell: Chichester; 2006

[27] Mba C, Afza R, Shu QY. Mutagenic radiations: X-rays, ionizing particles and ultraviolet. Plant mutation breeding and biotechnology. 2012:83-90. DOI: 10.1079/9781780640853.0083

[28] Mba C. Induced mutations unleash the potentials of plant genetic resources for food and agriculture. Agronomy. 2013;**3**(1):200-231. DOI: 10.3390/agronomy3010200

[29] Liu L, Lee GA, Jiang L, Zhang J. Evidence for the early beginning (c. 9000 cal. BP) of rice domestication in China: A response. The Holocene. 2007;**17**(8):1059-1068. DOI: 10.1177/0959683607085121

[30] Mei M, Qiu Y, Sun Y, Huang R, Yao J, Zhang Q, et al. Morphological and molecular changes of maize plants after seeds been flown on recoverablf satellite. Advances in Space Research. 1998;**22**(12):1691-1697. DOI: 10.1016/S0273-1177(99)00034-4

[31] IAEA. 2011. IAEA mutant database. International Atomic Energy Agency, Vienna; URL: http:// mvgs.iaea.org/Search.aspx.

[32] Wu JL, Wu C, Lei C, Baraoidan M, Bordeos A, Madamba MR, et al. Chemical-and irradiation-induced mutants of indica rice IR64 for forward and reverse genetics. Plant molecular biology. 2005;**59**(1):85-97. DOI: 10.1007/s11103-004-5112-0

[33] Medina FI, Amano E, Tano S. FNCA mutation breeding manual. InForum for Nuclear Cooperation in Asia 2011 May (Vol. 10).

[34] Auerbach C, Robson JM. Mutation from mustard and related substances. Nature. 1946; 157:302. DOI10.1038/157302a0

[35] Wani MR, Kozgar MI, Tomlekova N, Khan S, Kazi AG, Sheikh SA, et al. Mutation breeding: A novel technique for genetic improvement of pulse crops particularly chickpea (*Cicer arietinum* L.). InImprovement of crops in the era of climatic changes 2014 (pp. 217-248). Springer, New York. NY. . DOI: 10.1007/978-1-4614-8824-8_9

[36] Viana VE, Pegoraro C, Busanello C, de Oliveira AC. Mutagenesis in rice: The basis for breeding a new super plant. Frontiers in plant science. 2019;**10**. DOI: 10.3389/fpls.2019.01326

[37] Till BJ, Cooper J, Tai TH, Colowit P, Greene EA, Henikoff S, et al. Discovery of chemically induced mutations in rice by TILLING. BMC plant biology. 2007;7(1):19. DOI: 10.1186/1471-2229-7-19

[38] Joint FA. Manual on mutation breeding. International Atomic Energy Agency (IAEA); 1977.

[39] Talebi AB, Talebi AB, Shahrokhifar B. Ethyl methane sulphonate (EMS) induced mutagenesis in Malaysian rice (cv. MR219) for lethal dose determination. DOI 10.4236/ajps.2012.312202

[40] Bhat RS, Upadhyaya NM, Chaudhury A, Raghavan C, Qiu F, Wang H, et al. Chemical-and irradiation-induced mutants and TILLING. InRice functional genomics 2007 (pp. 148-180). Springer, New York. NY. . DOI: 10.1007/0-387-48914-2_8

[41] McCallum CM, Comai L, Greene EA, Henikoff S. Targeted screening for induced mutations. Nature biotechnology. 2000;**18**(4):455-457. DOI: 10.1038/74542

[42] McCallum CM, Comai L, Greene EA, Henikoff S. Targeting induced locallesions in genomes (TILLING) for plant functional genomics. Plant Physiology. 2000;**123**(2):439-442. DOI: 10.1104/pp.123.2.439

[43] Bentley A, MacLennan B, Calvo J, Dearolf CR. Targeted recovery of mutations in drosophila. Genetics. 2000;**156**(3):1169-1173

[44] Henikoff S, Comai L. Single-nucleotide mutations for plant functional genomics. Annual Review of Plant Biology. 2003; 54(1):375-401.DOI 10.1146/annurev.arplant.54.031902.135009

[45] Jain SM. Mutagenesis in crop improvement under the climate change. Romanian biotechnological letters. 2010;**15**(2):88-106

[46] Vasline A. Chlorophyll and viable mutations in rice (*Oryza sativa* L.). Plant Archives. 2013;**13**:531-533

[47] Rajarajan D, Saraswathi R, Sassikumar D. Determination of lethal dose and effect of gamma ray on germination percentage and seedling parameters in ADT (R) 47 rice. International Journal of Advanced Biological Research. 2016;**6**(2):328-332

[48] Oldach KH. Mutagenesis. p. 208-219. In: Pratap A, Kumar J, editors. Biology and Breeding of Food Legumes. Wallingford: CABI; 2011

[49] Comai L, Henikoff S. TILLING: Practical single-nucleotide mutation discovery. The Plant Journal. 2006;**45**(4):684-694. DOI: 10.1111/j.1365-313X.2006.02670.x

[50] Lo SF, Fan MJ, Hsing YI, Chen LJ, Chen S, Wen IC, et al. Genetic resources offer efficient tools for rice functional genomics research. Plant, cell & environment. 2016;**39**(5):998-1013. DOI: 10.1111/pce.12632

[51] Bibikova M, Beumer K, Trautman JK, Carroll D. Enhancing gene targeting with designed zinc finger nucleases. Science. 2003;300(5620):764- DOI.10.1126/science.1079512

[52] Bogdanove AJ, Voytas DF. TAL effectors: Customizable proteins for DNA targeting. Science. 2011 Sep 30;**333**(6051):1843-1846. DOI: 10.1126/science.1204094

[53] Jinek M. Krzysztof; Chylinski, Ines; Fonfara, M; Hauer, Jennifer; Doudna, and Emmanuelle Charpentier. "A programmable dual-RNA-guided DNA endonuclease in adaptive bacterial immunity." Science 2012; 337:816-821. DOI 10.1126/science.1225829

[54] Schaart JG, van de Wiel CC, Lotz LA, Smulders MJ. Opportunities for products of new plant breeding techniques. Trends in Plant Science. 2016; 21(5):438-49.DOI 10.1016/j.tplants.2015.11.006

[55] Piatek A, Ali Z, Baazim H, Li L, Abulfaraj A, Al-Shareef S, et al. RNA-guided transcriptional regulation in planta via synthetic dCas9-based transcription factors. Plant Biotechnology Journal. 2014;**13**(1):578-589

[56] Mojica FJ, Díez-Villaseñor C, García-Martínez J, Almendros C. Short motif sequences determines the targets of the prokaryotic CRISPR defence system. Microbiology. 2009;**155**:733-740

[57] Brouns SJ, Jore MM, Lundgren M, Westra ER, Silikhuis RJ, Snijders AP, et al. Small CRISPR RNAs guide antiviral defense in prokaryotes. Science. 2008;**32**:960-964

[58] Jia H, Wang N. Targeted genome editing of sweet orange using Cas9/sgRNA. PLoS One. 2014;**9**(4):93806

[59] Mali P, Yang L, Esvelt KM, Aach J, Guell M, DiCarlo JE, et al. RNA-guided human genome engineering via Cas9. Science. 2013;**339**(6121):823-826. DOI: 10.1126/science.1232033

[60] Voytas DF. Plant genome engineering with sequence-specific nucleases. Annual Review of Plant Biology. 2013;**64**:327-350

[61] Shan Q, Voytas DF. Editing plant genes one base at a time. Nature plants. 2018;**4**(7):412-413. DOI: 10.1038/s41477-018-0177-y

[62] Zong Y, Wang Y, Li C, Zhang R, Chen K, Ran Y, Qiu JL, Wang D, Gao C. Precise base editing in rice, wheat and maize with a Cas9-cytidine deaminase fusion. Nature biotechnology. 2017; 35(5):438.DOI 10.1038/nbt.3811

[63] Li C, Zong Y, Wang Y, Jin S, Zhang D, Song Q, et al. Expanded base editing in rice and wheat using a Cas9-adenosine deaminase fusion. Genome biology. 2018;**19**(1):59. DOI: 10.1186/s13059-018-1443-z

[64] Biswas S, Tian J, Li R, Chen X, Luo Z, Chen M, et al. Investigation of CRISPR/Cas9-induced SD1 rice mutants highlights the importance of molecular characterization in plant molecular breeding. Journal of Genetics and Genomics. 2020 May;**21**

[65] Zheng M, Zhang L, Tang M, Liu J, Liu H, Yang H, et al. Knockout of two Bna MAX 1 homologs by CRISPR/Cas9-targeted mutagenesis improves plant architecture and increases yield in rapeseed (*Brassica napus* L.). Plant biotechnology journal. 2020;**18**(3):644-654

[66] Cai Y, Wang L, Chen L, Wu T, Liu L, Sun S, et al. Mutagenesis of GmFT2a and GmFT5a mediated by CRISPR/Cas9 contributes for expanding the regional adaptability of soybean. Plant biotechnology journal. 2020;**18**(1):298-309. DOI: 10.1111/pbi.13199

[67] Zhang A, Liu Y, Wang F, Li T, Chen Z, Kong D, Bi J, Zhang F, Luo X, Wang J, Tang J. Enhanced rice salinity tolerance via CRISPR/Cas9-targeted mutagenesis of the OsRR22 gene. Molecular Breeding. 2019; 39(3):47. DOI.org/10.1007/s11032-019-0954-y

[68] Shao X, Wu S, Dou T, Zhu H, Hu C, Huo H, He W, Deng G, Sheng O, Bi F, Gao H. Using CRISPR/Cas9 genome editing system to create MaGA20ox2 gene-modified semi-dwarf banana. Plant Biotechnology Journal. 2020; 18(1):17-9. DOI.org/10.1111/pbi.13216

[69] Karunarathna NL. Wang H. Jiang L, Jung C. Elevating seed oil content in a polyploid crop by induced mutations in SEED FATTY ACID REDUCER genes. Plant Biotechnology Journal: Harloff HJ; 2020 Mar 26

[70] Zhang N, Roberts HM, Van Eck J, Martin GB. Generation and molecular characterization of CRISPR/Cas9-induced mutations in 63 immunity-associated genes in tomato reveals specificity and a range of gene modifications. Frontiers in plant science. 2020;**11**

[71] Wan DY, Guo Y, Cheng Y, Hu Y, Xiao S, Wang Y, Wen YQ. CRISPR/Cas9-mediated mutagenesis of VvMLO3 results in enhanced resistance to powdery mildew in grapevine (Vitis vinifera). Horticulture Research. 2020;7(1):1-4. DOI org/10.1038/s41438-020-0339-8

[72] Vlčko T, Ohnoutkova L. Allelic variants of CRISPR/Cas9 induced mutation in an inositol trisphosphate 5/6 kinase gene manifest different phenotypes in barley. Plants. 2020;**9**(2):195. DOI: 10.3390/plants9020195

[73] Zhang M, Liu Q, Yang X, Xu J, Liu G, Yao X, Ren R, Xu J, Lou L. CRISPR/Cas9-mediated mutagenesis of Clpsk1 in watermelon to confer resistance to *Fusarium oxysporum f. sp. niveum*. Plant Cell Reports. 2020:1-7. e DOI org/10.1007/s00299-020-02516-0

[74] Qin L, Li J, Wang Q, Xu Z, Sun L, Alariqi M, et al. High-efficient and precise base editing of C• G to T• a in the allotetraploid cotton (*Gossypium hirsutum*) genome using a modified CRISPR/Cas9 system. Plant biotechnology journal. 2020;**18**(1):45-56. DOI: 10.1111/pbi.13168

[75] Divya K. Study of genetic diversity in Karnataka rice (*Oryza Sativa*) landraces using trait specific simple sequence repeat (SSR) markers. Int. J. Thesis Proj. Diss. 2013;**1**:45-70

[76] Desta ZA, Ortiz R. Genomic selection: genome-wide prediction in plant improvement. Trends in plant science. 2014;19(9):592-601.DOI 10.1016/j.tplants.2014.05.006

[77] Bajracharya J, Steele KA, Jarvis DI, Sthapit BR, Witcombe JR. Rice landrace diversity in Nepal: Variability of agro-morphological traits and SSR markers in landraces from a high-altitude site. Field Crops Research. 2006;**95**(2-3):327-335. DOI: 10.1016/j.fcr.2005.04.014

[78] Cross RJ. A proposed revision of the IBPGR barley descriptor list. Theoretical and Applied Genetics. 1992;**84**(3-4):501-507. DOI: 10.1007/BF00229513

[79] Ayana A, Bekele E. Multivariate analysis of morphological variation in sorghum (*Sorghum bicolor* (L.) Moench) germplasm from Ethiopia and

Eritrea. Genetic Resources and Crop Evolution. 1999;**46**(3):273-284. DOI: 10.1023/A:1008657120946

[80] Hailu F, Merker A, Belay G, Johansson E. Multivariate analysis of diversity of tetraploid wheat germplasm from Ethiopia. Genetic Resources and Crop Evolution. 2006;**53**(6):1089-1098. DOI: 10.1007/s10722-005-9776-3

[81] Upadhyaya HD, Reddy LJ, Dwivedi SL, Gowda CL, Singh S. Phenotypic diversity in cold-tolerant peanut (Arachis hypogaea L.) germplasm. Euphytica. 2009;**165**(2):279-291. DOI: 10.1007/s10681-008-9786-2

[82] Chakravorty A, Ghosh PD, Sahu PK. Multivariate analysis of phenotypic diversity of landraces of rice of West Bengal. Journal of Experimental Agriculture International. 2013:110-123. DOI: 10.9734/AJEA/2013/2303

[83] Virk PS, Newbury HJ, Jackson MT, Ford-Lloyd BV. Are mapped markers more useful for assessing genetic diversity? Theoretical and Applied Genetics. 2000;**100**(3-4):607-613. DOI: 10.1007/s001220050080

[84] Edwards D, Batley J. Plant bioinformatics: From genome to phenome. TRENDS in Biotechnology. 2004;**22**(5):232-237. DOI: 10.1016/j.tibtech.2004.03.002

[85] Henry RJ, editor. Plant Genotyping. II: SNP technology. CABI; 2008. DOI: 10.1017/S0014479709990378

[86] Sonah H, Deshmukh RK, Sharma A, Singh VP, Gupta DK, Gacche RN, Rana JC, Singh NK, Sharma TR. Genome-wide distribution and organization of microsatellites in plants: an insight into marker development in Brachypodium. Plos one. 2011; 6(6):e21298. DOI10.1371/journal.pone.0021298

[87] Botstein D, White RL, Skolnick M, Davis RW. Construction of a genetic linkage map in man using restriction fragment length polymorphisms. American journal of human genetics. 1980;**32**(3):314. DOI: 10.1007/BF00292654

[88] Williams JG, Kubelik AR, Livak KJ, Rafalski JA, Tingey SV. DNA polymorphisms amplified by arbitrary primers are useful as genetic markers. Nucleic acids research. 1990;**18**(22):6531-6535. DOI: 10.1093/nar/18.22.6531

[89] Vos P, Hogers R, Bleeker M, Reijans M, Lee TV, Hornes M, et al. AFLP: A new technique for DNA fingerprinting. Nucleic acids research. 1995;**23**(21):4407-4414. DOI: 10.1093/nar/23.21.4407

[90] Joshi SP, Gupta VS, Aggarwal RK, Ranjekar PK, Brar DS. Genetic diversity and phylogenetic relationship as revealed by inter simple sequence repeat (ISSR) polymorphism in the genus Oryza. Theoretical and Applied Genetics. 2000;**100**(8):1311-1320. DOI: 10.1007/s001220051440

[91] McCouch SR, Teytelman L, Xu Y, Lobos KB, Clare K, Walton M, Fu B, Maghirang R, Li Z, Xing Y, Zhang Q. Development and mapping of 2240 new SSR markers for rice (*Oryza sativa* L.). DNA research. 2002;9(6):199-207.DOI 10.1093/dnares/9.6.199

[92] Ganal MW, Altmann T, Röder MS. SNP identification in crop plants. Current opinion in plant biology. 2009;**12**(2):211-217. DOI: 10.1016/j.pbi.2008.12.009

[93] Ijaz S. Microsatellite markers: An important fingerprinting tool for characterization of crop plants. African Journal of Biotechnology. 2011;**10**(40):7723-7726. DOI: 10.5897/AJBx10.021

[94] Chen X, Liu X, Wu D, Shu QY. Recent Progress of Rice Mutation Breeding and Germplasm Enhancement in China. Plant Mutation Reports 1: 4-6. International Atomic Energy Agency, Vienna, Austria. 2006.

[95] Ma H, Yin Y, Guo Z, Chen L, Zhang L, Zhong M, et al. Establishment of DNA fingerprinting of Liaojing series of japonica rice. Middle East Journal of Scientific Research. 2011;**8**(2):384-392

[96] Ni J, Colowit PM, Mackill DJ. Evaluation of genetic diversity in rice subspecies using microsatellite markers. Crop science. 2002;**42**(2):601-607. DOI: 10.2135/cropsci2002.6010

[97] McCouch SR, Chen X, Panaud O, Temnykh S, Xu Y, Cho YG, et al. Microsatellite marker development, mapping and applications in rice genetics and breeding. InOryza: From molecule to plant 1997 (pp. 89-99). Springer. Dordrecht. . DOI: 10.1023/A:1005711431474

[98] Joshi RK, Kuanar A, Mohanty S, Subudhi E, Nayak S. Mining and characterization of EST derived microsatellites in Curcuma longa L. Bioinformation. 2010;**5**(3):128. DOI: 10.6026/97320630005128

[99] Rafalski A. Applications of single nucleotide polymorphisms in crop genetics. Current opinion in plant biology. 2002 Apr 1;5(2):94-100.10.1016/S1369-5266(02)00240-6

[100] Agarwal M, Shrivastava N, Padh H. Advances in molecular marker techniques and their applications in plant sciences. Plant cell reports. 2008; 27(4):617-31. DOI10.1007/s00299-008-0507-z

[101] Zhu YL, Song QJ, Hyten DL, Van Tassell CP, Matukumalli LK, Grimm DR, et al. Single-nucleotide polymorphisms in soybean. Genetics. 2003;**163**(3):1123-1134

[102] Batley J, Edwards D. SNP applications in plants. InAssociation mapping in plants 2007 (pp. 95-102). Springer, New York, NY.DOI 10.1007/978-0-387-36011-9_6

[103] Varshney RK, Thiel T, Sretenovic-Rajicic T, Baum M, Valkoun J, Guo P, Grando S, Ceccarelli S, Graner A. Identification and validation of a core set of informative genic SSR and SNP markers for assaying functional diversity in barley. Molecular Breeding. 2008; 22(1):1-3.DOI 10.1007/s11032-007-9151-5

[104] Duran C, Appleby N, Clark T, Wood D, Imelfort M, Batley J, Edwards D. AutoSNPdb: an annotated single nucleotide polymorphism database for crop plants. Nucleic Acids Research. 2009; 37(suppl_1):D951-3. 10.1093/nar/gkn650

[105] Caicedo AL, Williamson SH, Hernandez RD, Boyko A, Fledel-Alon A, York TL, Polato NR, Olsen KM, Nielsen R, McCouch SR, Bustamante CD. Genome-wide patterns of nucleotide polymorphism in domesticated rice. PLoS Genet. 2007; 3(9):e163.DOI 10.1371/journal.pgen.0030163

[106] Jannink JL, Lorenz AJ, Iwata H. Genomic selection in plant breeding: from theory to practice. Briefings in functional genomics. 2010; 9(2):DOI 166-77.10.1093/bfgp/elq001

[107] Bao JS, Corke H, Sun M. Microsatellites, single nucleotide polymorphisms and a sequence tagged site in starch-synthesizing genes in relation to starch physicochemical properties in nonwaxy rice (*Oryza sativa* L.). Theoretical and Applied Genetics. 2006;113(7):1185-96.DOI 10.1007/s00122-006-0394-z

[108] Shirasawa K, Fukuoka H, Matsunaga H, Kobayashi Y, Kobayashi I, Hirakawa H, Isobe S, Tabata S. Genome-wide association studies using single nucleotide polymorphism markers developed by re-sequencing of the genomes of cultivated tomato. DNA research. 2013; 20(6):593-603.DOI 10.1093/dnares/dst033

[109] He J, Zhao X, Laroche A, Lu ZX, Liu H, Li Z. Genotyping-by-sequencing (GBS), an ultimate marker-assisted selection (MAS) tool to accelerate plant breeding. Frontiers in plant science. 2014;**5**:484. DOI: 10.3389/fpls.2014.00484

[110] McNally KL, Bruskiewich R, Mackill D, Buell CR, Leach JE, Leung H. Sequencing multiple and diverse rice varieties. Connecting whole-genome variation with phenotypes. Plant Physiology. 2006;141(1):26-31.DOI 10.1104/pp.106.077313

[111] Imelfort M, Duran C, Batley J, Edwards D. Discovering genetic polymorphisms in next-generation sequencing data. Plant biotechnology journal. 2009;**7**(4):312-317. DOI: 10.1111/j.1467-7652.2009.00406.x

[112] Deschamps S, la Rota M, Ratashak JP, Biddle P, Thureen D, Farmer A, Luck S, Beatty M, Nagasawa N, Michael L, Llaca V. Rapid genome-wide single nucleotide polymorphism discovery in soybean and rice via deep resequencing of reduced representation libraries with the Illumina genome analyzer. The Plant Genome. 2010; 3(1):53-68.DOI 10.3835/plantgenome2009.09.0026

[113] Subbaiyan GK, Waters DL, Katiyar SK, Sadananda AR, Vaddadi S, Henry RJ. Genome-wide DNA polymorphisms in elite indica rice inbreds discovered by whole-genome sequencing. Plant Biotechnology Journal. 2012;**10**(6):623-634. DOI: 10.1111/j.1467-7652.2011.00676.x

[114] Yamamoto T, Nagasaki H, Yonemaru JI, Ebana K, Nakajima M, Shibaya T, Yano M. Fine definition of the pedigree haplotypes of closely related rice cultivars by means of genome-wide discovery of single-nucleotide polymorphisms. BMC genomics. 2010; 11(1):1-4.DOI 10.1186/1471-2164-11-267

[115] Ossowski S, Schneeberger K, Clark RM, Lanz C, Warthmann N, Weigel D. Sequencing of natural strains of Arabidopsis thaliana with short reads. Genome Research. 2008;**18**(12):2024-2033. DOI: 10.1101/gr.080200.108

[116] Lai J, Li R, Xu X, Jin W, Xu M, Zhao H, et al. Genome-wide patterns of genetic variation among elite maize inbred lines. Nature genetics. 2010;**42**(11):1027-1030. DOI: 10.1038/ng.684

[117] Lam HM, Xu X, Liu X, Chen W, Yang G, Wong FL, Li MW, He W, Qin N, Wang B, Li J. Resequencing of 31 wild and cultivated soybean genomes identifies patterns of genetic diversity and selection. Nature genetics. 2010; 42(12):1053-9.DOI 10.1038/ng.715

[118] Habib M, Awan FS, Sadia B, Zia MA. Genome-Wide Association Mapping for Stripe Rust Resistance in Pakistani Spring Wheat Genotypes. Plants. 2020; 9(9):1056.DOI 10.3390/plants9091056

[119] Sasaki T. The map-based sequence of the rice genome. Nature. 2005 Aug; 436(7052):793-800.10. DOI 1038/nature03895

[120] Feltus FA, Wan J, Schulze SR, Estill JC, Jiang N, Paterson AH. An SNP resource for rice genetics and breeding based on subspecies indica and japonica genome alignments. Genome Research. 2004;**14**(9):1812-1819. DOI: 10.1101/gr.2479404

[121] Guo LB, Ye GY. Use of major quantitative trait loci to improve grain yield of rice. Rice Science. 2014;**21**(2):65-82. DOI: 10.1016/S1672-6308(13)60174-2

[122] Li JY, Wang J, Zeigler RS. The 3,000 rice genomes project: new opportunities and challenges for future rice research. Gigascience. 2014; 3(1):2047-217X. DOI 10.1186/2047-217X-3-8

[123] Alexandrov N, Tai S, Wang W, Mansueto L, Palis K, Fuentes RR, et al. SNP-seek database of SNPs derived from 3000 rice genomes. Nucleic acids research. 2015;**43**(D1):D1023-D1027. DOI: 10.1093/nar/gku1039

Chapter 3

Adaptive Evolution and Addressing the Relevance for Genetic Improvement of Sago Palm Commodity

Barahima Abbas

Abstract

Adaptive evolution implies evolutionary shifts within an organism which make it suitable and adaptable for its environment. Genetic resources of sago palm (*Metroxylon sagu* Rottb.) populations in Indonesia were explicated as follows: (1) Characters of sago palm in Indonesia were shown varied based on cpDNA markers and large variation based on RAPD markers. (2) Variation of starch production of sago palm correlated with *Wx* genes variation, (3) Distances barrier and geographies isolation in line of sago palm dispersions in Indonesia (4) Characteristics of genetic were observed does not related with vernacular names those were given by local people (5) Papua islands, Indonesia territorial is proposed the center of sago palm diversities, (6) Papua islands, Sulawesi islands and Kalimantan islands will be the provenance of the diversities (7) Genetic improvement of sago palm might enhanced using molecular marker that link to interesting genes by developing marker-assisted breeding.

Keywords: breeding, DNA, genetic, marker, *Metroxylon sagu*

1. Introduction

Information on adaptive evolution and genetic diversity of an organism are very important in supporting genetic improvement and germplasm conservation. Adaptive evolution implies evolutionary shifts within an organism which make it suitable for its environment. The improvements lead to improved chances of survival and reproduction. In order to conserve germplasm of an organism, information on genetic diversity is needed so that it can capture germplasm as a whole and efficiently in the implementation of germplasm conservation activities. In addition, information on the diversity of organisms needs to be documented to maintain information on the wealth and existence of certain types of an organism, including sago palm.

Several markers that can be used for accessing the diversity of an organism are morphology, protein, and DNA marker. Morphological and protein markers are not sufficiently used as indicators for measuring genetic characteristic because they are heavily affected by the surrounding factors. One of the markers that is not influenced by the surrounding factors is a molecular marker. Thereby, in expressing adaptive evolution and genetic characteristics, it is necessary to be based on molecular markers. Disclosure of the genetic characteristics of organism such as plant in Indonesia will be better focused on molecular-based markers.

Several DNA markers that can be used for accessing adaptive evolution of an organism are: Simple Sequence Repeat (SSR) in the nuclear genome and chloroplast genome (cpSSR), Random Amplified Polymorphism DNA (RAPD), functional gene such as Waxy gene in sago palm, 5S, Restriction Fragment Length Polymorphism (RFLP), and Amplified Fragment Length Polymorphism (AFLP), chloroplast DNA (cpDNA) such as *mat*K gene, and mitochondrial DNA (mtDNA) such *nad* gene. These molecular markers are widely used as markers to express adaptive evolution of plant.

SSR markers have been shown to have high polymorphisms in soybean and in apples [1–4], thereby, can be used for revealing the adaptive evolution of an organism. SSR is composed of 1–6 base pairs (bp) of repeated DNA sequences with varying amounts [5]. The polymorphic fragments (alleles) are produced from variations in the length of the SSR repeats which can be separated by electrophoresis to display the genetic profile of the genome and the organelle genome. SSR alleles are codominant monogenic inherited and can be distinguished between homozygous and heterozygous in segregated populations [1].

The advantages of SSR DNA markers or microsatellite markers in genome analysis are that SSR sequences are found in many eukaryotic genomes, high diversity, stable inheritance, co-dominant markers and high accuracy detection [6]. The RAPD marker is a technique that is widely used for genetic characterization because the RAPD technique is simpler than other techniques. Molecular markers related to the expression of certain genes are interesting molecular markers because it can be seen the variation of genes encoding certain characters, making it easier to trace genes that have specific expressions and are desired for the improvement of certain gene of organisms.

The *Wx* gene molecular marker is a marker related to the starch biosynthesis process and amplifies the plant DNA sequences that linked to the starch formation. The Waxy (Wx) gene in cereals and *amf* in potato is called isoform gene, Granule-bound starch synthase I (GBSS I) that it encodes starch synthesis [7, 8]. Furthermore, starch synthesis process is regulated by one of the key genes, those the *Wx* gene [9]. Starch from rice plants consists of amylopectin and amylose [10]. Furthermore, it was stated that the *Wx* gene regulates the level of amylose content in starch-producing plants such as wheat and rice [10–12]. The motive structure of the *Wx* gene was reported that it has a very conservative sequence [8] so it fulfills the requirements to be used as a marker. The *Wx* gene marker have been used in various types of crops, i.e. rice [13], barley [9], wheat [14, 15], and sago palm [16, 17].

Large numbers of insertions and deletions in the genome can be detected using agaros gel separation techniques. A technique that is more suitable for small changes in DNA sequences, such as mutations or small deletions or insertions, is fragment analysis using sequencer tools. The technique can detect a change in the size of one base in a DNA fragment. The use of a separation technique that is able to distinguish the differences of one base pair makes it possible to detect the genetic diversity of sago palm that occur at the individual and population. The estimation of adaptive evolution that occurs over a long period of time (hundreds to thousands of years) can be determined based on the chloroplast Simple Sequence Repeat (cpSSR) marker and barcode *mat*K gene in the cpDNA genome. The barcode *mat*K gene was commonly use in the vascular plant, such as Dipterocarpaceae [18], Arecaceae [19] and in the species of sago palm also [20]. The variation that occurs in a relatively short period of time can be determined based on RAPD markers and other markers used to investigate the nucleus genome.

2. Adaptive evolution of sago palm

Diversity is a reflection adaptive evolution in an organism. Variations within a population and inter species that are affected by the occurrence of adaptive evolution. Adaptive evolution of sago palm can be measured by using various markers. The characteristics of sago palm in Indonesia were shown widely varies in morphological phenotypic. It was reported that around Sentani, Jayapura there are 15 varieties [21]. These varieties show variation in a broad sense, not only in morphological characters, but also in their adaptation to the environment (tolerant to fire and waterlogging). Furthermore, the variation of sago palm in Papua is very large based on morphological phenotypic, there are 96 varieties based on vernacular name [22]. The variation base on morphological phenotypic may differences from another population and location because morphological characters are strongly influenced by environmental factors. Observing the variation of sago palm need a marker that are not influenced by the environment so that they can reflect the actual state of plant variation. Markers developed in a wide variety of organisms including plants, namely chloroplast genome molecular markers (cpDNA) and nuclear genome molecular markers (RAPD, Wx gene expression, and others).

The cpDNA molecular marker is a very conservative molecular marker, so it is very suitable to be used to estimate long-term adaptive evolution for a particular organism. The cpDNA locus mutation rates was estimated between 3.2 x 10–5 and 7.9 x 10–5 [23]. Apart from this, cpDNA sequences are conservative in comparison to nuclear genome because they do not undergo recombination in the genome and uniparental inherited [24, 25]. Based on the information found in the chloroplast genome, it is a difference that occurred hundreds or thousands of years ago.

The cpDNA markers were developed in plants showed that the cpDNA of sago palm varied, the total 10 haplotypes were found throughout Indonesia territorials [26]. Seven haplotypes were found on the island of Papua and three haplotypes were found apart from the island of Papua and two haplotypes were found on several islands (sharing haplotypes). Based on highly conservative cpDNA criteria, the variations in cpDNA detection were reflect conditions hundreds or thousands of years ago. It is hypothetically that gene flow of sago palm since ancient times moving from one island to another in various ways. It was found that only two haplotypes experienced displacement. This phenomenon was corresponded of *Pinus silvestris* L. and *Abies alba* Mill referred to as the refugee population [27, 28].

Base on the largest number of haplotypes were found on several islands where sago samples were taken, the island of Papua is the center of sago diversity because the island of Papua has the highest number of cpDNA haplotypes. Large amount of diversity is found in natural populations [29]. Based on this statement, it can be said that the sago palm in Papua is a natural population (not refers to a migrant population). When talking about the source of diversity, the islands of Papua, Sulawesi and Kalimantan are the sources of diversity of sago palm because it has a specific haplotype. Large number of haplotypes reflects the high variation or diversity in a population [28] and differences in cpDNA haplotypes in each population reflect differences in genetic entities (sources of variation) [29].

Based on the developed molecular markers of the chloroplast genome (cpDNA) and nucleus genomes, it was revealed that individuals with different local names within and between populations were generally not different. This indicates that the environmental influence on the appearance of the morphological phenotype is very large because the local name given by the local community is based on morphological phenotypic and local language. In Papua alone, there are a lot of regional languages which make the local names for the sago palm too many. People in Jayapura (West, Central, and East Sentani) give local names for one type of sago palm which differs from one another [30]. If the grouping and naming of sago palm varieties is based on local names, there will be a very large number of vernacular names comparing from the real thing. It was documented that in Papua there are 96 vernacular name of sago palm [31]. Furthermore, the farmers indicated that there are 21 varieties in Sentani and Scientist only recognized 15 varieties out of 21 varieties based on morphological phenotypic [21]. Based on this information, it reflects confusion and there is an overlap in the naming of varieties, which makes the classification and number of varieties recorded larger than the real thing. Cases like these are make molecular markers play an important role for clarification as well as correction of varieties number.

Molecular markers of the chloroplast genome and nucleus genome developed on sago palm detected that sago palm in various islands in Indonesia experienced high diversities as seen from the varying values of genetic diversity: $\sum$H, HE, S, G, $\hat{H}$, V$\hat{H}$, π, πn, and P. This means that in a population there are individuals who are very different from one another. In general, it can be interpreted that the sago palm scattered in various islands in Indonesia, even though the samples from the island of Java with the Wx gene marker and samples from the islands of Ambon and Java with the nucleus genomic SSR markers are not differentiated. This is probably due to the discriminatory focus of each molecular marker that is different from one another. The Wx gene marker focuses its discrimination on genes encoding the biosynthesis of amylose. If the DNA sequence of the *Wx* gene in the population sample did not vary like the population sample from Bogor, then the amylose content did not vary either. Various *Wx* gene alleles determine the amylose content in starch-producing plants [10, 12, 32].

Based on the codominant molecular markers (*Wx* genes and nucleus genomic SSR) used, it shows that the level of heterozygosity of sago palm in various populations in Indonesia varies in terms of the ratio of heterozygous and homozygous values. Based on the *Wx* gene marker, it shows that the samples from the Palopo and Bogor populations are all heterozygous, in contrast the SSR markers of the nuclear genome of the individual samples from the Ambon and Bogor populations are all homozygous. This phenomenon reflects the degree of individual heterozygosity depending on the particular character observed. The heterozygous diversity of the *Wx* gene was relevant to the quality and quantity of plant starch production which also varied. Starch content of sago palm varied as well as the accumulated dry matter [21]. Variations in the *Wx* gene in wheat caused variations in the viscosity of the resulting starch production [15]. The heterozygosity values based on the nucleus genomic SSR markers also varied, although they were not as high as the heterozygosity values of the *Wx* gene markers [16]. SSR markers when designed based on SSR sequences information of the plant genome under study will produce high levels of polymorphism. Previous studies on various types of plants have shown that SSR markers are commonly used to measure adaptive evolution because of their high rates of polymorphism [33–36].

Genetic hierarchy and genetic differentiation based on chloroplast genome markers and nucleus genome indicate that sago samples with cpDNA markers and *Wx* genes differentiate at individual and population levels [16, 19, 26, 37]. Furthermore, samples with RAPD markers experience differentiation at the individual and population levels [16, 26]. The levels of genetic hierarchy observed were individual, population, and island levels [38]. On the other hand, the SSR marker of the nucleus genome was only a sample between populations from the island of Papua which experienced differentiation. This difference is strongly influnced by the nature and the degree of polymorphism of the genetic markers used. The conservative genetic markers such as *mat*K gene markers and mitochondrial *nad2* gene markers tend to show lower levels of polymorphism and only at lower levels of genetic hierarchy are significantly different [20, 37]. Low levels of polymorphism between populations and did not experience genetic differentiation in Pinaceae using the cpSSR marker, but with the RAPD marker, high polymorphism and genetic differentiation were found [39]. Furthermore, the cpDNA characters that evolved in the cotton genus were low [40].

Genetic relatedness of the population based on phylogenetic constructs shows that the SSR molecular marker of the nucleus genome divides the sample into two groups, the cpDNA and RAPD molecular markers divide the samples into three groups, and the *Wx* gene molecular marker divides the sample into four groups [17]. The variations that occur may be due to the different nature and focus of discrimination for each molecular marker used. This case is something that is often encountered in various kinds of molecular markers. Previous studies have shown that different molecular markers infer variability, genetic relatedness, and adaptive evolution of individual or population variations [40–42]. Furthermore, the genetic variation in *Pseudotsuga menziesii* (Mirb.) Franco. using univarentally inherited (cpSSR) and biparental inherited (isozyme and RAPD) molecular markers concluded that the level of polymorphism and differentiation of cpSSR markers was lower than that of isozyme and RAPD markers [39].

Based on molecular markers of cpDNA, RAPD, Wx genes, SSR nucleus genome, cpDNA *mat*K gene, and mitochondrial *nad*2 gene, it shows that sago palm in Indonesia are diverse [17, 19, 37]. The relevance of genetic diversity generated by molecular markers of the chloroplast genome and nucleus genome with the morphological diversity that has been revealed by sago plant researchers is that they both reveal that sago palm in Indonesia are diverse, but the level of diversity based on genetic markers is lower than that based on morphological markers [43]. The variation of sago palm in Papua is very large based on morphological phenotypics, namely that in total there are 96 varieties found from eight locations (Salawati, Waropen, Sentani, Kaureh, Wasior, Inanwatan, Onggari, and Windesi) in Papua and west Papua Province [22]. It was reported three varieties of sago palm in Kendari, Southeast Sulawesi [21]. Furthermore, it was documented 11 varieties of sago palm in Southeast Sulawesi, North Sulawesi and North Ambon based on morphological characteristics [44]. Genetic diversity based on the molecular markers that have been disclosed classifies sago palm in Indonesia from two to four groups. It was reported that sago palm is divided into two clusters and two sub-clusters [45]. The Morphological performance of Sago palm forest is shown on **Figure 1** and the morphological performance at the russet growth is shown on **Figure 2**.

Based on the molecular markers that have been used on sago palms, nothing has been associated with the morphological characters. The same thing was also that spineless and spiny of sago palm was not related to genetic distance based on RAPD markers [45]. It is believed that spine and spineless in sago palm is controlled

Figure 1.
Morphological performance of sago palm forest.

Figure 2.
Morphological performance of sago palm in the russet growth. Spineless with purple color of the young leaf (A), spineless with green color of the young leaf (B), spiny with purple color of the young leaf (C), and spiny with green color of the young leaf (D).

by certain genes, so that there are certain nucleotide sequences in the sago palm genome that undergo transcription and translation processes in the spine formation process. Molecular markers encoding the characteristics of sago palm can be designed if desired by reverse transcription of sequences encoding protein synthesis for spiny formation in sago.

Genetic relatedness based on the phylogenetic constructs of each tested molecular marker shows that the distribution of the level of sample similarity according to the size of genetic distance is not limited by location and geographical isolation because samples from one island to others islands blending with each other [17]. The blending of Stylosanthes sp. obsessions from various regions in the dendrogram construction indicates that the obsessions have geographic distribution [46]. It was reported that, if there is distance and geographic isolation in the long term, the population from one region to another will experience differentiation or adaptive evolution as happened in Brassicaceae [47]. Furthermore, Scientist documented that population differentiation of O. rifipogon affected by distance or geographic isolation [36].

3. Genetic assessment of sago palm

3.1 Random amplified polymorphism DNA (RAPD) marker

RAPD polymorphisms amplified on the PCR machine produced polymorphic fragments and the number of genotypes of each population. RAPD polymorphisms and high number of genotypes are a reflection of plant genetic diversity and adaptive evolution based on RAPD markers [38]. This result is in line with the diversities of sago palm revealed by using RAPD markers on several samples from Indonesia and Malaysia [48].

Population genetic diversity shows that the population samples from Papua have the highest of polymorphic sites number (S), the moderate of pairwise differences values (π), and the highest percentage of haplotype polymorphic compared to other populations from several islands in Indonesia [38]. Genotype diversity equal to one means that no identical genotype is found in a population sample. The value of Ĥ of individual samples at the island level all shows number of one, which means that one sample of individuals with another sample of individuals differs from one another based on the RAPD markers. Sago progenies obtained from semi-cultivated sago populations showed genetic differences among the progeny tested [43]. The varying values of S, π, and Ĥ indicate the genetic variation values of sago palm population in Indonesia. In the previous studies of sago palm by using RAPD markers showed that the sago palm diversities among individual was recorded also high [48] and other scientist reported 15 varieties of sago palm based on morphological characters in around Sentani lake [21].

The genetic hierarchy was estimated based on Analyses Molecular of Variance (AMOVA) calculations. The AMOVA calculation value shows that 89.35% of the total variety of samples is contributed by individuals with very significantly different with probability (P) values, 6.58% and 8.4% variance is contributed among populations [17]. The rates of diversities and adaptive evolution were detected in sago palm, those related to the genetic diversity of *M. sativa* L. by using RAPDs marker [47] as well as *Cynara scolymus* L. [49]. The statistical test method used to reveal the differentiation that occurs at the population and island level is also found to be used to reveal the differentiation that occurs at the population level in various types of plants, such as in *M. sativa* L [47], in *Acacia radiana* [50], and in *Primula elatior* (L.) Oxlip [51].

Genetic relatedness among individual shows that the sago palm were classified into three groups based on the dendrogram construction. Group I include sago palm from all the populations as well as group III, while Group II includes sago palm from Jayapura, Serui, Manokwari and Ambon. This is related to the grouping of sago palm that it was reported in the previously study and divided sago palm from Indonesia and Malaysia into two groups and subgroups based on RAPD markers [48]. Previous sago genetic studies that focused on the Indonesian archipelago showed that sample individuals were divided into four groups based on RAPD markers [52]. Grouping of individuals in a dendrogram is largely determined by the genetic distance used, the method of grouping, and the desired bootstrap coefficient or rate. The differences between the groupings based on the cpDNA markers and the RAPD markers observed in previous studies are common in genetic relatedness studies [24, 39, 41].

Genetic relatedness among population shows a clustering pattern similar among individual. Genetic relatedness based on the dendrogram sample construction at the island level shows that the samples from the island of Papua are more closely related to the samples from Sumatra and Kalimantan, the samples from the island of Sulawesi are closely related to the samples from Ambon, and the samples from the island of Java are separate from other islands based on the RAPD marker [38]. Here there is something interesting to observe because the sample at the island level forms a group together with samples from other islands that are far away, such as the sample from the island of Papua which forms a group together with the sample from the island of Sumatra. When examined from the migration side, it is possible that individual sago palm from Papua population have mingled with sago palm from the island of Sumatra. This phenomenon is possible because the molecular markers (RAPD) used are not as conservative as the cpDNA molecular markers that are uni-parental inherited [24, 25]. The RAPD marker is a nucleus genomic molecular marker associated with the DNA recombination process and is biparental inherited [39] so that the RAPD marker is a molecular marker that has a relatively short conservative time (one generation) compared to the cpDNA molecular marker. Previous studies suggest that higher variation is found using nucleus genomic markers rather than cpDNA markers [39–41, 53].

3.2 Gene encoding starch biosynthesis (waxy genes)

Polymorphisms of *Wx* gene markers that were found of 8 polymorphisms alleles and 14 genotypes of the *Wx* genes [16]. The polymorphism detected in sago palm was in line of the polymorphism in *Triticum aestivum* L. by using the *Wx* (SunI) gene markers [14]. The number of alleles and genotypes of sago palm at the level populations and islands varies as well as their frequency [17]. The Wx gene variations found in sago are similar with the *Wx* gene variations on wheat [15]. Furthermore, Scientist were reported a high Wx gene polymorphism in barley [9] and in rice [13]. This phenomenon indicates that the source of the *Wx* genes diversity is the Papua islands Papua and Sulawesi islands because these islands are found genotypes that are not found on other islands [16]. If the center of diversity is the object of attention, then the island of Papua is the center of diversity of the *Wx* genes because the most genotypes of *Wx* genes are found on the island of Papua [17].

The genetic diversities of *Wx* genes that was observed to the sago palm from various islands were shown varied. The genetic diversities calculation results showed varying values except for samples from Jawa [17]. The sago palm variations were detected, those a reflection of sago palm variations that it occurs in the several islands in Indonesia [16]. The *Wx* gene is one of many genes that it is regulated biosynthesis process for resulting starch of plants, including sago palm. If the *Wx* gene has high variations that will be resulting various quantity and quality of starch. In the previous studies were reported the quantities of starch accumulation of sago palm range from 28 to 710 kg trunk^{-1} [45] and starch accumulation of sago palm trunk^{-1} will be depend on the varieties [21]. The *Wx* gene was one of the genes that influenced of starch synthesis in rice endosperm [54]. Two alleles of the Wx gene that is Wxa and Wxb gene were reported regulating to increase *Wx* protein and amylose content [10]. *Wx* allelic pulp in wheat showed a significantly different reduction in amylose content [12] and recombinant inbred line (RIL) of wheat that has integrated three *Wx* genes in their genome was reported resulting high quality starch than wheat RIL which did not contain the three *Wx* genes [32].

The heterozygote values of sago palm in the populations that was observed were shown variation from 0.52–1.00 with a low standard deviation of 0.0000–0.0014 [17]. The heterozygosity variations were indicated variations in the *Wx* gene in the genome of sago palm. The key gene that influences starch synthesis in rice endosperm is the *Wx* gene [54]. Variation of the *Wx* gene causes a variation in the viscosity of starch production in wheat (Boggini et al. 2001). The *Wx*a and

*Wx*b alleles were found to regulate quantitative levels of *Wx* protein as well as amylose content [10].

The genetic hierarchy calculation using AMOVA shows that individuals and populations was estimated significantly different [17]. The differentiation values based on the chi-square test at the population and island level were found sago palm differentiated that occurs at the population and island level [17]. The detected variance is an indicator that the *Wx* gene varies both at the individual level and at the population level. Previous studies of sago palm using different markers also showed that sago palm varied both in terms of quantity and quality of production [21, 45]. The allelic levels of *Wx* genes and their interactions in starch-producing plants were reported increasing quality and quantity of starch production [10, 11, 32], and [55]. It is predicted that the *Wx* gene variation in sago palm is one of the genes that determines the variation in the quantity and quality of sago starch yields [16]. The sample diversity at the inter-island population was not significantly different based on the AMOVA value as was the sample at the inter-island population. This phenomenon indicates that the variation of the *Wx* gene in sago palm is more caused by variations at the individual and populations, not due to the isolation of different distances and geographic differences due to the low FCT value of 0.06044 [17].

Genetic relationship among individual shows that sago palm are grouped into four groups based on dendrogram construction [17]. The division into four groups was strengthened by the MDS test which showed the sample was distributed in four quadrants. The data illustrates that certain individuals are not grouped based on population origin but rather mixed with each other with different population origins and different local names [16]. This description implies that local names are not appropriate when used as a reference for determining the number of species or varieties of sago palm without the support of other data such as molecular data. In the vicinity of Sentani Lake, the local community revealed that there were 21 types of sago palm based on morphology and scientist found only 15 species based on the same marker [21].

Genetic relationship of sago palm in the population level shows that sago palm from the populations of Jayapura, Serui, Sorong, and Pontianak are closely related and form group I, samples from populations from Manokwari, Palopo, and Selat Panjang cluster to form group II, then groups III and IV only formed from one population. The grouping of the population into four groups is also strengthened by the MDS test which shows the population sample is distributed in four quadrants [16]. Previous studies have discussed the genetic relationships of populations using various markers [39, 47, 49, 51]. Populations contained in one group are closely related, on the other hand, populations in different groups are not closely related. The differences in a population is thought to be caused by outbreeding so that the population experiences differentiation. Population differentiation can be caused by pollen migration [56]. In general, it can be interpreted that there is a tendency for sago palm in Indonesia to be differentiated inter-island and among island based on the *Wx* gene marker. Differentiation can be caused by evolutionary processes, georaphic isolation, distance isolation, genetic drift and gene flow. Population differentiation is caused by evolution, natural selection, migration, and genetic drift [57] and the differentiation of Cruciferae due to gene flow [58].

3.3 Chloroplast DNA (cpDNA) marker

Based on cpDNA markers, various polymorphic and haplotypic alleles were found in sago palm. Studies related to the use of the NTCP21 and NTCP22 markers in potato have also demonstrated allele polymorphisms in potato [59]. Locus rpl1671, NTCP21, and NTCP22 on sago were detected in three haplotypes out of 10 haplotypes which were specific haplotypes in populations from Jayapura and one specific haplotype each for populations from Serui, Palopo, and Pontianak [26]. The specific haplotype phenomenon is also found in several types of plants i.e. Cunninghamia spp. [60], *Pinus sylvestris* L. [27], and *Alyssum* spp. [29]. The specific haplotypes were found in a population, those indicated the source of diversities in a population. The specific haplotypes of sago palm were found in the populations of Papua, Sulawesi, and Kalimantan indicated the provenance of the diversities, while the most haplotypes of sago palm diverse is the population from Jayapura then followed by the sago palm population from Serui [17]. The large number of haplotypes reflects the high variation in a population in line of the *Abies alba* Mill population [28]. The differences in chloroplast haplotypes in each population reflect differences in genetic entities or sources of variation [29]. The number of haplotypes that were found to be present together in each population is an indication that genetic similarities among individual in a population. It is hypothetically that the sago palm migration by carrying of people. Four haplotypes of 10 haplotypes of sago palm were found in to two or more populations, which means that only four haplotypes were found migration through various kinds of intermediaries. The same thing was also found in *P. sylvestris* L. and *A. alba* Mill. referred to as the refugial population [27, 28].

Population genetic diversity shows that the population from Papua has the highest number of haplotypes ($\sum$H), the number of polymorphic sites (S), and the highest percentage of haplotype polymorphics compared to other populations. A value (HE) equal to one means that no haplotype numbers are the same in individual samples in a population (single haplotype individuals) as happened in the population from Bogor. This is similar with individual haplotype on *P. sylvestris* L [27]. Previous studies on sago palm using RAPD markers showed that sago plant diversity at the individual level was also high [48, 52].

The genetic hierarchy based on cpDNA was estimated by using analysis of molecular variance (AMOVA) was calculated of differentiation level of population samples at the inter-island level (−3.88% and FCT = −0.03884), between populations within islands (8.49% and FSC = 0.08177), and Papua and others (5.05% and FCT 0.05054) which is low with the probability value not significantly different. High percentage values of variance were observed at the level among individuals (95.39% and FST = 0.04610) and between populations (5.91% and FST = 0.05914) with significantly different probability values [26]. The same thing was also found in *P. sylvestris* L., namely the percentage value of variance between populations (3.24%) with a significantly different probability [27]. The negative value observed at the inter-island level indicates that the sample island level does not contribute to the total measured variance. This phenomenon resembles the tetraploid alfalfa population [47]. Negative correlation coefficients have a biological significance in that the samples at the inter-island level are more closely related than those at the island level [61]. Based on this, it indicates that island or geographic differences do not cause variations in the chloroplast genome, even though the distance between one island and another is far (hundreds to thousands of kilometers). The variation between individuals and between populations contributed 95.39% and 5.91% to the total variety and was significantly different [18]. The results observed were similar with Abies species that was only a small variance value between populations (6.10%), high proportion of variance within the population or between individuals (74.66%) [62]. A low trend of genetic variability between populations is also found in Pinaceae [39] and species other than pine [63].

Genetic differentiation based on the Fst value shows that among the populations being compared, only the population from Jayapura is significantly different from the population from Palopo and Pontianak [26]. These populations based on cpDNA markers each have a genetic entity, which means that the diversity that occurs in this population has appeared separately since ancient times (thousands of years ago). The genetic differentiation of samples at the population level based on the X^2 test shows that the population originating from Jayapura is different from the population originating from Serui, Manokwari, Sorong, Palopo, Pontianak, and Selat Panjang and the population originating from Serui is different from the population originating from Pontianak but not different from other populations [17]. Based on the X^2 test at the population level, it indicates that the source of sago plant diversity at the population level is the population from Jayapura, Serui, and Pontianak. Genetic differentiation of samples at the island level based on the X^2 test shows that samples originating from the island of Papua are different from samples from the islands of Sulawesi and Kalimantan [26]. This is in line with the specific haplotypes found on the three islands. For this reason, it is suggested that the sources of sago palm diversity based on the samples tested are the islands of Papua, Sulawesi and Kalimantan. This data is also consistent with the grouping of sago palm samples through phylogenetic construction at the island level which divides the samples into three groups, namely the Papua group, the Sulawesi group and the Kalimantan group. This indicates that the source of the diversity of sago palm in Indonesia, apart from being on the island of Papua, is also found in other islands, namely Sulawesi and Kalimantan [26].

The genetic relationship of the samples at the individual level shows that the samples are classified into three groups based on the phylogenetic construction. Sago palm relationship studies previously show that sago palm originating from the Malay Archipelago and several samples of sago from Indonesia clustered into two groups and two sub-groups based on the RAPD markers [48]. The sago relationship study focused on the Indonesian archipelago, but with a larger number of samples, showed that the sample individuals were divided into four groups based on RAPD markers [52]. The discrepancy in the division of the number of groups (groups) between the groupings based on cpDNA markers and RAPD markers that was observed in previous studies is something that is often found in studies of genetic relationship using molecular markers. Previous genetic related studies which showed that different molecular markers led to different groupings of certain plants by using cpDNA, RAPD and isozyme markers in Pseudotsuga spp. (Pinaceae) [39], cpDNA and inter-SSR (ISSR) markers in the nucleus genome on *Brassica oleraceae* L. plants using [41], and using cpDNA and mitochondrial DNA on apple plants [24].

Based on cpDNA, the sources of sago diversity in Indonesia are predicted to come from three islands, namely Papua, Sulawesi and Kalimantan. It is suspected that from these three islands, individual sago palm experienced migration in line with migration and population mobilization in Indonesia that had occurred hundreds of years ago. This assumption is reinforced by haplotype data, phylogenetic analysis, and genetic hierarchies which show that samples at the inter-population level and between individuals are significantly different, which means that there are one or more different individuals or populations. Although the source of diversity is found in three islands, if the number of haplotypes is the size of the center of diversity, then the island of Papua is the center of diversity of sago palm in Indonesia because that island is found in the largest number of haplotypes compared to other islands. Apart from the highest number of haplotypes, on the island of Papua, the wild relatives of sago palm are found. If the data obtained is linked to the incidence of sago distribution in Indonesia, it is strongly suspected that only four haplotypes experienced migration from one population to another, which were then given different local names. The sago population with a specific name for the origin of the Papua region which groups together with populations from other places with other names is also a reflection that in the past these sago populations were only one then experienced joint migration with the migration of people from one island to another or from one population to another. If population migration events have occurred hundreds of years ago and are thought to have caused sago palm to spread from sources of diversity to form new populations or join old populations on islands that are sources of diversity, it is still possible because the measure of similarity is cpDNA, which has very conservative sequences [64], a very low mutation rate of between 3.2 x 10^{-5} and 7.9 x 10^{-5} [23], is not recombinant [24, 64] and are inherited uniparental [25, 39].

3.4 Genetic improvement by using marker-assisted breeding (MAB)

The development of genetics and technology molecular has facilitated our understanding of the genetics underlying the traits sought by plant breeding. The development of molecular markers allows plant breeding to develop faster and more advanced in producing superior organisms. The benefits of DNA markers are for germplasm characterization, selection of desired traits from genomic regions involved in the expression of traits of interest, and single gene transfer. The application of selection using efficient and effective markers to improve polygenic properties certainly requires new technology. Genetic improvement of sago palm may use transformation agrobacterium-mediated and particle bombardment. Successfully introgression *bar* and *gus* gene into sago palm genome [65]. The embryogenic callus was the most appropriate transformation material compared to the via callus, the embryoid stage and the shoots initiated by using *Agrobacterium*-mediated. The transformation of the gene gun demonstrated greater efficiency of transformation than those transformed with Agrobacterium when targets were bombarded once or twice with 280 psi helium pressure at a distance of 6 to 8 cm [65]. Therefore, economics interesting genes may introgression into sago palm genome in the future.

The purpose of MAB is to enhance certain characteristics in plant or animal breeding programs. Strategy for rapidly integrating a targeted gene into a wheat genotype in only two generations and restoring 97% or more of the recurrent genotype of the parent by using MAB [66]. Deconvolution of ancestry offers a first step towards selection of suitable admixture profiles at the seed or seedling level, which will support marker-assisted breeding aimed at introgressing wild Vitis species while maintaining the desirable characteristics of elite *V. vinifera* cultivars [67]. Marker-assisted backcrossing can be used in plant breeding to integrate traits into elite cultivars while minimizing the transfer of unwanted alleles from the donor genome [68]. This method includes the selection of foreground as well as context. Foreground selection refers to offspring screening and selection based on the presence or absence of a particular allele associated with a feature of interest. Conversely, selection of offspring on the basis of genomic ancestry estimates is the history selection.

The MAB needs to be developed to accelerate and increase the success of the breeders to produce superior seeds. Recently, breeders were developed abundant MAB linked with specific characters of plant genetics. Simple sequence repeat (SSR), namely Md-PG1$_{SSR}$10kd tightly linked with fruits texture of apple [69] and microsatellites RM5926 and AP5659–5 were developed for detecting rice blast resistance genes, those markers tightly linked with *Pi*-1 and *Piz*-5 genes respectively [70]. Marker-Assisted Introgression of b-carotene hydroxylase was developed for detecting b-Carotene Rich in maize hybrid [71]. Furthermore, Muthusamy et al. (2014) stated that B-carotene concentration among crtRB1-introgressed inbred ranged from 8.6 to 17.5 mg/g - a maximum increase up to 12.6 times over recurrent parent. In comparison to 2.6 mg/g in the original hybrid, the reconstituted hybrids formed from improved parental inbred also showed enhanced kernel *b-carotene* as high as 21.7 mg/g [71]. This study may use as a model for increasing quality starch that resulting of sago palm and other plant in the current time and in future time.

4. Conclusions

Genetic resources of sago palm in Indonesia were explicated as follows: (1) Characters of sago palm in Indonesia were shown varied based on cpDNA markers and large variation based on RAPD markers. (2) Variation of starch production of sago palm correlated with *Wx* genes variation, (3) Distances barrier and geographies isolation in line of sago palm dispersions in Indonesia (4) Characteristics of genetic were observed does not related with vernacular names those were given by local people (5) Papua islands, Indonesia territorial is proposed the center of sago palm diversities, (6) Papua islands, Sulawesi islands and Kalimantan islands will be the provenance of the diversities (7) Genetic improvement of sago palm might enhanced using molecular marker that link to interesting genes by developing marker-assisted breeding.

Acknowledgements

This work is supported by Research Development Project No.198/SP2H/LT/DRPM/2020 from the Directorate General of Strengthen Research and Community Service (DRPM-DIKTI), Republic of Indonesia.

Author details

Barahima Abbas
Faculty of Agriculture, Post Graduate Program, University of Papua (UNIPA), Manokwari, Indonesia

*Address all correspondence to: b.abbas@unipa.ac.id

References

[1] Narvel JM, Fehr WR, Chu WC, et al (2000) Simple sequence repeat diversity among soybean plant introduction and elite genotype. Crop Sci. 40: 1452-1458.

[2] Akkaya MS, Bhagwat AA, and Cregan PB (1992) Length polymorphism of simple sequence repeat DNA in soybean. Genetics 132:11331-1139.

[3] Powel W, Morgante M, Andre C, Hanafey M, et al (1996) The comparison of RFLP, RAPD, AFLP, and SSR (microsatellite) markers for germplasm analysis. Molecular Breeding 2:225-238.

[4] Diwan N, and Cregan PB (1997) Automatic sizing of fluorescent-labeled simple sequence repeat (SSR) markers to assay genetic variation in soybean. Theor Appl Genet. 95:723-733.

[5] Tautz D (1993) Notes on definition and nomenclature of tandemly repeat DNA sequence. In S.D. Pena et al. (eds.) DNA Fingerprinting. CRC Press. USA. Pp. 1-20.

[6] Mba C, Fregene M, and Tohme J. 2001. Simple sequence repeat markers as efficient and robust molecular tools in cassava varietal development. Biotechnology Research Unit, International Center for Tropical Agriculture (CIAT), KM17 Recta Cali-Palmira, AA 6713 Cali, Colombia. m.fregene @cgiar.org.

[7] Edwards A, Marshall J, Denyer K, et al (1996) Evidence that a 77-kilo Dalton protein from the starch of pea embryos is an isoform of starch synthase that is both soluble and granule bound. Plant Physiol. 112:89-97.

[8] Gamer RJM, Weil CF, Kellogg EA (1998) Granule-bound starch synthase: structure, function, and phylogenetic utility. Mol biol Evol 15(12):1658-1673.

[9] Domon E, Yanagisawa T, Saito A, Takeda K (2004) Single nucleotide polymorphism genotyping of the barley Waxy gene by polymerase chain reaction with confronting two-pair primers. Plant Breed 123(3):225-228

[10] Wanchana S, Toojinda T, Tragoonrung S, Vanavichit A (2003) Duplicated coding sequence in thewaxy allele of tropical glutinous rice (Oryza sativa L.) Plant Sci 16(6):1193-1199.

[11] Wickramasinghe HAM and Miura H (2003) Gene dosage effect of wheat Wx alleles and their interaction on amylase synthesis in the endosperm. Euphytica 132(3):303-310.

[12] Sharma R, Sison MJ, Rathjen AJ, Jenner CF (2002) The null allele at the waxy locus in durum wheat affect pasta cooking quality. J Cereal Sci 35(3):287-297.

[13] Bao JS, Corke H, Sun M (2002) Microsatellites in starch-synthesizing genes in relation to starch physicochemical properties in waxy rice (Oriza sativa L.) Theor Appl Genet 105:898-905

[14] Shariflou MR, Sharp PJ (1999) A polymorphic microsatellite in the 3′ end waxy genes of wheat, Triticum aestivum. Plant Breed 118:275-277.

[15] Boggini G, Cttaneo M, Paganoni P, Vaccino P (2001) Genetic variation for waxy proteins and starch properties in Italian wheat germplasm. Euphytica 119:111-114

[16] Abbas B, Ehara H (2012) Assessment genetic variation and relationship of sago palm (Metroxylon sagu Rottb.) in Indonesia based on specifi expression gene (Wx genes) markers. African J Plant Sci 6(12):314-320

[17] Abbas B (2018) Sago palm genetic resource diversity in Indonesia. In: Ehara H, Toyoda Y, Johnson D (eds.).

Sago Palm: Multiple Contributions to Food Security and Sustainable Livelihoods. Springer, Singapore. DOI:10.1007/978-981-10-5269-95.

[18] Harnelly E, Thomy Z, Fathiya N (2018) Phylogenetic analysis of Dipterocarpaceae in Ketambe Research Station, Gunung Leuser National Park (Sumatra, Indonesia) based on rbcL and matK genes. Biodiversitas 19: 1074-1080. DOI: 10.13057/biodiv/d190340.

[19] Abbas B, Tjolli I, Munarti (2020). Genetic diversity of sago palm (Metroxylon sagu) accessions based on plastid cpDNA matK gene as DNA barcoding. Biodiversitas 21: 219-225.

[20] Abbas B, Kabes RJ, Mawikere NL, et al (2020). DNA barcode of Metroxylon sagu and other palm species using matK gene. Biodiversitas 21: 4047-4057.

[21] Yamamoto Y, Yoshida T, Miyazaki A et al (2005) Biodiversity and productivity of several sago palm varieties in Indonesia. Eighth International Sago Symposium in Jayapura, Indonesia. Japan Society for the Promotion Science, pp 35-40

[22] Matanubun H, Santoso B, Nauw M, Rochani A, et al (2005) Feasibility study of the natural sago forest for the establishment of the commercial sago palm plantation at Kaureh District, Jayapura, Papua, Indonesia. The Eight International Sago Symposium in Jayapura, Indonesia. Japan Society for the Promotion Science.

[23] Provan J, Soranzo N, Wilson NJ, et al (1999) A low mutation rate for chloroplast microsatellites. Genetics 153:943-947.

[24] Ishikawa S, Kato S, Imakawa S et al (1992) Organelle DNA polymorphism in apple cultivars and rootstocks. Theor Appl Genet:963-967.

[25] Savolainen V, Corbaz R, Moncousin C, et al (1995) Chloroplast DNA variation and parentage analysis in 55 apples. Theor Appl Genet 90:1138-1141.

[26] Abbas B, Renwarin Y, Bintoro MH et al (2010) Genetic diversity of sago palm in Indonesia based on chloroplast DNA (cpDNA) markers. J Biol Div 11(3):112-117

[27] Provan J, Soranzo N, Wilson NJ, et al (1998) Gene-pool variation in Caledonian and European Scots pine (Pinus sylvestris L.) revealed by chloroplast simple-sequence repeats. Proc Biol Sci 265(1407):1697-1705.

[28] Vendramin GG, Degen B, Petit RJ (1999) High level of variation at Abies alba chloroplast microsatellite loci in Europe. Mol Ecol 8:1117-1126

[29] Mengoni A, Gonelli C, Brocchini C, et al (2003) Chloroplast genetic diversity biogeography in the serpentine endemic Ni-hyperaccumulator Alyssum bertolonii. New Phytol 157:349-356.

[30] Renwarin Y, Dedaida HT, Matanubun H, Abbas B (1998) Identification, collection, and evaluation of sago palm cultivars in Irian Jaya for supporting commercial and plantation sago palm in Indonesia. In Indonesian Language. Competitive research Grant Report. 91p.

[31] Matanubun H, Maturbongs L (2005) Sago palm potential, biodiversity and socio-cultural considerations for industrial sago palm development in Papua, Indonesia. Abstracts of The Eight International Sago Symposium in Jayapura, Indonesia. Japan Society for the Promotion Science.

[32] Epstein J, Morris CF, Huber KC (2002) Instrumental texture of white salted noodles prepared from recombinant inbred lines of wheat differing in the three granules bound

starch synthase (waxy) genes. J Cereal Sci 35(1):51-63

[33] Rossetto M, McNally J, Henry RJ (2002) Evaluating the potential of SSR flanking regions for examining taxonomic relationships in the Vitaceae. Theor Appl Genet. 104:61-66.

[34] Anthony F, Combes MC, Astorga C, et al (2002) The origin of cultivated Coffea arabica L. varieties revealed by AFLP and SSR markers. Theor Appl Genet. 104:894-900.

[35] Grassi F, Labra M, Imazio S, et al (2003) Evidence of a secondary grapevine domestication centre detected by SSR analysis. Theor Appl Genet. 107:1315-1320.

[36] Song ZP, Xu X, Wang B. Chen JK, Lu CB (2003) Genetic diversity in the northern most Oriza rufipogon population estimated by SSR markers. Ther Appl Genet. 107:1492-1499.

[37] Abbas B, Tjolli I, Dailami M, Munarti (2019) Phylogenetic of sago palm (Metroxylon sagu) and others monocotyledon based on mitochondrial nad2 gene markers. Biodiversitas 20: 2249-2256

[38] Abbas B, Bintoro MH, Sudarsono S, et al (2009) Genetic relationship of sago palm (Metroxylon sagu Rottb.) in Indonesia based on RAPD markers. J Biol Div 10(4):168-174

[39] Viard F, Kassaby YAE, Ritland K (2001) Diversity and genetic structure in populations of Pseudotsuga menziesii (Pinaceae) at chloroplast microsatellite loci. Genome 44:336-344

[40] Cronn RC, Small RL, Haselkorn T, Wendel JF (2002). Rapid diversification of the cotton genus (Gossypium: Malvaceae) revealed by analysis of sixteen nuclear and chloroplast genes. American Journal of Botany 89(4):707-725.

[41] Panda S, Martin JP, Agunagalde I (2003) Chloroplast and nuclear DNA studies in a few members of the Brassica oleracea L. group using PCR-RFLP and ISSR-PCR markers: a population genetic analysis. Theor Appl Genet. 106:1122-1128

[42] Russel JR, Booth A, Fuller JD, Baum M, et al (2003) Patterns of polymorphism detected in the chloroplast and nuclear genomes of barley landrases sampled from Syria Jordan. Theor appl Genet. 107:413-421.

[43] Riyanto R, Widodo I, Abbas B (2018) Morphology, growth and genetic variations of sago palm (Metroxylon sagu) seedlings derived from seeds. Biodiversitas 19: 602-608. DOI: 10.13057/biodiv/d190241.

[44] Ehara H, Susanto S, Mizota C, Hirose S, and Matsuno T. 2000. Sago palm (Metroxylon sagu, Arecaceae) production in the eastern archipelago of Indonmesia : Variation in Morphological characters and pith. Economic Botany 54(2):197-206.

[45] Ehara H, Naito H, Mizota C (2005) Environment factors limiting sago production and genetic variation in Metroxylon sagu Rottb. Eighth International Sago Symposium in Jayapura, Indonesia. Japan Society for the Promotion Science, pp 93-104

[46] Liu CJ (1997) Geographical distribution of genetic variation in Stylosanthes scabra revealed by RAPD analysis. Euphytica 98:21-27.

[47] Mengoni A, Gori A, Bazzcalupo M (2000) Use of RAPD and micro satellite (SSR) variation to assess genetic relationships among populations of tetraploid alfalfa, Medicago sativa. Plant Breeding 119: 311-317

[48] Ehara H, Kosaka S, Shimura N, et al (2003) Relationship between geographical distribution and genetic

distance of sago palm in Malay Archipelago. Sago Palm 11:8-13

[49] Lanteri S, Leo ID, Ledda L, et al (2001) RAPD variation within and among population of globe artichoke cultivar 'Spinoso sardo'. Plant Breeding 120: 243-246.

[50] Shrestha MK, Goldhirsh AG, Ward D (2002) Population genetic structure and the conservation of isolated population of Acacia raddiana in the Negev Desert. Biological Conservation 108:119-127.

[51] Jacquemyn H, Honnay O, Galbusera P, Ruiz IR (2004) Genetic structure of forest herb Primula elatior in a changing landscape. Molecular Ecology. 13:211-219.

[52] Abbas B, Ehara H, Bintoro MH, Sudarsono, et al (2004) Genetic Relationship among Sago Palms in Indonesia Based on RAPD and SSR-cpDNA Markers. Proceeding of Japanese Society Sago Palm 53-58.

[53] Hultquist SJ, Vogel KP, Lee DJ, Arumuganathan K, and Kaeppler S (1996) Chloroplast DNA and nuclear DNA content variations among cultivars of Switchgrass, Panicum virgatum L. Crop Science 36:1049-1052.

[54] Larkin PD, Park WD (2003) Association of waxy gene single nucleotide polymorphisms with starch characteristics in rice (Oriza sativa L.) Mol Breed 12(4):335-339.

[55] Miura H, Araki E, and Tarui S (1999) Amylose synthesis capacity of the Three Wx genes of wheat cv. Chinese Spring. Euphytica 108:91-95.

[56] Latta RG. and Mitton JB. 1997. A comparison of population differentiation across four classes of gene marker in limber pine (Pinus flexilis James). Genetics (146):1153-1163.

[57] Hartl DL and Clark AG. 1989. Principle of Population Genetics, Second Edition. Pub. Sinauer Associates Inc. Sunder land Massachusetts. 682p.

[58] Mayer MS, Soltis PS, and Soltis DE. 1994. The evolution of the Streptanthus glandulosus complex (Cruciferae): Genetic divergence and gene flow in serpentine endemics. American Journal of Botany 81:221-231.

[59] Bryan GJ, McNicol J, Ramsay G, et al (1999) Polymorphic simple sequence repeat markers in chloroplast genomes of solanaceous plants. Theor Appl Genet 99:859-867

[60] Hwang SY, Lin TP, Ma CS, et al (2003) Postglacial population grow.th of Cunninghamia konishii (Cupressaceae) inferred from phylogeographical and mismatch analysis of chloroplast DNA variation. Mol Ecol 12:2689-2695

[61] Schneider S, Roessli D, Excoffier L (2000) Arlequin: A Software for population genetics data analysis. Ver 2.000. Genetics and Biometry Laboratory, Department of Anthropology, University of Geneva.

[62] Parducci L, Szmidt AE, Madaghiele A, Anzidei M, and Vendramin GG. 2001. Genetic variation at chloroplast microsatellites (cpSSRs) in Abies nebrodensis (lojac.) Mattei and three neighboring Abies Sp. Theor appl Genet. 102:733-740.

[63] Austerlitz F, Mariette S, Machon N, Gouyon PH, and Godelle B. 2000. Effect of colonization processes of genetic diversity: differences between annual plants and tree species. Genetics 154:1309-1321.

[64] Provan J, Powel W, Hollingsworth M (2001) Chloroplast microsatellites: new tools for studies in plant ecology and evolution. Trends Ecol Evol 16(3):142-147.

[65] Ibrahim ER, Hossain MA, Roslan HA (2014) Genetic Transformation of Metroxylon sagu (Rottb.) Cultures via Agrobacterium-Mediated and Particle Bombardment. BioMed Research International, http://dx.doi.org/10.1155/2014/ 348140

[66] Randhawa HS, Mutti JS, Kidwell K, Morris CF, Chen X, et al. (2009) Rapid and Targeted Introgression of Genes into Popular Wheat Cultivars Using MarkerAssisted Background Selection. PLoS ONE 4(6): e5752. doi:10.1371/journal.pone.0005752

[67] Sawler J, Reisch B, Aradhya MK, Prins B, Zhong G-Y, et al. (2013) Genomics Assisted Ancestry Deconvolution in Grape. PLoS ONE 8(11): e80791. doi:10.1371/journal.pone.0080791.

[68] Collard BCY, Mackill DJ (2008) Marker-assisted selection: an approach for precision plant breeding in the twenty-first century. Philos Trans R Soc Lond B: Biol Sci 363: 557-572. doi:10.1098/rstb.2007.2170. PubMed: 17715053

[69] Longhi S, Cappellin L, Guerra W, Costa F (2013) Validation of a functional molecular marker suitable for marker-assisted breeding for fruit texture in apple (Malus × domestica Borkh.). Molecular Breeding 32: 841-852.

[70] Gouda PK, Saikumar S, Varma CMK, Nagesh K, et al (2013) Marker-assisted breeding of Pi-1 and Piz-5 genes imparting resistance to rice blast in PRR78, restorer line of Pusa RH-10 Basmati rice hybrid. Plant Breeding 132(1):61-69

[71] Muthusamy V, Hossain F, Thirunavukkarasu N, Choudhary M, Saha S, et al. (2014) Development of b-Carotene Rich Maize Hybrids through Marker-Assisted Introgression of b -carotene hydroxylase Allele. PLoS ONE 9(12): e113583. doi:10.1371/journal. pone. 0113583

Chapter 4

The Sensitiveness of Expected Heterozygosity and Allelic Richness Estimates for Analyzing Population Genetic Diversity

María Eugenia Barrandeguy and María Victoria García

Abstract

Genetic diversity comprises the total of genetic variability contained in a population and it represents the fundamental component of changes since it determines the microevolutionary potential of populations. There are several measures for quantifying the genetic diversity, most notably measures based on heterozygosity and measures based on allelic richness, i.e. the expected number of alleles in populations of same size. These measures differ in their theoretical background and, in consequence, they differ in their ecological and evolutionary interpretations. Therefore, in the present chapter these measures of genetic diversity were jointly analyzed, highlighting the changes expected as consequence of gene flow and genetic drift. To develop this analysis, computational simulations of extreme scenarios combining changes in the levels of gene flow and population size were performed.

Keywords: allelic richness, computational simulations, gene diversity, molecular markers, population genetics

1. Introduction

Genetic diversity comprises the total of genetic variability contained in a population and it represents the row material for evolutionary changes since it determines the microevolutionary potential of populations.

The most popular measure of genetic variation is the average heterozygosity expected in Hardy–Weinberg equilibrium. Nei [1] called this measure as gene diversity index, and defined it as either the average proportion of heterozygotes per locus in a randomly mating population or the probability that two alleles randomly and independently selected from a gene pool will represent different alleles. Expected heterozygosity at n loci within a population is calculated, as:

$$H_e = 1 - \sum_{i=1}^{n} p_i^2 \quad (1)$$

Being p_i the allele frequency. Since this index has been formulated entirely in terms of alleles and genotypic frequencies, its treatment is biologically the most direct [2]. Expected heterozygosity can be applied to any population of all organisms (sexual or asexual, diploid or non-diploid) independently of the number of alleles at a given locus or the pattern of evolutionary forces [1].

The total number of alleles at a locus has also been used as a measure of genetic variation and is an important measure of the long-term evolutionary potential of populations [3]. The major drawback of the number of alleles is that, unlike heterozygosity, it is highly dependent on sample size. Therefore, samples sizes must be equal in order to obtain meaningful comparisons between samples because of the presence of many alleles at low frequencies in natural populations. In this way, the allelic richness estimator (*r*) can avoid this problem owing to this estimator represents a measure of allelic diversity that takes into account the sample size [4]. By means of rarefaction method, the *r* estimator calculates the expected number of alleles at a locus for a fixed sample size, considering generally the smallest sample size in a series of sampled populations [5].

1.1 Loss of genetic diversity in reduced sized populations

The starting question for analyzing the effect of reduced sized populations on genetic diversity levels is how population size (N) influence on the allele and genotype frequencies. In case that Hardy–Weinberg principle assumption of infinite population size being violated, genetic drift will occur in populations. Genetic drift is a stochastic sampling process that determines what alleles will constitute the gene pool in the next generation. Fragmentation and isolation due to habitat loss and landscape modification can reduce the population size of many species of plants and animals throughout the world hence understand genetic drift and its effects is extremely important for biodiversity conservation [3].

The implementation of molecular biology techniques for differentiation of individuals directly at DNA level allows inferring genetic diversity parameters in real populations even these parameters were defined prior to the development of DNA-based molecular markers. In addition, technological development of capillary electrophoresis has improved the resolution power for allele identification and advances in computer power has allowed the analysis of a huge number of highly polymorphic loci simultaneously in a simply and quickly manner.

1.2 Molecular markers as workhorses for genetic diversity studies

A molecular marker is known as any specific DNA fragment that may or may not correspond to coding regions of the genome [6] and is representative of differences at the genomic level [7]. In case that a molecular marker shows segregation according to the Mendelian laws of inheritance, it can also be defined as a genetic marker and it provides genetic information [6]. Molecular markers offer advantages over conventional alternatives based on phenotype, since contrary to morphological data, molecular data are stable and detectable in all tissues without being related to the development, differentiation, growth, or defense state of the cell and they are not influenced by environmental effects [7, 8].

Although there are several type of molecular markers the ideal genetic marker must be reliably measurable, exhibit highly variable loci, be codominant, and be densely distributed throughout the genome. The microsatellite markers also called Simple Sequence Repeat (SSRs) meet all these requirements [9]. SSRs are monotonous repeats of short nucleotide motifs of 1 to 6 base pairs (e.g., cgtcgtcgtcgtcgt, which can be represented by $(cgt)_n$ where n = 5). These repetitive elements can be found interspersed in the three eukaryotic genomes: nucleus (SSRs), mitochondria (mtSSRs) and chloroplasts (cpSSRs) [10]. The different SSRs alleles are mainly generated through simple repeat addition and subtraction mechanisms that occur with equal probability [11], and they are rarely found in coding regions [9]. SSRs are informative and practical markers because of they provide information about the amount and distribution of genetic diversity and the processes that determine the genetic structure and variation within and between natural populations [12]. Regarding methodological concerns, they present high stability with high intra- and inter-laboratory repeatability and they can be implemented in low complexity laboratories using external sequencing services. A limitation for SSRs implementation is that the sequence of repetitive flanking region is required to the development of specific primers although the cross transference of primers between closely related species is usually successful. SSRs have become the most widely used DNA marker in population genetics for genome mapping, molecular ecology, and conservation studies [3]. Despite the fact that massive sequencing methods to identify single nucleotide polymorphisms (SNPs) have gained prominence, microsatellites continue to be widely used tool because the analysis of generated data is simple and easily comparable with previous studies.

1.3 Simulations as a tool for predicting what is expected under certain conditions

Simulations help to recreate the stochastic process that accompanies the transmission of genes from parents to offspring because they recreate the movement of alleles under a model with same conditions several times. In addition, using different model conditions can help to disentangle sampling effects and scale dependencies, as well as historical influences of gene flow.

Any model (analytical, simulation, and otherwise) makes simplifying assumptions, excepting that it be “an entire reconstruction of the actual system—whereupon it ceases to be a model” [13].

The focus of this chapter is define the simplest model that show the effects of population size and gene flow on contemporary levels of genetic diversity, attending to the influence that multiplicity and abundance play on the classic genetic diversity estimators.

2. Materials and methods

2.1 Simulations

In order to test the effect of population size and gene flow on the magnitude of genetic diversity parameters simulated genetic data were obtained using IBDsim program [14]. This program simulates genetic data under isolation by distance model using a backward simulation strategy at population level. Stepping Stone Model was considered which assumes discrete populations, discrete number of generations, genetic drift within each population, and migration between adjacent or spatially proximal population [15–17] being m the total dispersal rate in one dimension [18]. Four different scenarios were simulated considering a population composed by a square grid of 6 x 6 subpopulations. Those scenarios combine two subpopulation sizes (n): 100 or 20 diploid individuals and two migration rates (m): 0.5 or 0.005, respectively (**Table 1**). The four combinations of n and m allowed to obtain scenarios that show expected genetic diversity with low or high levels of gene flow in population of small or large populations. Scenarios A-C and A-D allowed to evaluate the consequences of high or low levels of gene flow on the diversity parameters in populations of high size, respectively while scenarios B-C and B-D allowed to evaluate the consequences of high or low levels of gene flow on diversity

Population size (n)	Migration rate (m)	
	0.5	0.005
100	A - C	A - D
20	B - C	B - D

Table 1.
Four simulated scenarios combining population size (n) *and migration rate* (m).

parameters in populations of small size, respectively. Each data set was composed by 180 diploid individuals sampled from nine subpopulations. To avoid edge effects, a two-dimensional lattice was represented in a torus [18]. At grid edges, we used 'absorbing' boundaries in IBDSim whereby 'the probability mass of going outside the lattice is equally shared on all movements inside the lattice' [19]. The total simulated population was kept constant, but samples were taken from a smaller area of 3 x 3 subpopulations with 20 individuals per node. This sampling strategy was implemented in order to restrict the sampling design to a relatively small geographical area in order to work at a local geographical scale [19]. Each individual was characterized by a multilocus genotype defined by ten nuclear microsatellite loci of a two base pair repeated motif with a mutation rate (μ) of 10^{-3} with two to 20 alleles per locus. From each scenario, 10 data sets were simulated.

2.2 Analysis of simulated data

Expected heterozygosity (*He*) was estimated using Nei's gene diversity index (1) [1] and allelic richness (*r*) was estimated using a rarefaction method. Both estimators were calculated for each subpopulation (nine in each data set) under each scenario (four) and for each repetition (10 in each scenario) obtaining as result 360 estimations of each genetic diversity measures. These estimations were developed using FSTAT software [20]. Means of *He* and *r* were estimated for each scenario. In order to determine if differences between means were statistically significant a standard *t*-test of means was implemented. Differences between means was considered statistically significant if the chance occurrence of such statistic was 5 percent or less ($p < 0.05$). This test was implemented using Microsoft Excel software.

In addition, the spread and skew of both estimated parameters in all simulations by each scenario was shown using box and whisker plots that display a five-number summary: minimum, maximum, median, upper and lower quartiles. The central rectangle spans the first quartile to the third quartile, or the interquartile range (IQR). A segment inside the rectangle shows the median while whisker to the left and to the right show the locations of the minimum and maximum. These estimations were calculated using Microsoft Excel software.

3. Results

Combination of *n* and *m* allowed analyze the effect of population size and genetic isolation among population on genetic diversity estimators based on all differences between scenarios parameters estimations were statistically significant (**Table 2**). Scenarios A-C and A-D which consider large population size the allelic richness and the expected heterozygosity were higher than scenarios B-C and B-D which consider small population size (**Figure 1**). However, allelic richness showed lower values than heterozygosity in smaller populations comparing with large

	A-C	A-D	B-C	B-D
A-C	—	9.05511E-11	2.27959E-75	2.20212E-68
A-D	6.23453E-15	—	3.10501E-68	3.01124E-66
B-C	4.77563E-87	1.35851E-69	—	8.60895E-19
B-D	9.19086E-97	1.10061E-81	4.24449E-15	—

Table 2.
Pairwise t-test *results between scenarios. Below diagonal* p *values of* t-test *applied for allelic richness* (r) *means and above diagonal* p *values of* t-test *applied for expected heterozygosity* (He) *means.*

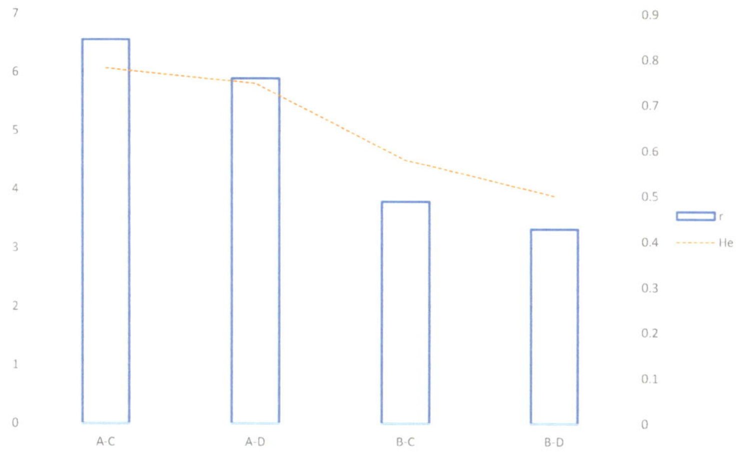

Figure 1.
Allelic richness (r) *and expected heterozygosity* (He) *means by scenario.*

populations with the same migration rate (A-C vs. B-C and A-D vs. B-D, respectively) (**Figure 1**). **Figure 2** shows box and whisker plots of *r* and *He* parameters for all simulated populations in the fourth scenarios. Despite the overlapping in simulated data from same population size and differences in the migration rates (A-C vs. A-D and B-C vs. B-D, respectively) differences in median values among all scenarios were detected. In addition, these plots show higher spread of *r* than *He* (**Figure 2**). In the comparison of means and median values between scenarios considering high levels of gene flow (m = 0.5) with differences in population size (A-C vs. B-C) and low levels of gene flow with differences in population size (A-D vs. B-D) *r* showed higher reduction than *He* (**Table 3**). Furthermore, the reduction was higher for *r* than the reduction for *He* between scenarios considering large population size with differences in migration rates (A-C vs. A-D). However, the reduction was higher for *He* than the reduction for *r* between scenarios considering small population size with differences in migration rates (B-C vs. B-D) (**Table 4** and **Figure 3**).

4. Discussion

Genetic diversity is a pre requisite for population adaptation to environmental changes [12]. Large populations of naturally outbreeding species usually have extensive genetic diversity, but genetic diversity is usually reduced in populations and species of conservation concern [12]. Theoretical analyses based on simulations give information for understanding empirical results.

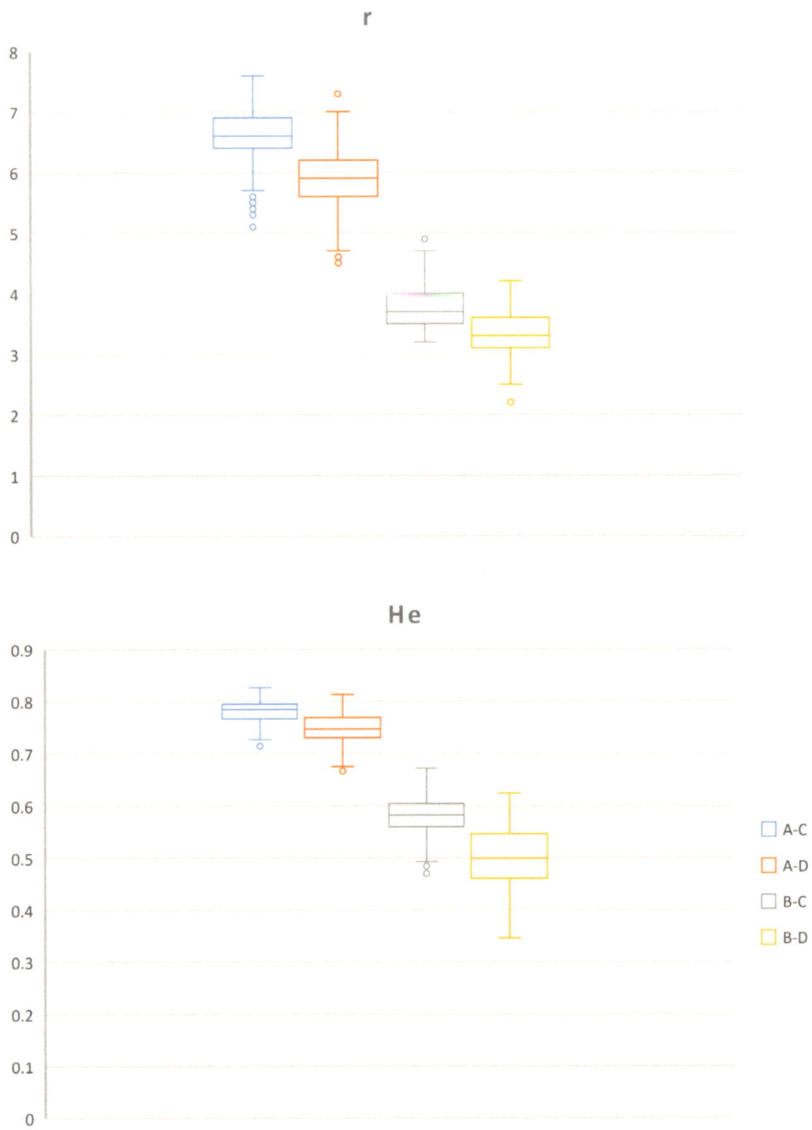

Figure 2.
Box and whisker plots for allelic richness (r) *and expected heterozygosity* (He) *by scenario.*

Parameter	Statistic	A-C vs B-C	A-D vs B-D
r	Mean	2.769 (42.24%)	2.575 (43.69%)
	Median	2.900 (43.94%)	2.600 (54.93%)
He	Mean	0.201 (25.77%)	0.246 (32.98%)
	Median	0.202 (26.77%)	0.248 (33.20%)

Table 3.
Reduction of allelic richness (r) *and expected heterozygosity* (He) *as consequence of changes in population size with high levels of gene flow* (m = 0.5) *(A-C vs. B-C) and in populations with low levels of gene flow* (m = 0.005) *(A-D vs. B-D). Reduction percentage are showed between brackets.*

The total allele number by locus is a complementary measure of genetic diversity because it is more sensitive to loss of genetic variation as consequence of small population size than heterozygosity. In this way, *r* becomes in an important measure

Parameter	Statistics	A-C vs A-D	B-C vs B-D
r	Mean	0.662 (10.10%)	0.468 (12.36%)
	Median	0.700 (11.31%)	0.400 (10.81)
He	Mean	0.034 (4.35%)	0.079 (13.64%)
	Median	0.037 (4.72%)	0.083 (14.26%)

Table 4.
Reduction of allelic richness (r) and expected heterozygosity (He) as consequence of changes in gene flow levels in large populations (n = 100) (A-C vs. A-D) and in small populations (n = 20) (B-C vs. B-D). Reduction percentage are showed between brackets.

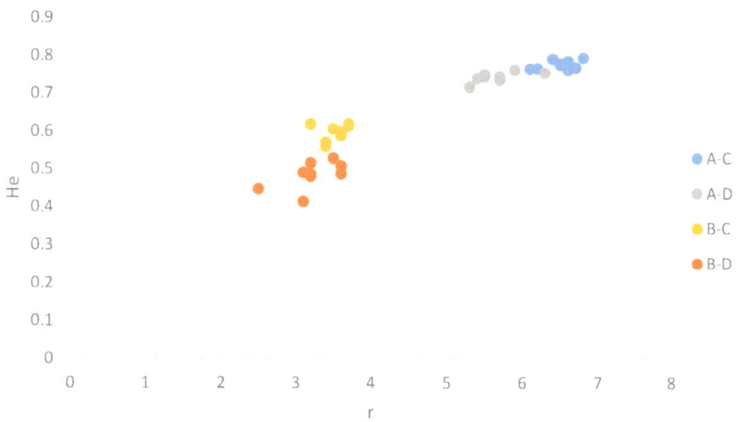

Figure 3.
Plot of allelic richness (r) and expected heterozygosity (He) of nine populations at one simulation for each scenario.

for long-term evolutionary population potential [3]. We will represent this statement using a hypothetical situation: population A (n = 100) and population B (n = 10) (**Figure 4**). There, population B is a random sample from population A. Population B shows three out of eight alleles from population A because of the reduction in population size, which cause that only alleles present in a high frequency remain in the small population. It means that by chance the more frequent alleles have a highest probability to being contained in the gene pool of small population while the rare alleles shows low frequency and as consequence they have high probability to be lost. In this way, the genetic drift is operating and as consequence of this microevolutionary process, not all alleles of a population will be present in the next generation producing a sampling error. As results of this sampling error, the change in the allelic frequencies is at random and the action of genetic drift does not have pre-established direction. However, in the analyzed example (**Figure 4**) the estimated value of *He* changes from 0.719 to 0.620 as consequence of 10 times reduction of population size. This change could indicate that *He* is less sensitive to rare allele lost as consequence of population size reduction. We can explain it by means of other hypothetical situation: We consider four pairs of small populations that contain between eight and 10 alleles (**Figure 5**). At left side of **Figure 5**, four populations show one allele at high frequency and rare alleles increase successively their number step by step (a, b, c and d) while at right side in the same Figure, four populations show alleles at equal frequency that increase successively their number step by step (a, b, c and d). For each population *r* and *He* were estimated. In the step (a) both populations show two alleles (r = 2) but *He* was lower in the population at

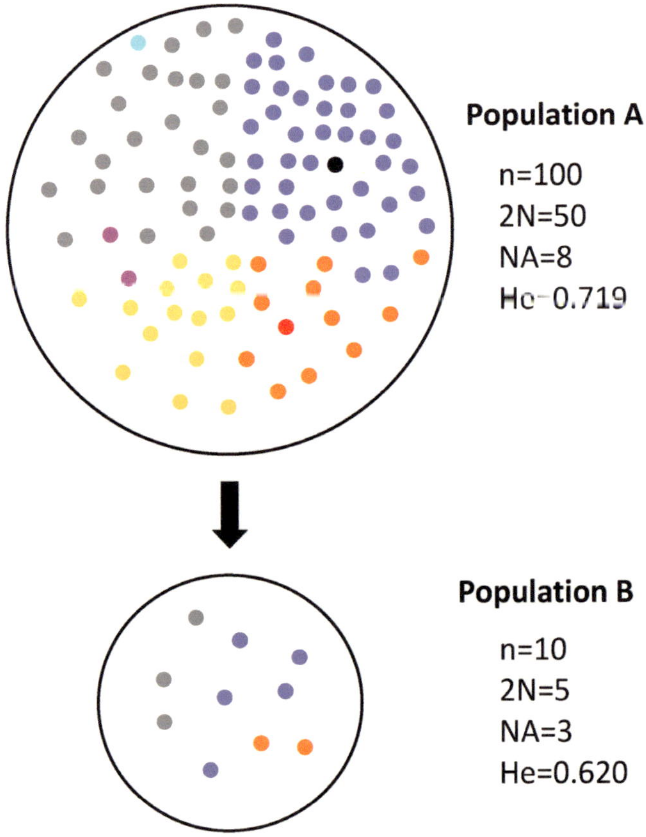

Figure 4.
Changes in number of alleles (NA) *and expected heterozygosity* (He) *as consequence of population size reduction.*

left side than population at right side (0.18 vs. 0.50, respectively), being the alleles frequencies the unique difference between both populations. Successively, in the following steps (b, c and d) while the number of different alleles increases, *He* also increases in populations at both sides. However, in populations at the right side, since the alleles are equally frequent in all steps, *He* reaches the maximum values, while in the populations at left side, the new alleles show low frequencies (rare alleles) and *He* increases little by little. Finally, in the step (e) *He* reaches the maximum value although all alleles are rare because of they show the same frequency. Hence, the estimation of *He* is highly dependent on allele frequencies and its value will be determined in a greater extent by the presence of alleles at high frequency which usually show high probability to be proportionally maintained when population reduce its size.

The effects of changes in population size on genetic diversity estimators considering different gene flow levels were studied in the present chapter by means of simulations (A-C vs. B-C and A-D vs. B-D, respectively). As expected, reductions in *r* and *He* values were obtained between large and small populations. In case that *r* and *He* are used for detecting genetic diversity reduction, *r* is more sensitive than *He* to detect genetic diversity reduction independently gene flow levels (**Table 3**).

The effects of gene flow levels on genetic diversity estimators considering different population sizes were studied in the present chapter by means of simulations

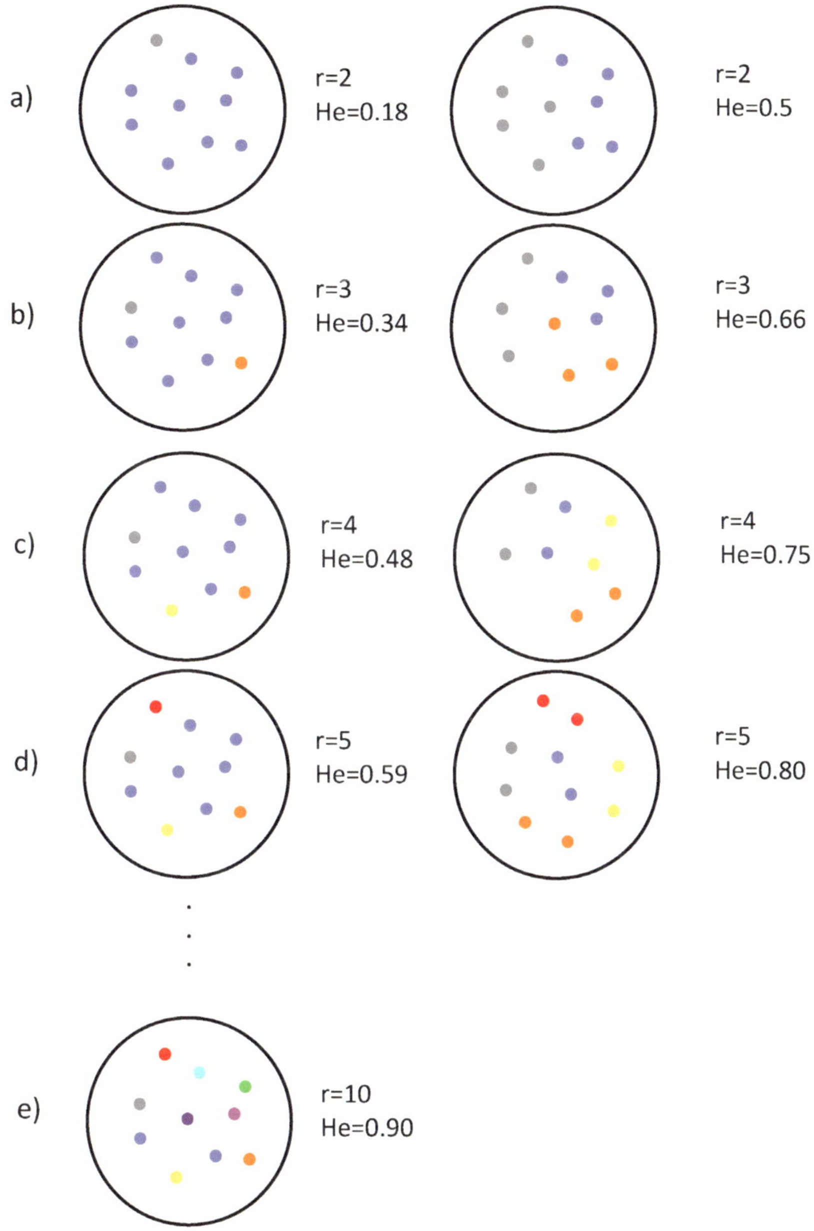

Figure 5.
Changes in allelic richness (r) *and expected heterozygosity* (He) *in small populations with increasing in number of different alleles: two, three, four, five and ten (a, b, c, d and e, respectively).*

(A-C vs. A-D and B-C vs. B-D, respectively). In large populations, *r* is more sensitive than *He* to detect genetic diversity reduction as consequence of low gene flow level. On the other hand, in small populations *He* is more sensitive than *r* to detect genetic diversity reduction as consequence of low gene flow level (**Table 4**).

Gene flow is a microevolutionary process that maintain the genetic exchange among local populations increasing population genetic diversity [21]. Gene flow can be quantified by the parameter *m*, which describes the movement of each gamete or individual independently of population size [22]. As microevolutionary process, gene flow counteracts the genetic drift effect and the balance between gene flow and genetic drift determine genetic diversity levels for neutral alleles. Genetic diversity is the basis for local adaptation and genetic drift could be understood as a threat for biodiversity because of it causes genetic diversity loss in natural populations. Current climate change and fragmentation of natural populations as consequence of anthropic impacts are calling to urgent collective and interdisciplinary actions from researchers. The study of genetic diversity levels is especially important for the management of endangered and valuable species. The focus in conservation biology is the maintenance of genetic diversity because of inbreeding and reduction in reproductive fitness is often associated with loss of genetic diversity [12]. Although the International Union for Conservation of Nature (IUCN) recognizes the need to conserve genetic diversity as one of three global conservation priorities [23] the genetic factors are not currently considered to assign the conservation status of species [24].

5. Conclusion

The comprehensive quantification of genetic diversity levels demand the estimation of *r* and *He* because of the sensitiveness of both estimators depends on allele multiplicity and frequencies. In this way, the estimation of *r* and *He* is recommended for genetics studies in populations that inhabit disturbed environments.

Acknowledgements

The authors wish to thank National Council of Scientific and Technical Research (CONICET, Argentina).

Conflict of interest

The authors declare no conflict of interest.

Author details

María Eugenia Barrandeguy[1,2]* and María Victoria García[1,2]

1 Population and Quantitative Genetics, Department of Genetics, Faculty of Exact, Chemicals and Natural Sciences, National University of Misiones (FCEQyN – UNaM), Argentina

2 Institute of Subtropical Biology – Posadas Node (UNaM – CONICET), National Council of Scientific and Technical Research, Argentina

*Address all correspondence to: ebarran@fceqyn.unam.edu.ar

References

[1] Nei M. Mint: Analysis of Gene Diversity in Subdivided Populations. Proceedings of the National Academy of Sciences.1973; 70 (12) 3321-3323. DOI: 10.1073/pnas.70.12.3321

[2] Nagylaki T. Mint: The expected number of heterozygous sites in a subdivided population. Genetics.1998; 149: 1599-1604.

[3] Allendorf FW, Luikart GH. Conservation and the Genetics of Populations. Blackwell Publishing; 2007. 642 p.

[4] El Mousadik A, Petit RJ. Mint: High level of genetic differentiation for allelic richness among populations of the argan tree [Argania spinosa (L.) Skeels] endemic to Morocco. Theoretical and Applied Genetics. 1996; 92: 832-839. DOI: 10.1007/BF00221895

[5] Petit R, El Mousadik A, Pons O. Mint: Identifying Populations for Conservation on the Basis of Genetic Markers. Conservation Biology. 1998; 12(4): 844-855.

[6] Ferreira M, Grattapaglia D. Introducao ao uso de marcadores moleculares em análise genética. EMBRAPA-CENARGEN; 1996. 220 p.

[7] Agarwal M, Shrivastava N, Padh H. Advances in molecular marker techniques and their applications in plant sciences. Plant Cell Reports 2008; 27:617-631.

[8] Marcucci Poltri S. Marcadores Moleculares aplicados a Programas de Mejoramiento Genético de Eucalyptus. In: Secretaría de Agricultura. Ganadería, Pesca y Alimentos editors. Mejores árboles para más forestadores, 2005. 241 p.

[9] Karhu A. Evolution and applications of pine microsatellites. [thesis]. Faculty of Science. University of Oulu. Oulu. 52 p.

[10] Tautz D, Renz M. Mint: Simple sequences are ubiquitous repetitive components of eukaryotic genomes. Nucleic Acid Research. 1984; 12(10):4127-4138.

[11] Schlötterer C, Tautz D. Mint: Slippage synthesis of simple sequence DNA. Nucleic Acids Research. 1992; 20: 211-215.

[12] Frankham R, Ballou JD, Briscoe D.A. Introduction to Conservation Genetics. Cambridge University Press, 2002. 617 p.

[13] Epperson BK, Mcrae BH, Scribner K, Cushman SA, Rosenberg MS, Fortin MJ, James PM, Murphy M, Manel S, Legendre P, Dale MR. Mint: Utility of computer simulations in landscape genetics. Molecular Ecology. 2010; 19: 3549-3564.

[14] Leblois R, Estoup A, Rousset F IBDSim: a computer program to simulate genotypic data under isolation by distance. 2008; Molecular Ecology Resources 9(1): 107-109. DOI: 10.1111/j.1755-0998.2008.02417.x

[15] Kimura M. Mint: "Stepping stone" model of population. Annu Rep Natio Inst Genet. 1953; 3: 62-63.

[16] Kimura M, Weiss GH. Mint: The stepping stone model of population structure and the decrease of genetic correlation with distance. Genetics. 1964; 49: 561-576.

[17] Weiss G H, Kimura M. A mathematical analysis of the stepping stone model of genetic correlation. Appl Probab. 1965; 2: 129-149. DOI: 10.2307/3211879

[18] Leblois R, Beeravolu C R, Rousset F. IBDSim version 2.0 User manual.

[19] Leblois R, Estoup A, Rousset F. Mint: Influence of mutational and sampling factors on the estimation of demographic parameters in a "continuous" population under isolation by distance. Mol Biol Evol. 2003; 20(4): 491-502. DOI: 10.1093/molbev/msg034

[20] Goudet J. Mint: FSTAT (vers.2.9.3.2): a computer program to calculate F statistics. Heredity. 1995; 86:485-486.

[21] Hartl, DL, Clark AG. Principles of population genetics. Sinauer Associates, Inc Publishers; 2007. 652 p.

[22] Slatkin M, Barton NH. A comparison of three indirect methods for estimating average levels of gene flow. Evolution. 1989; 43(7):1349-1368

[23] McNeely JA, Miller KR, Reid WV, Mittermeier RA, Werner TB. Conserving the world's biological diversity. IUCN, World Resources Institute, Conservation International, WWF-US, and the World Bank, 1990.

[24] Garner BA, Hoban S, Luikart G. Mint: IUCN Red List and the value of integrating genetics. Conservation Genetics. 2020; 21: 795-801. DOI: 10.1007/s10592-020-01301-6

Chapter 5

Genetic Diversity of *Coffea arabica*

Juliano Lino Ferreira, Eveline Teixeira Caixeta, Fernanda Fatima Caniato, Tesfahun Setotaw, Gustavo César Sant'Ana and Leila Maria Ferreira

Abstract

Coffea arabica L. is a native coffee species probably originated in Abyssinia, now Ethiopia. The genetic diversity of *C. arabica* has economic implications directly related to profits by breeding for developing new varieties to a global market. The economic value of *C. arabica* genetic resources are estimated at US$ 420 million, considered a 10% discount rate. Understanding the extent of traits variability and genetic diversity is essential to guide crosses between genotypes, targeting the development of new varieties with high economic value. This chapter will present the *C. arabica* economic importance, primarily to Brazil, the most significant world producer; we will outline the origin and dispersion of arabica coffee and briefly show the leading germplasm banks. We will also point out contribution of genetic diversity studies based on morphological, agronomic traits, and molecular markers supporting the development of new varieties. Finally, we present an outline for the future.

Keywords: economic importance, genetic resources, molecular markers

1. Introduction

Coffee is an everyday beverage and consumed enthusiastically throughout the world. This popular beverage is a primary source of annual income and employment, contributing economically, on four continents, as well as too many emerging nations. In the second half of the nineteenth-century coffee was transformed into an industrial product as a consequence of the accelerated expansion of coffee production in Brazil, which in turns, nurtured the growth of a mass consumer market in the United States [1].

Coffee crop, in current times, spread in over 10 million hectares grown in more than 80 tropical and sub-tropical nations. On a social basis, it plays a relevant role notably for the subsistence of nearly 20 million coffee-farming families in underdeveloped countries of Asia, Africa and Latin America [2]. In the world, coffee places in the second-largest export commodity position only behind to the petroleum products [3].

According to the USDA 2020/21 Forecast Overview [4], the world coffee production is estimated at approximately 9 million bags (60 kilograms) superior to the past year record of 176.1 million. The forecast is that Brazil accounts for most considerable of the increment because its arabica coffee crops start the on-year of the biennial production cycle and robusta coffee is achieving record output. Brazil is the leading supporter of the forecast for the expansion in world exports. Arabica output in Brazil is forecast to achieve 6.8 million bags above the preceding season to 47.8 million. Most favorable climate conditions prevailed in the majority coffee regions, promoting coffee fruit setting and development and filling, thus succeeding in high yields.

The genus *Coffea* includes approximately 124 well-identified species. *Coffea canephora* P. and *Coffea arabica* L. are commercially highpoints species [5]. *C. arabica* is a member of the family Rubiaceae and is a single polyploid species inside the genus *Coffea*. This true allotetraploid has 2n = 4x = 44 chromosomes. It is the old and most cultivated species of a coffee plant [6, 7]. Concerning floral biology, each species of the *Coffea* genus has its particularities and *C. arabica* can be characterized as self-fertile, it means that reproduction occur through self-fertilization, with an allogamy index of about 10%, on average. One research carried out at the Instituto Agronômico de Campinas concluded that insects (bees in particular), wind and gravity, are the main responsible for the pollination of coffee [8].

Genetic diversity is a prerequisite component of biodiversity, obligatory for species reproduction, and essential for adapting species to a dynamic environment [9]. Besides, assessment of genetic diversity directly impacts the development of new varieties through breeding. So, valuable genetic traits can be transferred to existing plant cultivars to achieve goals towards increasing crop yields, characteristics related to the quality of crops, resistance to disease and pest, etc. Plant breeding focused on wild races genetic information is usual for most global crops and has driven an essential contribution to increasing global food security [10]. Notwithstanding, genetic information wild races have been reduced at an alarming rate, specifically for tropical crop species, including *C. arabica* [11, 12]. Among other expert scholars, Labouisse et al. [13] cite deforestation as a noticeable contributing factor affecting the genetic erosion of coffee in Ethiopia. McNeely et al. [14] states some issues like land use conversion, overexploitation, and introduction of exotics species as factors contributing to native populations decimate.

In plant breeding, it is decisive to identify the most critical phenotypic traits to increase plant production. Therefore, the assessment of trait occurrences and dissimilarities in a population is the key to defining possibly useful crosses among accessions. The first line of attack is to understand the extent of the variability of some species. To do that, many countries around the world strategically centered money and human capital on collecting, assessing, and keeping the genetic resources available on germplasm banks. Although many studies emphasize the genetic diversity with molecular markers, it is also useful for plant breeders to contemplate the morphological and agronomical diversity of interest traits. In this context, we briefly show the coffee chain's budget value, summarily point out the leading germplasm banks, and concisely demonstrate the employment of genetic diversity assessed on morphological and agronomical traits, along with molecular markers approaches.

2. Economic importance of *Coffea arabica*

Coffee represents an agricultural commodity that has stood out in international trade and domestic supply in terms of quantity and value [15]. Developing countries correspond to the leading suppliers, while the main buyers are developed countries, in which coffee consumption is full-bodied. The soil characteristics of the intertropical and equatorial regions of the world play a fundamental role in the coffee marketing chain worldwide [16].

Approximately 170 countries are coffee producers, and almost all countries are consumers, highlighting its commercial importance, which has grown steadily over the last 150 years [17]. Even with the crop distribution range, it has not represented a barrier to the growing production concentration in some nations. Currently, 70% of the consumed coffee worldwide comes from Brazil, Vietnam, Colombia, and Indonesia. In contrast, the primary consumer countries are the United States, the European Union, Brazil, and Japan, which account for two-thirds of the global demand for coffee [16].

Agricultural products usually have limited extended storage to avoid severe losses of quality. However, in coffee, the beans can be stored for decades, since observed aspects regarding the limits of humidity, light, temperature, and the latter keeping with reasonable consumption conditions. The coffee profile allows coffee growers to use the harvest with a strategic vision of economics. Many of them prefer store in bags instead of selling them immediately, hoping they will reach better prices [16].

Coffee was introduced in Brazil in 1727 through French Guiana and spread from northern Brazil to the southeast states, mainly in the mountain regions. Coffee developed in these areas due to favorable climate conditions for its grown, such as mild temperature, heavy rains, and distinct dry season [18].

The cultivation of coffee has evolved significantly and contributed to economic development throughout the history of Brazilian regions, particularly during early times and locations where the crop implantation occurred. The establishment of farming dates back to the 18th century in the northern land, precisely at the State of Pará. Later, it moved to the states of Rio de Janeiro and São Paulo (which corresponds to the Paraíba Valley). In 1850, cultivation spread rapidly towards Serra da Mantiqueira and Santos. In the 20th century, coffee cultivation continued its expansion in the states of São Paulo, south of Minas Gerais, Espírito Santo, Paraná and also to the northern region of Brazil, in the State of Rondônia. During this period of growth, the Brazilian economy, in general, was strongly associated with the coffee market, and the Brazilian Federal Government heavily regulated the coffee market until the mid-1990s [19].

Twelve states represent the primary coffee-producing regions in Brazil, and there are about 300,000 coffee plantations in the country, spread over 1950 cities is estimated. The state of Minas Gerais holds about 50% of the total coffee production in Brazil. Minas Gerais state offers topography and mountain climate ideal for the cultivation of coffee that along with the low-cost land, and abundance of cheap labor, may contribute to their outstanding position [20].

Minas Gerais accounts for approximately 50% of coffee cultivated in Brazil, 98% of this occupancy with *C. arabica* species that is the most economically relevant species. Some of the most economically outstanding cultivars of *C. arabica* in Brazil are Mundo Novo, Bourbon, Catuai Vermelho, and Catuai Amarelo [21].

At world scenario, the International Coffee Organization estimated that in 2014, the consumption of coffee was 150.3 million bags of 60 kg, in 2015; it rose to 152.1 million bags. In the last four years, the annual increase has remained an average of 2%. There was a significant increase in consumption in the Asia region, with rates of growth in the range between 4.5 and 9% in Indonesia, the Philippines, India, and Thailand. World coffee production in 2015 was 143.4 million 60 kg bags [22].

In April 2020, analyzing world coffee exports, it appoints an estimate to the 10.82 million bags, whereas, in April 2019, this number was 11.17 million. In the first seven months of the coffee year from 2019 to 2020 (the period between October/19 to April/20), they decreased 3.8% concerning exports in the same period from 2018 to 2019, totaling 72.78 bags, against 75.67 million. Shipping of beans of *C. arabica* species in the 12 months ending in April 2020 totaled 81.30 million bags, against 80.75 million bags the previous year [23]. In this context, Brazil has been responsible for 20% of coffee exports in the world. Due to the exponential growth of global consumption and the capacity to produce in large quantities, Brazil has become one of the largest coffee beans exporters. In numbers, it represents more than 34 thousand bags, which corresponds to the US $ 5.4 billion in revenue, 15% of which consists of *Specialty* coffee. The United States and Germany are the major importing countries [24]. The coffee tree farmland employs approximately 26 million people, many of whom are small farmers, dependent mainly on coffee for their livelihood [25].

In the budgetary part, the International Coffee Organization's composite indicator fell 4.1% in May 2020, registering an average of 104.45 US cents per pound, which represented a second consecutive month of decline. The price trend curve for all *C. arabica* groups was bearish. From October 2019 to April 2020, shipments from Africa increased 7% to 7.66 million bags, and those from Asia and Oceania increased by 0.6%, to 23.62 million bags. In the same period, shipments from Central America and Mexico fell 4.9% to 8.77 million bags, and those from South America fell 8.6% to 32.74 million [23].

A series of research recognizes the economic value of genetic diversity [26]. However, these authors confirm the market failure in the case of conservation of coffee genetic resources, especially in Ethiopian highland forests, alerting that in 10 years, the coffee forest will disappear if the current devastation rates persists, which is alarming. This study addressed Ethiopian genetic coffee resources, the primary centre of diversity, revealing the potential economic importance of amounts to nearly US$1458 million, considering a 5% discount rate and US$420 million for a 10% discount rate. A good explanation of this outsized discount rate impact may be the expressive time lag between the required cost of coffee breeding programmes and the gains resulting from enhanced cultivars development.

3. Origin and distribution of *Coffea arabica*

The study of plant domestication, beyond its role in man's cultural evolution, is an excellent experimental system for the study of biological evolution. Numerous dissimilarities in the middle of wild and domesticated types are related to essential features and basic plant biology processes, such as adaptation, development, and reproduction [27].

The *C. arabica* had its origin in the highlands of tropical forests located in southwestern Ethiopia. Under the specter of the biological structure, the genetic basis of the world's coffee plantations is considerably small, as are most commercial coffee varieties to date, derived from a limited number of accessions from Ethiopia's forests [26].

C. arabica is one of the most favorite beverage crops globally that accounts for about 70% of the total international coffee market. This crop species is the most valuable globally due to their high beverage quality and taken every day by a million people worldwide. The *C. arabica* was assumed to be originated in the Southwestern part of Ethiopia in specifically called the Keffa area [28]. It is also considered the possibility that *C. arabica* was originated in the Boma plateau in Sudan and Mount Marsabit of Kenya. Ethiopia is recognized strongly substantiated as a primary centre of diversity for coffee arabica [29–31].

In ancient times, coffee was first noticed by the Arab merchants in Ethiopia and taken to Yemen [32]. The origin of *C. arabica* has been subject to both molecular and archeological studies, confirming the Ethiopian origin of *C. arabica* [28, 33, 34]. *C. arabica* is a true allotetraploid species with 2n = 4x = 44 that considered as originated from the interspecific hybridization of *C. canephora* and *C. eugenioides* [35, 36].

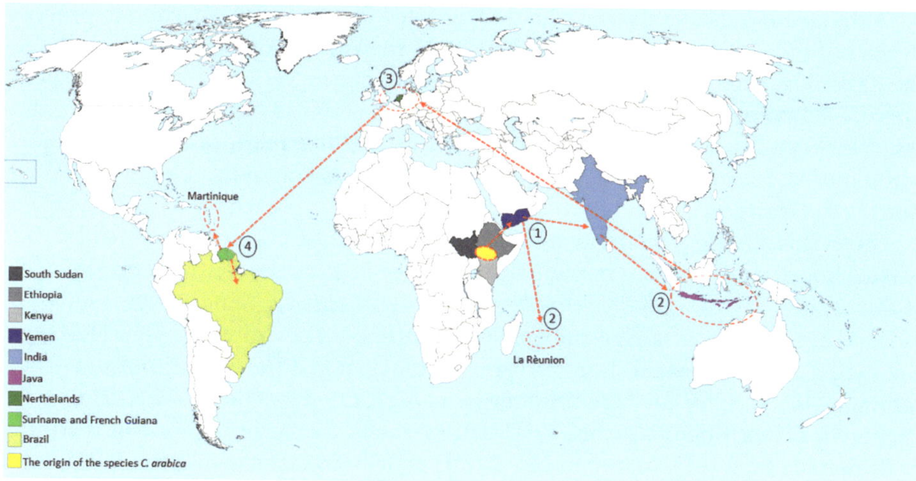

Figure 1.
Origin and dispersion course of Coffea arabica *all the way through the world. Image reproduced from reference [43] with permission from the Royal Society of Chemistry (RSC).*

C. arabica cultivation was started after the wild coffee introduced from Ethiopia to Yemen as early as 575 AD [32]. The cultivated coffee arabica divided in to *C. arabica* var. typica and *C. arabica* var. Bourbon [37]. After its introduction to Yemen, the coffee arabica was distributed worldwide and became the most popular beverage crop. The crop distributed to Reunions Island from Yemen and then introduced to India and Java (Indonesia) [38, 39]. The coffee crop was then distributed from Java to Europe (Amsterdam botanical garden) in 1710 [28, 40]. After that, the coffee plant was taken to South America in 1718 from Europe. It was introduced to Martinique Island in 1720 or 1723 and Brazil via French Guiana in 1727 [40–42]. Finally, the coffee was spread throughout the world from South America. Ferreira et al. 2019 [43] precisely illustrate the origin and dispersion of *C. arabica* (**Figure 1**).

4. *Coffea arabica* genetic resources

The efficient use of available germplasm for breeding purposes requires detailed information on the relationship of genetic relatedness among accessions that compose it, primarily affected by the domestication process. The prospect of coffee improvement in all desirable aspects depends on the availability and use of the mostly untapped genes found in the wild, in farmers' fields and *in* and *ex-situ* germplasm collections [3].

Conservation *in-situ* of plant species make possible the maintenance a greater diversity of species and genepools in a dynamic environment, supporting populations that continue to evolve [44]. Understory trees in the tropical forests of Africa are the range where wild coffee grows spontaneously. It covers a wide geographic area from Guinea in West Africa through Central to eastern Africa, with additional centres of diversity add the Mascarene Islands (La Réunion and Mauritius) in the Indian Ocean, Madagascar, and the Comoros Islands [45].

From 1971 and 1997, the deforestation took place in around 235,400 ha of closed and slightly disturbed forests in the highland plateau of southwest Ethiopia. Numerous international organizations have outlined proposals for *in-situ* conservation of *C. arabica*, but regrettably, implementation has been lagging as a result of financial constraints [46].

An effort to preserve the last remaining coffee forests in Ethiopia and to prevent the loss of biodiversity resulted in a creation of the Yayu Biosphere Reserve and the Kafa Biosphere Reserve, in 2010. At that time, due to the sustainable strategic interest, it became component of the United Nations World Network of Biosphere reserves. Yayu Coffee Forest Biosphere plays a crucial role in the *in-situ* conservation being the last remaining montane rainforest fragments with wild *C. arabica* populations in the world [47].

Given this alarming scenario, in the past, the strategic importance of wild *C. arabica* boosted exploration missions guided to in its primary centre of origin (Ethiopia and Kenya) and the secondary centre of diversity, Yemen. In this sense, in 1964–1965, a Food and Agriculture Organization of the United Nations (FAO) conducted collecting expedition of coffee germplasm in different locations in Ethiopia [48]. In 1966, an expedition mission performed by ORSTOM (Office de la Recherche Scientifique et Technique Outre-Mer; a formerly designation of Institute de Recherche pour le Développement [IRD]) collected germplasm from 70 different origins. Despite the original purpose, most accessions were collected from cultivated coffee being only some native of the understory of tropical forest [49].

The accelerated devastation of the tropical forest ecosystems in Africa, Madagascar, the Comoros and Mascarene islands drove collecting mission for other *Coffea* species. The result of those collecting expedition yielded a total of 20,000 wild coffee trees collected, representing more than 70 species and also the identification of 300 wild coffee populations [49].

According to Bramel et al. [31], is consensus in the majority of institutions worldwide indicates the conservation of the collection is secure due to the adherence and engagement of the institutes and their team. In most institutions, everyone is challenged, to some degree, to cover the yearly cost for everyday conservation operations. One critical study concerning costing for Centro Agronómico Tropical de Investigación y Enseñanza (CATIE) confirms the long-term implications of negligence if the fund is insufficient, which is quite alarming.

Comprehensively, the conservation tactics applied to *C. arabica* accessions may be *in-situ* sites or both *ex-situ* and *in-situ*. *In-situ* involves the maintenance of genetic material in the arrangement of native populations by implementing ecosystem reserves such as national parks and refuges. On the order hand, *ex-situ* that deals maintenance of a species out its original habitat. In this approach, farmed and natural plant species are collected and transferred to a specific site aiming to conserve the genetic information. Furthermore, the accessions are maintained locally in the forms seeding, seeds or *in vitro* culture [31].

In this sense, the chief way of knowing and measuring the size of species variability is to carry out collection expeditions to acquire materials in a vast natural geographic occurrence. After that, each accession must be documented, and subsequently, the measurement of its phenotype must be carried out. In germplasm conservation *ex-situ*, the most common scheme used in coffee, this surveying must be made with suitable statistical designs, plot sizes suitably reliable, an adequate number of repetitions and field locations.

According to Giomo et al. [50], the first and critical step in a breeding programme is the presence and understanding of genetic diversity. In this sense, the knowledge of a series of desirable traits is required to develop a new cultivar of coffee such as adaptability, architecture, fruit color, longevity, maturation, precocity, productivity, resistance to pests and diseases, size, type of grain, quality of coffee cupping, vigor, among others. Therefore, it is imperative to know the distinguished accessions selecting particular interest traits, including agronomic characterization of plants up to the beans' chemical composition and sensory quality, to meet the specific coffee production chain demands.

In coffee species, a significant marketable crop, the research on genetic improvement carried out by a renowned research center around the world has in the germplasm banks its primary source of raw material, essentially in *C. arabica* and *C. canephora*. Germplasm banks guard and preserve an extensive collection of genetic resources used in breeding research and biotechnology to obtain increasingly adapted and productive cultivars.

Among the world-leading significant germplasm resources and conservation of the *Coffea* genus, we highlight the following research institute: Centre National de Recherche Agronomique (CNRA), United States Department of Agriculture - National Plant Germplasm System, CATIE, Centro de Cooperación Internacional de Investigación Agricola para el Desarrollo (CIRAD), Ethiopian Institute of Agricultural Research, Jimma Agricultural Research Center (JARC), Institute of Biodiversity Conservation, Instituto Agronômico do Paraná (IAPAR), and Instituto Agronômico de Campinas (IAC). Those institutes enable the acquisition, exchange, conservation, duplication, and documentation of this crop's valuable genetic resources, aiming the world food security. These organizations also performs phenotypic, cytogenetic, and molecular evaluation seeking elite accessions looking for specific attractive traits, primarily due to the already known low variability of *Coffea arabica* species, allowing in this way, putative well successful crosses.

The genebanks around the world have a collection of *C. arabica* which stands out with the most significant number of accessions (11,415), immediately succeeded by *C. canephora* (625), *C. liberica* (94), *C. eugenioides* (81) and other *Coffea* species (7756) [31].

CNRA was founded in 1998 and headquartered in Abidjan, Ivory Coast. According to Labouisse [13], CNRA has the most extensive genebank field collection of coffee in the world with 8003 accessions that resulted of prospecting conducted in eight African countries: Cote d'Ivoire, Guinea, Cameroon, Tanzania, Kenya, Madagascar and the Democratic Republic of the Congo.

Currently, the United States Department of Agriculture (USDA) comes again developing a *Coffea* collection as part of the National Plant Germplasm System, with approximately 300 accessions. In the past, this governmental department used to maintain 500 accessions of arabica coffee [31, 51].

Established in 1942, CATIE botanical garden and germplasm collection inaugurated its headquarter in Turrialba, Costa Rica. In 1948, the field collections of rubber, cocoa and coffee launched the germplasm preservation in Turrialba [52]. The CATIE International Coffee Germplasm Center is one organization in the public domain because of its designation to the International Institute *ex-situ* collections network under the auspices of FAO [3]. Their field genebank of coffee places the third in the world [52], and include to an ample range the entire genetic diversity of *C. arabica* recording 1987 accessions and above 9000 coffee trees. Also, the genetic diversity of a couple of other *Coffea* species is represented to a minor extent, covering 68 introductions of *C. canephora* and 24 introductions of *C. liberica* [49]. The *C. arabica* germplasm bank of CATIE possess 880 wild and semi-wild genotypes, 581 accessions of them acquired from collecting expedition performed by FAO and ORSTOM in Ethiopia - the known biodiversity hotspots; 923 belongs cultivars, mutants and selections section; 19 interspecific hybrids; and 165 intraspecific hybrids [3]. Considering that field collections maintenance is very costly to maintain, and the conserved genetic material is continuously endangered to biotic and abiotic stress, the research team of CATIE, from this point of view, developed a methodology for cryopreservation in liquid Nitrogen for long-term germplasm conservation of coffee seeds. Lately, CATIE maintains a core subset of 63 accessions from Ethiopia cryopreserved and thus establishing the first world cryobank [3, 53].

CIRAD commenced does collecting mission since the 1960s, being some of these collecting expeditions occurred in association with other institutions - viz., ORSTOM, International Plant Genetic Resources Institute (IPGRI), and IRD [3]. In 1977, an ORSTOM/CIRAD mission arrived in Kenya where they collected eighty different accessions of *C. arabica* at Mount Marsabitan, along with samples of *C. eugenioides*, *C. zanguebariae*, and *C. fadenii*. Subsequently, in 1989, samples from coffee plantation arising from 22 different origins were collected by an IPGRI/CIRAD mission-focused in Yemen. Besides that, the mission recognized six morphologically different types of coffee plants [3]. According to FAO-WIEW database 1990–2001, CIRAD maintain in Guyana a total 3800 accession *ex-situ* of coffee [31].

In Ethiopia, the Jimma Agricultural Research Center (JARC) has the commitment to be a leading centre of excellence research for arabica coffee on the planet, operating ten research stations located strategically in the main coffee production areas. The Jimma Research Station initiated variety development and germplasm conservation activity in 1966–1967. From 1966 to nowadays, the field collection has assembled 5853 accessions of *C. arabica* grouped in the following program/type: National collection - 1431, Exotic collection - 78, Coffee Berry Disease (CBD) resistance collection - 825, and Local landrace - 3519. To date, JARC has launched 42 coffee varieties. In Ethiopia, JARC is the unique public institution that has taken the initiative of multiplying and providing basic coffee seeds, primarily, coffee adapted varieties and CBD resistant material. Furthermore, this research institute plays a considerable role in dissemination and adoption of improved coffee technologies by innovative farmers, private and state-owned farms throughout the countryside [13, 54, 55]. Other important genetic resources organization in Ethiopia is the Institute of Biodiversity Conservation established in Choche (Limu) field genebank with 5196 accessions conserved [13].

In Brazil, the IAPAR was founded in 1972 and headquartered in Londrina, in the state of Paraná. The IAPAR operates in a 300 ha-farm, of which 40 ha are cultivated with coffee. In 1975 was established the field genebank of coffee that were primarily composed by IAC accessions with posterior inclusions of accessions from the FAO/IBPGR collection. Also, they have a partnership with five farmers to test the F3/F4 generations. Several cultivars have been released by IAPAR improved to achieve high yield, drought tolerance, resistance to rust, nematodes, bacterial blight, and leaf miner; and also, different ripening cycles. The IAPAR combine testing, seed production and demonstration to farmers in the F_6 generation, speeding, in this way, the time to release genetic material. The IAPAR institution has a good reputation among of coffee farmer's producers in Paraná [31].

The Instituto Agronômico de Campinas (IAC), institution of the Brazilian Coffee Consortium, maintains the largest and the oldest coffee germplasm bank in the country, with 5451 records. Supported by the framework of this diversity, the active germplasm bank of the "Instituto Agronômico" has contributed for 87 years with significant results in the Brazilian coffee research. IAC also perform a series of research in collaboration with other research institutions within the Brazilian Coffee Consortium [50, 56].

IAC continuously performs morphological, agronomic, chemical and molecular characterization of the genetic materials maintained in its germplasm banks. This is essential for the definition and identification of the most genetic promising materials, with better productivity and other attributes considered according to each survey. To achieve an desired coffee cultivar is required a long-term due to the time demanded to advance the genetic material from generation to generation. In Brazil, the two most adopted cultivars in coffee plantations, Mundo Novo and Catuaí, are the results of improvement research conducted by the "Instituto Agronômico" from its germplasm bank. They are planted in about 80% of Brazilian coffee crops area today [56].

Besides, the germplasm bank of the IAPAR and IAC, there are other five coffee germplasm banks in Brazil: Empresa de Pesquisa Agropecuária de Minas Gerais (EPAMIG), Universidade Federal de Viçosa (UFV), Instituto Capixaba de Pesquisa, Assistência Técnica e Extensão Rural (INCAPER), Fundação Procafé, and Embrapa Rondônia. According to Bramel [31], the collection of these germplasm banks has an estimation of about 13,856 accessions; however, the number of accessions may be inconsistent across reports.

As stated by Bramel [31], it is estimated 21,026 accessions in a compilation of world coffee collections that account 52 holding coffee germplasm collection with at least ten accessions.

5. Breeding and genetic diversity based on morphological and agronomic traits

In plant breeding, it is crucial to identify the most critical phenotypic traits to boost plant production. Consequently, the evaluation of trait occurrences and differences in a population is a key to determining probably valuable crosses among accessions. Although most studies focus on genetic diversity with molecular markers, it is also useful for plant breeders to recognize the morphological diversity of traits of interest [57].

Around the world, the arabica breeding programmes has the primary purpose of developing new cultivars taking into account the economic benefits to be returned to coffee growers. The target characteristics in the desired arabica cultivar are productivity, mainly focused on bean size as well as cup quality and resistance to major diseases and pests. On the other hand, each breeding programme has its own particularities that establish the priorities of selection criteria usually defined based on multifactorial variations in specific circumstances of weather conditions, soil, biotic and abiotic stresses, cropping systems, socio-economic factors, market dynamics and consumer preferences. In arabica coffee, typically, four primary methods of breeding and selection are used: 1- Pure line selection; 2- Pedigree selection after hybridization (sometimes also backcrossing); 3 - Intraspecific F1 hybrids; 4 - Interspecific hybridization (arabica x robusta), backcrossing and pedigree selection. The comprehensive overview of selection criteria and outcomes from each breeding method is presented in detail by Van der Vossen [58].

Gathering a series of studies, Monge and Guevara [52] make the compilation of the critical phenotypic markers for evaluation of coffee and suggests a list of appropriated traits evaluation markers: morphological descriptors - viz.: architectural (ramification degree, number of internodes, and length of plagiotropic branches) and physical (dimensions and color of leaves, flowers and fruits, flush color, stem diameter, *etc.*); phenological descriptor (flowering dates, fructification cycle duration); ecological adaptation descriptors (altitude, dry or humid regions, resistance to pest and diseases); productive descriptors (productivity level and early or late flowering, and fruit set); technological descriptors (coffee quality, the weight of 100 beans, caracoli rate, etc).

Monge and Guevara [52] in a review also outlined a compilation result of two studies concerning the phenotypic evaluation of 300 wild *C. arabica* collected in eight Ethiopia area, those accessions were added into CATIE collections in 1985. It highlighted the high variability in fruit maturation length (ranging from 130 to 258 days), a caracoli rate (varying from 1 to 71%), size of leaves, internode length and bean size. Furthermore, there was a detected correlation concerning morphologic variables - viz.: the lower ramification of the tree, the bigger the leaves and the bean that produces.

Cilas et al. [59], in a study concerning genetic value prediction for *C. arabica* production through evaluation of morpho-agronomic traits, having the yield registered throughout the first four years of production. They concluded that better coffee yield may be increased by the addition of the medium level of heterozygosity, once the hybrid present immense superiority in comparison to the parental line. Furthermore, these authors also affirm that the prediction of yield may also be fully achieved by combining morphological traits, for instance, stem diameter, number of primary branches and tree height.

Bertrand et al. [60], addressing efforts towards sustainability, performed a study in three Central American countries comprising of 15 trials between 2000 to 2006 aiming to assess F1 hybrids of *C. arabica* in the agroforestry system (shade) compared to full-sun (unshade) crop system. The experiment involved thirteen lines and twenty-one F1 hybrids that were measured to average production throughout the first production cycle earlier than pruning and coppicing. The results point out that the green coffee per tree yield was higher among F1 hybrids in contrast to traditional cultivar in 58%, aggregating to 170 g in agroforestry, whereas in the full-sun system this increment was 34%, accumulating 190 g. In this respect, the economic outcomes of both systems look quite similar. This study also discussed the economic advantage in the agroforestry system renovation with hybrids, indicating that after six years of replacing the traditional cultivar by hybrids could earn up 5000 USD/ha. They were also pointing to the facilitation of credit policies and the opportunity of reaching new market niches with differentiated prices.

The first original phenotypic structure within *C. arabica* was present by Montagnon and Bouharmont [61]. The authors observed eighteen morphological and agronomic characteristics in a field collection of 148 accessions used the analyzed by multivariate approach. Interestingly, the result allowed identifying a sharp structure split into two main groups, comprise respectively 53 and 76 accessions. The other six groups are composed of less than five entries. The principal component analysis explained 77% of the accumulated variation within the first two axes, which is reasonably good. Also, the authors believe that the arrangement of the two main structured groups combined with the historical evidence of those accessions infers that group 1 has not been engaged within the domestication pathway of *C. arabica*. The traits modified by the course of domestication partly explained the well-defined separation of those two main groups.

The genetic diversity study conduced in Tepi National Spices Agricultural Research Center on 93 *C. arabica* accessions based 22 quantitative characteristics was able to detect five clusters by using multivariate techniques of hierarchical cluster and principal component analysis [62]. According to Klief [62], the significant inter-cluster distances between clusters point out that there is a high probability for obtaining transgressive segregates and maximize heterosis by crossing germplasm accessions across distinct clusters.

An study carried out in southwestern Saudi Arabia evaluated the genetic variation of accessions of *C. arabica* conserved *in-situ* in 19 localities, where stressful conditions prevail. Multivariate approach applied on 17 quantitative traits detected five groups. Interestingly, four accessions from the same place were grouped in four different clusters, supporting the importance of *in-situ* conservation strategy. All cluster showed significant inter-cluster distance, where two clusters present highest cluster distance. Therefore, Tounekti et al. [63] affirms that from these findings, it is suitable to explore this variability in breeding programmes to overcome environmental stresses.

The biochemical aspect of coffee liquor is highly essential. From this point of view, it was made a study addressing the genetic diversity based on caffeine content level concurrently with physical aspects of green bean characteristics and coffee cup quality. The examination of dissimilarities involved cluster analysis based on unweighted pair group arithmetic average (UPGMA), together with correlation among those variables analyzed. The outcome results consisted of two main groups were distinguished. The first cluster formed by 11 accessions distinguished by high caffeine content, undesirable physical characteristics of green bean and poor coffee cup quality. The other cluster split into two subgroups: the first with 26 accessions with caffeine content varying from low to average level and cup quality; the next subgroup with five accessions characterized by a medium level of caffeine content, desirable physical qualities of green coffee bean and high-grade cup quality. The authors also identify negative and significant associations linking caffeine content and all other variables related to cup quality. From that perspective, it is possible a simultaneous improvement of desirable cup quality plus low caffeine content [64].

A research, performed in IAC, evaluated the effectiveness of a minimum set of descriptors established for the conduct of test for distinctness, uniformity and stability in *C. arabica*. Twenty-nine cultivars were scattered in 11 groups when assessed by 35 morphological characteristics and three agronomic traits during three years. The results demonstrate that those descriptors were skilled in discriminating cultivar groups but a minor role in the identification of cultivars within each group. Therefore, the authors recommend the adoption of molecular markers and biochemical descriptors to identify cultivars to be protected more accurately [65].

Weldemichael et al. [66] conducted one well-designed study estimating genetic parameters in 49 accessions of *C. arabica*. It was used 26 carefully chosen appropriated quantitative traits aiming to estimate the phenotypic variation. The statistical analyses approach consisted of a series of adequate genetic parameters estimation. The findings exhibited the occurrence of variability for some morphological traits among coffee germplasm accessions. Interestingly, coffee berry disease recorded a pronounced genetic gain per population mean (88.8%); this point draws particular attention, once in arabica coffee disease resistance is a breeding objective of the chief priority to plant breeders. The detected low genetic advance as per cent mean and/or low genotypic coefficients of variation exhibited in most traits indicating these characteristics could not be developed through simple section rather heterosis breeding. Conversely, they advise that high morphological variation is not a guarantee of pronounced genetic variation; in this viewpoint, it is helpful to take into consideration the molecular and biochemical studies as a complementary approach.

6. Genetic diversity based on molecular markers

The progress achieved in plant breeding programmes culminated in reduced genetic variability in the improved populations [36, 67–69]. This problem may be worse in species with a narrow genetic base, such as Arabica coffee (*C. arabica*). The narrow genetic base of this species is associated with its autogamy, the low number of plants that were initially distributed worldwide, and the recent evolution of the species [30, 36, 70]. Thus, genotype discrimination based on differences in phenotypic characteristics may be difficult because individuals who are genetically distinct may be phenotypically similar, which reduces the selective efficiency. To overcome this difficulty, molecular markers have been used as an important tool in the accurate discrimination of genotypes [71, 72].

DNA markers allow the detection of variations in DNA sequences between individuals of the same species. Because they identify variations in DNA, they are stable and are unaffected by the environment or by pleiotropic or epistatic effects [73]. Thus, molecular markers have been used in breeding programmes as an efficient tool for the discrimination of genotypes and the analysis of genetic variability, as their analysis is a precise association strategy between phenotypic and genotypic variability.

Genetic diversity assisted by molecular markers has been used in several stages of Arabica coffee breeding programmes. The molecular characterization of coffee accessions is an accurate tool for the conservation and more efficient use of genetic resources by breeders. This molecular information is useful in evaluating the redundancies and deficiencies of the germplasm and generates information on the efficiency of the collection, maintenance, and expansion of a germplasm bank. In addition, the study of molecular diversity provides fundamental information to help breeders choose parents to integrate into cross-breeding schemes, as well as in directing the improvement of the genetic base during the course of a breeding programme.

Different molecular markers, such as simple sequence repeats (SSRs), sequence-characterized amplified regions (SCARs), and single-nucleotide polymorphisms (SNPs), have been identified and made available for coffee [71, 72, 74–82]. These species-specific markers combined with random markers, such as inter-simple sequence repeats (ISSRs), random amplified polymorphic DNA (RAPD) and amplified fragment length polymorphisms (AFLPs); support the genetic breeding of this crop.

Genetic studies and analyses of diversity and molecular characterizations of different germplasm banks and cultivars of *C. arabica* have benefited from molecular marker technology. Coffee plants belonging to the group of the Híbrido de Timor (HdT) from the Brazilian germplasm bank of the Universidade Federal de Viçosa (UFV) in partnership with Empresa de Pesquisa Agropecuária de Minas Gerais (EPAMIG) and Empresa Brasileira de Pesquisa Agropecuária (Embrapa Café) have been studied in detail using AFLP and SSR markers [83]. HdT coffee plants are the result of natural hybridization between *C. arabica* and *C. canephora* and are one of the main sources of resistance genes to coffee diseases and pests [84–86]. Through molecular markers, redundancy was observed in the core collection of the HdT, so that two plants with different identifications corresponded to the same genotype. One of them was eliminated, resulting in a core collection containing 151 unique and properly discriminated HdTs. The data obtained allowed fingerprinting of the accessions [83]. The fingerprinting of each genotype allows the identification of individuals through a unique code. This information will provide reliability to breeders for germplasm maintenance, preservation, and exchange.

With 52 alleles from 22 SSRs, it was possible to access the diversity of the Core Collection of HdT [83]. Considerable variability was observed between the accessions, which were separated into 21 groups. This grouping result was analyzed together with the resistance data obtained for the main coffee diseases, rust and coffee berry disease. The concentration of individuals resistant to both diseases was verified in eight groups. Through this analysis, it was possible to identify HdT coffee plants belonging to distinct genetic diversity groups that have not yet been used in genetic breeding. This made it possible to select genotypes in the obtained dendrogram that were as distinct as possible from the sources already explored to date and that have different disease resistance genes. The selected HdT accessions consist of potential parents for breeding aiming resistance to multiple diseases [83].

Molecular markers were also analyzed in the HdT to understand the introgression of the genomes from the coffee species of their origin (*C. arabica* and *C. canephora*), as well as their potential impact on the cup quality on the *C. arabica* cultivars. HdT has the largest portion of the genome corresponding to *C. arabica* [87]; however, the small portion of *C. canephora* provides disease resistance genes. This portion, even though small, raises concern about the possibility of *C. canephora* affect the cup quality, since the beverage quality of *C. canephora* is known to be lower. Thus, the effect of introgression of *C. canephora* on HdT derivatives were evaluated [88, 89]. The study also demonstrated the presence of disease-resistant genotypes combined with good cup quality typical of *C. arabica* cultivars. The genetic diversity analysis showed high genetic similarity between HdT with *C. arabica* and clear differentiation among coffee species. The introgression of *C. canephora* in the HdT accessions did not reach 30%. The sensory analysis of the coffee genotypes showed no significant difference in the beverage quality parameters between *C. arabica* cv. Bourbon and HdT-derived cultivars, which demonstrated the possibility of developing *C. arabica* cultivars without affecting beverage quality [89].

Accessions of different species and interspecific hybrids from the germplasm bank of UFV/EPAMIG/Embrapa were also analyzed with genomic SSRs and expressed sequence tag–SSR markers. The combination of these two types of markers allowed discriminating all accessions, including genotypes traditionally of *C. arabica*, genotypes containing introgression of HdT, *C. canephora*, HdT, *C. racemosa*, and triploids of *C. arabica* and *C. racemosa*. This study also identified unique alleles that are useful for accession discriminating in breeding programmes and for cultivar fingerprinting [90, 91].

Using the currently available large-scale genotyping technology, genetic diversity between and within Brazilian coffee breeding progenies was assessed by 49,567 SNPs. The significant number of SNP molecular markers distributed throughout *C. arabica* genome was efficient in discriminating all evaluated accessions by grouping them according to their genealogies. Mixtures within the families were identified. New parents to be introduced in the ongoing breeding were identified, and the parents currently used were analyzed in detail. The population structure and its effect on obtaining the improved varieties of *C. arabica* were discussed [72].

Accessions from the germplasm bank and cultivars launched by the breeding programme of the Instituto Agronômico de Campinas were analyzed with RAPD, AFLP, and SSR markers [92]. The variability observed between accessions was small, and only two groups were formed, one containing genotypes that included most cultivars and the other containing accessions/cultivars derived from interspecific crosses.

A more comprehensive analysis of Brazilian coffee plants was performed in 34 cultivars belonging to the Brazilian Cultivar Trial, using SSR markers [93]. The molecular pattern obtained allowed the discrimination of all cultivars and the creation of a fingerprinting data of the main cultivars of the country. The ability of markers to detect varietal mixtures and the diversity between and within cultivars was demonstrated.

The genetic variability of *C. arabica* accessions from other countries, such as Costa Rica [94], Mexico [95], Nicaragua [96], India [97–99], Indonesia [100], China [101], Kenya [102] and Ethiopia [34, 103–105], has also been analyzed using markers such as ISSRs, SSRs, sequence-related amplified polymorphisms (SRAPs), AFLPs, and SNPs. In Ethiopia, different studies have shown the presence of great genetic variability in coffee plants. This variability has been attributed to the particular ecological characteristics of the country, such as its rainfall amplitude and its different altitudes, temperatures, and soil fertility, which are suitable for the crop. The presence of indigenous coffee production methods in the country has also contributed to this diversity [5, 106]. Greater genetic diversity has been reported among wild coffee populations than cultivated genotypes [103].

A broader study of the diversity and fingerprinting of Arabica coffee accessions from various producing regions of the world was done in 2533 genotypes [107]. These genotypes corresponding to the Core Collection of the germplasm of the Tropical Agricultural Research and Higher Education Center, accessions from Southern Sudan, and cultivars/germplasm from North, Central, and South America as well as Africa and Asia. The obtained fingerprinting was efficient. Based on this tool, farmers can verify and trust the identity of the cultivars being planted, and coffee roasters can rely on marketing related to the cultivars they are growing and selling. The seed and nursery sector can become more professional and reliable by using this new monitoring tool to establish and verify the genetic purity of the seed and seedling stock.

Currently, SNP markers are using for genome-wide investigation [72, 82, 108]. In an original work of genome-wide association, candidate genes associated with lipids and diterpenes contents in *C. arabica* were identified [108]. This study detects the domestication and breeding process in *C. arabica*, pointing out the switch in allele frequency, revealing high allelic richness in wild accessions. In this regard, the identification of these candidate genes outlining potential targets for improving beverage cup quality in a coffee breeding programme.

7. Conclusion

Genetic resources commendably provide the basis of genetics solution to solve numerous problems of coffee growing areas throughout the world. The experimental schemes that lead to the introgression of new agronomic traits are known and have previously been validated with large populations. This approach has allowed the combination of several desirable traits in a single coffee cultivar. Also, plant breeders currently can count on the employment of molecular genetics to enhance the competence to introduce the desirable characteristics in the new cultivar. Molecular marker approach in association with morpho-agronomic characterization and diversity study helps to efficiently maintain the germplasm bank and facilitated its use by the breeder. Molecular tools are also useful to detect genetic structure and divergent breeding subpopulation. Application of genomics as a supplementary approach to conventional coffee breeding is highly recommended, improve the productivity of the breeding programme by reducing time to variety development as well as assure selection of desirable traits on the course of the breeding process, this is specifically relevant for the coffee crop that is perennial and has a narrow genetic base. Furthermore, molecular and morphological diversity approach provides nurseries, farmers and the whole coffee industry an opportunity to increase knowledge about the genetic identity of the coffee tree planted or traded.

The highly-regarded line of attack in the coffee sector is the elaboration of a wide-ranging catalog on existing germplasm collections including the markers profile. In the world, the usage of genetic diversity available in germplasm collections faces two significant problems: limited access to the conserved genetic resources and the deficiencies of genetic evaluation. Anthropogenic disturbances have modified the natural habitats where wild coffee species have spontaneously evolved, and in consequence, much relevant germplasm is in the risk of destruction. So, efforts of the scientific community are essential to design and implement conservation strategies. The ongoing partnership between Latin America and the African countries involved in the conservation and evaluation of coffee genetic resources is a well-intentioned strategy. This network aims to revitalize and advance the research to boost the productivity and cup quality of the coffee.

Conflict of interest

The authors do not have conflict of interests.

Author details

Juliano Lino Ferreira[1]*, Eveline Teixeira Caixeta[2], Fernanda Fatima Caniato[3], Tesfahun Setotaw[4], Gustavo César Sant'Ana[5] and Leila Maria Ferreira[6]

1 Embrapa Pecuária Sul, Bagé, Brazil

2 Embrapa Café, Brasília, Brazil

3 Department of Fundamental Sciences and Agricultural Development, Federal University of Amazonas, Manaus, Brazil

4 Ethiopian Agricultural Research Institute, Addis Ababa, Ethiopia

5 Tropical Melhoramento & Genética, Cambé, Brazil

6 Department of Statistics, Federal University of Lavras, Lavras, Brazil

*Address all correspondence to: juliano.ferreira@embrapa.br

References

[1] Morris J. Coffee: A Global History. 1st ed. London: Reaktions Books; 2018. 176 p.

[2] Zhou L, Vega FE, Tan H, Lluch AER, Meinhardt LW, Fang W, et al. Developing Single Nucleotide Polymorphism (SNP) Markers for the Identification of Coffee Germplasm. Trop Plant Biol. 2016;9:82-95. DOI: 10.1007/s12042-016-9167-2

[3] Vega FE, Ebert AW, Ming R. Coffee germplasm resources, genomics, and breeding. Plant Breed Rev. 2008;30:415-447p.

[4] Agricultural USDA. Coffee: World Markets and Trade. Coffee: World Markets and Trade [Internet]. 2020;9. Available from: http://apps.fas.usda.gov/psdonline/circulars/coffee.pdf

[5] Legesse A. Assessment of coffee (*Coffea arabica* L.) genetic erosion and genetic resources management in Ethiopia. Int J Agric Ext. 2020;7:223-229. DOI: 10.33687/ijae.007.03.2972

[6] Pearl HM, Nagai C, Moore PH, Steiger DL, Osgood R V., Ming R. Construction of a genetic map for arabica coffee. Theor Appl Genet. 2004;108:829-835. DOI: 10.1007/s00122-003-1498-3

[7] Fazuoli LC. Metodologia, critérios e resultados da seleção em progênies do café Icatu com resistência a *Hemileia vastatrix* [thesis]. Campinas: Universidade Estadual de Campinas; 1991.

[8] de Melo B, de Sousa LB. Biologia da reprodução de *Coffea arabica* L. e *Coffea canephora* Pierre. Rev Verde Agroecol e Desenvolv Sustentável. 2011;6:1-7

[9] Mooney HA, Lubchenci J, Dirzo R, Sala OE. Biodiversity and ecosystem functioning: basic principles. In: Heywood V, H., Watson RT, editors. Global Biodiversity Assessment. 1st ed. New York: Cambridge University Press; 1995. p. 275-325.

[10] Morris ML, Heisey PW. Estimating the benefits of plant breeding research: Methodological issues and practical challenges. Agric Econ. 2003;29:241-252. DOI: 10.1016/S0169-5150(03)00051-3

[11] Hawkes JG. The diversity of crop plants. Cambridge: Harvard Univerity Press; 1983. 184 p.

[12] Fowler C, Mooney PR. Shattering: food, politics, and the loss of genetic diversity. 1st ed. Tucson: University of Arizona Press; 1990. 296 p.

[13] Labouisse JP, Bellachew B, Kotecha S, Bertrand B. Current status of coffee (*Coffea arabica* L.) genetic resources in Ethiopia: Implications for conservation. Genet Resour Crop Evol. 2008;55:1079-1093. DOI: 10.1007/s10722-008-9361-7

[14] McNeely JA, Gadgil M, Levèque C, Padoch C, Redford K. Human influences on biodiversity. In: Heywood VH, Watson RT, editors. Global Biodiversity Assessment. 1st ed. New York: Cambridge University Press; 1995. p. 711-821.

[15] Oliveira MNV, Santos TMA, Vale HMM, Delvaux JC, Cordero AP, Ferreira AB, et al. Endophytic microbial diversity in coffee cherries of *Coffea arabica* from southeastern Brazil. Can J Microbiol. 2013;59:221-30. DOI: 10.1139/cjm-2012-0674

[16] de Almeida LF, Spers EE. Coffee Consumption and Industry Strategies in Brazil: A Volume in the Consumer Science and Strategic Marketing Series. 1st ed. Sawston: Woodhead Publishing; 2019. 394 p.

[17] Silva JP, Mendez GL, Lombana J, Marrugo DG, Correa-Turizo R. Physicochemical Characterization of Spent Coffee Ground (*Coffea arabica* L) and its Antioxidant Evaluation. Adv J Food Sci Technol. 2018;16:220-225.

[18] Moura W de M, e Oliveira AMC, Gonçalves DR, de Carvalho CFM, de Oliveira RL, Cruz CD. Adaptability and stability of organic-grown Arabica coffee production using the modified centroid method. Crop Breed Appl Biotechnol. 2017;17:359-365.

[19] Volsi B, Telles TS, Caldarelli CE, Camara MRG da. The dynamics of coffee production in Brazil. Aldrich SP, editor. PLoS One. 2019;14:e0219742. DOI: 10.1371/journal.pone.0219742

[20] Brazilian Coffee Production and Trade [Internet]. 2020. Available from: https://www.arcgis.com/apps/MapJournal/index.html?

[21] Monteiro MC, Farah A. Chlorogenic acids in Brazilian *Coffea arabica* cultivars from various consecutive crops. Food Chem. 2012;134:611-614. DOI: 10.1016/j.foodchem.2012.02.118

[22] Ratton, Clarissa; Santos J. Embrapa. Consumo mundial de café foi de 152,1 milhões de sacas de 60kg em 2015, segundo relatório da Organização Internacional do Café – OIC. 2016 [Internet]. 2020. Available from: https://www.embrapa.br/busca-de-noticias/-/noticia/10619109/consumo-mundial-de-cafe-foi-de-1521-milhoes-de-sacas-de-60kg-em-2015-segundo-relatorio-da-organizacao-internacional-do-cafe--oic

[23] International Coffee Organization. 2020. Available from: http://www.ico.org/

[24] Boaventura PSM, Abdalla CC, Araújo CL, Arakelian JS. Value co-creation in the specialty coffee value chain: the third-wave coffee movement. Rev adm empress. 2018;58;254-266. DOI: 10.1590/s0034-759020180306

[25] Vellema W, Casanova AB, Gonzalez C, D'Haese M. The effect of specialty coffee certification on household livelihood strategies and specialisation. Food Policy. 2015;57:13-25. DOI: 10.1016/j.foodpol.2015.07.003

[26] Hein L, Gatzweiler F. The economic value of coffee (*Coffea arabica*) genetic resources. Ecol Econ. 2006;60:176-185. DOI: 10.1016/j.ecolecon.2005.11.022

[27] Gepts P. The contribution of genetic and genomic approaches to plant domestication studies. Curr Opin Plant Biol. 2014;18:51-59. DOI: 10.1016/j.pbi.2014.02.001

[28] Anthony F, Combes MC, Astorga C, Bertrand B, Graziosi G, Lashermes P. The origin of cultivated *Coffea arabica* L. varieties revealed by AFLP and SSR markers. Theor Appl Genet. 2002;104:894-900. DOI: 10.1007/s00122-001-0798-8

[29] Sylvain PG. Some observations on *Coffea arabica* L. in Ethiopia. Turrialba. 1955;5: 37-53

[30] Scalabrin S, Toniutti L, Di Gaspero G, Scaglione D, Magris G, Vidotto M, et al. A single polyploidization event at the origin of the tetraploid genome of *Coffea arabica* is responsible for the extremely low genetic variation in wild and cultivated germplasm. Sci Rep. 2020;10:4642. DOI: 10.1038/s41598-020-61216-7

[31] Bramel P, Krishnan S, Horna D, Lainoff B, Montagnon C, editors. Global Conservation Strategy for Coffee Genetic Resources. 1st ed. Bonn: Crop Trust; 2017. 72 p.

[32] Smith RF. A History of Coffee. In: Clifford, M.N.; Willson KC, editor. Coffee: Botany, Biochemistry and

Production of Beans and Beverage. 1st ed. Boston: Springer; 1985. p. 1-12.

[33] Montagnon C, Bouharmont P. Multivariate analysis of phenotypic diversity of *Coffea arabica*. Genet Resour Crop Evol. 1996;43:221-227. DOI: 10.1007/BF00123274

[34] Anthony F, Bertrand B, Quiros O, Wilches A, Lashermes P, Berthaud J, et al. Genetic diversity of wild coffee (*Coffea arabica* L.) using molecular markers. Euphytica. 2001;118:53-65. DOI: 10.1023/A:1004013815166

[35] Clarindo WR, Carvalho CR. Comparison of the *Coffea canephora* and *C. arabica* karyotype based on chromosomal DNA content. Plant Cell Rep. 2009;28:73-81. DOI: 10.1007/s00299-008-0621-y

[36] Setotaw TA, Caixeta ET, Pereira AA, de Oliveira ACB, Cruz CD, Zambolim EM, et al. Coefficient of parentage in *Coffea arabica* L. cultivars grown in Brazil. Crop Sci. 2013;53: 1237-1247. DOI: 10.2135/cropsci2012.09.0541

[37] Charrier A, Berthaud J. Botanical Classification of Coffee. In: Clifford, M.N.; Willson KC, editor. Botany, Biochemistry and Production of Beans and Beverage. Boston: Springer; 1985. p. 13-47.

[38] Anzueto F, Baumann TW, Graziosi G, Piccin CR, Söndahl MR, van der Vossen HAM. The plant. In: Illy, A.; Viani R, editors. Espresso Coffee [Internet]. 2nd ed. London: Elsevier; 2004. p. 21-86. DOI: 10.1016/B978-012370371-2/50003-2

[39] van der Vossen H, Bertrand B, Charrier A. Next generation variety development for sustainable production of arabica coffee (*Coffea arabica* L.): a review. Euphytica [Internet]. 2015;204:243-256. DOI: 10.1007/s10681-015-1398-z

[40] Lécolier A, Besse P, Charrier A, Tchakaloff TN, Noirot M. Unraveling the origin of *Coffea arabica* "Bourbon pointu" from la Réunion: A historical and scientific perspective. Euphytica. 2009;168:1-10. DOI: 10.1007/s10681-009-9886-7

[41] Candolle, A. The Origin of Cultivated Plants (Cambridge Library Collection - Botany and Horticulture). 1st ed. Cambridge: Cambridge University Press; 2011. 468 DOI:10.1017/CBO9781139107365

[42] Pendergrast M. Uncommon grounds: The history of coffee and how it transformed our world. New York: Basic Books; 2010. 424 p.

[43] Ferreira T, Shuler J, Guimarães R, Farah A. Introduction to Coffee Plant and Genetics. In: Farah A, editor. Coffee: Production, Quality and Chemistry. London: The Royal Society of Chemistry; 2019. p. 1-25. DOI: 10.1039/9781782622437-00001

[44] Engelmann F, Dulloo ME, Astorga C, Dussert S, Anthony F. editors Conserving coffee genetic resources : complementary strategies for *ex situ* conservation of coffee (*Coffea arabica* L.) genetic resources : a case study in Catie, Costa Rica. 1st ed. Rome: Bioversity international; 2007. 63 p.

[45] Davis AP, Govaerts R, Bridson DM, Stoffelen P. An annotated taxonomic conspectus of the genus *Coffea* (Rubiaceae). Bot J Linn Soc. 2006;152:465-512. DOI: 10.1111/j.1095-8339.2006.00584.x

[46] Gole TW, Denich M, Teketay D, Vlek PLG. Human Impacts on the *Coffea arabica* Genepool in Ethiopia and the Need for its *in situ* Conservation. In: Engels JMM, Rao VR, Brown AHD, Jackson MT, editors. Managing Plant Genetic Diversity. 1st ed. New York: CABI Publishing; 2002. p. 237-247. DOI: 10.1079/9780851995229.0237

[47] Gole TW. Conservation and use of coffee genetic resources in Ethiopia: challenges and opportunities in the context current global situations. In: Globalization and Equity. In: Proceedings of the 4th Annual Global Development Network Conference.19-21 January 2003; CAIRO; New Delhi: Global Development Network; 2003. p. 1-23

[48] Meyer FG, Fernie LM, Narasimhaswamy RL, Monaco LC, Greathead DJ. FAO coffee mission to Ethiopia 1964-1965. Rome: Food and Agriculture Organization of the United Nations; 1968. 200 p.

[49] Anthony F, Astorga C, Berthaud J. Los Recursos Genéticos: Las Bases De Una Solución Genética A Los Problemas De La Caficultura Latinoamericana. In: Bertrand B, Rapidel B, editors. Desafios de la caficultura en Centroamérica. 1st ed. San José: Centro de Cooperación Intemacional de Investigación Agricola para el Desarrollo (CIRAD) de Francia; 1999. p. 369-406.

[50] Giomo, Gerson Silva; Mistro, Julio Cesar; Pereira SP. Café dos Brasil - do IAC para o mundo. O Agronômico. 2017;69.

[51] Volk GM, Krishnan S. Case Study: Coffee Wild Species and Cultivars. In: Volk GM, Byrne P, editors. Crop Wild Relatives and their Use in Plant Breeding. 1st ed. Fort Collins: Colorado States University; 2020.

[52] Monge MA, Guevara R. Agriculture_in_alliance_with_nature. Série técnica. Informe Técnico 315. Turrialba, Costa Rica: CATIE; 2000.

[53] Dussert S, Vasquez N, Salazar K, Anthony F, Engelmann F. Cryopreservation of coffee genetic resources. In: Engelmann F, Dulloo ME, Astorga C, Dussert S, Anthony F, editors. Conserving coffee genetic resources. 1st ed. Rome: Bioversity international; 2007. p. 49-58.

[54] Benti T. Progress in Arabica Coffee Breeding in Ethiopia: Achievements, Challenges and Prospects. Int J Sci Basic Appl Res. 2017;33:15-25.

[55] Teferi D. Achievements and Prospects of Coffee Research in Ethiopia: A Review. Int J Res Stud Agric Sci. 2019;5:41-51. DOI: 10.20431/2454-6224.0511006

[56] Banco de Germoplasma de café do IAC contribui com pesquisas de melhoramento genético do Consórcio Pesquisa Café [Internet]. 2012. Available from: http://www.consorciopesquisacafe.com.br/index.php/imprensa/noticias/243-banco-de-germoplasma-de-cafe-do-iac-contribui-com-pesquisas-de-melhoramento-genetico-do-consorcio-pesquisa-cafe

[57] Ferreira JL, Gwinner R, Ferreira LM, Ferronato J, Leite LG, Ferreira KGG, et al. Understanding the extent of phenotypic variability in accessions of *Paspalum urvillei* Steud. From the USDA NPGS. Iheringia - Ser Bot. 2020;75:e2020006. DOI: 10.21826/2446-2312020v75e2020006

[58] Van der Vossen HAM. Agronomy I: Coffee Breeding Practices. In: Clarke RJ, Vitzthum OG, editors. Coffee: Recent Developments. 1st ed. London: Blackwell Science; 2001. p. 184-201.

[59] Cilas C, Bouharmont P, Boccara M, Eskes AB, Baradat P. Prediction of Genetic Value for Coffee Production in *Coffea arabica* from a Half-Diallel with Lines and Hybrids. Euphytica. 1998;104:49-59. DOI: 10.1023/A:1018635216182

[60] Bertrand B, Alpizar E, Lara L, SantaCreo R, Hidalgo M, Quijano JM, et al. Performance of *Coffea arabica* F1 hybrids in agroforestry and full-sun cropping systems in comparison with

American pure line cultivars. Euphytica. 2011 Sep;181:147-58. DOI: 10.1007/s10681-011-0372-7

[61] Montagnon C, Bouharmont P. Multivariate analysis of phenotypic diversity of *Coffea arabica*. Genet Resour Crop Evol. 1996;43:221-227. DOI: 10.1007/bf00123274

[62] Kifle AT, Ali HM, Ayano A. Genetic Diversity Analysis of Tepi Surroundings Coffee (*Coffea arabica* L.) Germplasm Accessions using Quantitative Traits in Ethiopia. Inter J Agri Biosci. 2018;7:76-80.

[63] Tounekti T, Mahdhi M, Al-Turki TA, Khemira H. Genetic Diversity Analysis of Coffee (*Coffea arabica* L.) Germplasm Accessions Growing in the Southwestern Saudi Arabia Using Quantitative Traits. Nat Resour. 2017;08:321-336. DOI: 10.4236/nr.2017.85020

[64] Dessalegn Y, Labuschagne MT, Osthoff G, Herselman L. Genetic diversity and correlation of bean caffeine content with cup quality and green bean physical characteristics in coffee (*Coffea arabica* L.). J Sci Food Agric. 2008;88:1726-30. DOI: 10.1002/jsfa.3271

[65] Aguiar AT da E, Guerreiro Filho O, Maluf MP, Gallo PB, Fazuoli LC. Caracterização de cultivares de *Coffea arabica* mediante a utilização de descritores mínimos. Bragantia. 2004;63:179-92. DOI: 10.1590/S0006-87052004000200003

[66] Weldemichael G, Alamerew S, Kufa T. Genetic variability, heritability and genetic advance for quantitative traits in coffee (*Coffea arabica* L.) accessions in Ethiopia. African J Agric Res. 2017;12:1824-1831. DOI: 10.5897/AJAR2016.12059

[67] Rodgers DM, Murphy JP, Frey KJ. Impact of Plant Breeding on the Grain Yield and Genetic Diversity of Spring Oats 1. Crop Sci. 1983;23:737-740. DOI: 10.2135/cropsci1983.0011183x002300040032x

[68] Ortiz R, Lund B, Andersen SB. Breeding gains and changes in morphotype of Nordic spring wheat (1901-1993) under contrasting environments. Genet Resour Crop Evol. 2003;50:455-459. DOI: 10.1023/A:1023902110224

[69] Smith JSC, Duvick DN, Smith OS, Cooper M, Feng L. Changes in pedigree backgrounds of pioneer brand maize hybrids widely grown from 1930 to 1999. Crop Sci. 2004;44:1935-1946. DOI: 10.2135/cropsci2004.1935

[70] Carvalho A, Krug CA. Agentes de polinização da flor do cafeeiro (*Coffea arabica* L. Bragantia. 1949;9:11-24. DOI: 10.1590/S0006-87051949000100002

[71] Ferrão LF V., Caixeta ET, Pena G, Zambolim EM, Cruz CD, Zambolim L, et al. New EST–SSR markers of *Coffea arabica*: transferability and application to studies of molecular characterization and genetic mapping. Mol Breed. 2015;35:31. DOI: 10.1007/s11032-015-0247-z

[72] Sousa TV, Caixeta ET, Alkimim ER, de Oliveira ACB, Pereira AA, Sakiyama NS, et al. Population structure and genetic diversity of coffee progenies derived from Catuaí and Híbrido de Timor revealed by genome-wide SNP marker. Tree Genet Genomes. 2017;13:124. DOI: 10.1007/s11295-017-1208-y

[73] Adhikari S, Saha S, Biswas A, Rana TS, Bandyopadhyay TK, Ghosh P. Application of molecular markers in plant genome analysis: a review. Nucl. 2017;60:283-97. DOI: 10.1007/s13237-017-0214-7

[74] Rovelli, P.; Mettulio, R.; Anthony, F.; Anzueto, F.; Lashermes, P.; Graziosi G. Microsatellites in *Coffea*

arabica L. In: Sera, T.; Soccol, C.R.; Pandey, A.; Roussos S, editor. Coffee Biotechnology and Quality. Springer Netherlands; 2000. p. 123-33. DOI: 10.1007/978-94-017-1068-8_9

[75] Combes MC, Andrzejewski S, Anthony F, Bertrand B, Rovelli P, Graziosi G, et al. Characterization of microsatellite loci in *Coffea arabica* and related coffee species. Mol Ecol. 2000;9:1178-1180. DOI: 10.1046/j.1365-294X.2000.00954-5.x

[76] Moncada P, McCouch S. Simple sequence repeat diversity in diploid and tetraploid *Coffea* species. Genome. 2004;47:501-509. DOI: 10.1139/g03-129

[77] Cubry P, Musoli P, Legnaté H, Pot D, De Bellis F, Poncet V, et al. Diversity in coffee assessed with SSR markers: Structure of the genus *Coffea* and perspectives for breeding. Genome. 2008;51:50-63. DOI: 10.1139/G07-096

[78] Missio RF, Caixeta ET, Zambolim EM, Zambolim L, Sakiyama NS. Development and validation of SSR markers for *Coffea arabica* L. Crop Breed Appl Biotechnol. 2009;9:361-371. DOI: 10.12702/1984-7033.v09n04a11

[79] Vieira ESN, von Pinho ÉVDR, Carvalho MGG, Esselink DG, Vosman B. Development of microsatellite markers for identifying Brazilian *Coffea arabica* varieties. Genet Mol Biol. 2010;33:507-514. DOI: 10.1590/S1415-47572010005000055.

[80] Diola V, de Brito GG, Caixeta ET, Maciel-Zambolim E, Sakiyama NS, Loureiro ME. High-density genetic mapping for coffee leaf rust resistance. Tree Genet Genomes. 2011;7:1199-1208. DOI: 10.1007/s11295-011-0406-2

[81] Ferrão LF V., Caixeta ET, Souza F de F, Zambolim EM, Cruz CD, Zambolim L, et al. Comparative study of different molecular markers for classifying and establishing genetic relationships in *Coffea canephora*. Plant Syst Evol. 2013;299:225-38. DOI: 10.1007/s00606-012-0717-2

[82] Alkimim ER, Caixeta ET, Sousa TV, Da Silva FL, Sakiyama NS, Zambolim L. High-throughput targeted genotyping using next-generation sequencing applied in *Coffea canephora* breeding. Euphytica. 2018;214:50. DOI: 10.1007/s10681-018-2126-2

[83] Silva RA, Zambolim L, Castro ISL, Rodrigues HS, Cruz CD, Caixeta ET. The Híbrido de Timor germplasm: identification of molecular diversity and resistance sources to coffee berry disease and leaf rust. Euphytica. 2018;214:153. DOI: 10.1007/s10681-018-2231-2

[84] Bettencourt, A.J.; Noronha-Wagner, M.; Lopes J. Factor genético que condiciona a resistência do clone 1343/269 ("Híbrido de Timor") à *Hemileia vastatrix* Berk. & Br. Brotéria Genética. 1980;1:53-58.

[85] Wallace, Gonçalves; Pereira AA. Resistência de cafeeiros a nematóides IV-reação de cafeeiros derivados de Híbridos de Timor a *Meloidogyne exigua*. Nematol Bras. 1998;22:39-50.

[86] Pereira AA, Sakiyama NS, Zambolim L, Moura WM, Zambolim EM, Caixeta ET. Identification and use of sources of durable resistance to coffee leaf rust in the UFV/EPAMIG breeding program. In: Zambolim, L.; Zambolim, E.M.; Várzea VMP, editors. Durable resistance to coffee leaf rust. Viçosa: UFV; 2005. p. 215-232.

[87] Lashermes P, Combes MC, Robert J, Trouslot P, D'Hont A, Anthony F, et al. Molecular characterisation and origin of the *Coffea arabica* L. Genome. Mol Gen Genet. 1999;261:259-266. DOI: 10.1007/s004380050965

[88] Bertrand B, Guyot B, Anthony F, Lasherme P. Impact of the *Coffea canephora* gene introgression on

beverage quality of *C. arabica*. Theor Appl Genet. 2003;107:387-94. DOI: 10.1007/s00122-003-1203-6

[89] Setotaw TA, Caixeta ET, Zambolim EM, Sousa TV, Pereira AA, Baião AC, et al. Genome Introgression of Híbrido de Timor and Its Potential to Develop High Cup Quality C. arabica Cultivars. J Agric Sci. 2020;12:64-76. DOI: 10.5539/jas.v12n4p64

[90] Missio RF, Caixeta ET, Zambolim EM, Pena GF, Ribeiro AP, Zambolim L, et al. Assessment of EST-SSR markers for genetic analisys on coffee. Bragantia. 2009;68:573-581. DOI: 10.1590/s0006-87052009000300003

[91] Missio RF, Caixeta ET, Zambolim EM, Pena GF, Zambolim L, Dias LAS, et al. Genetic characterization of an elite coffee germplasm assessed by gSSR and EST-SSR markers. Genet Mol Res. 2011;10:2366-2381. DOI: 10.4238/2011.October.6.2

[92] Maluf MP, Silvestrini M, Ruggiero LM de C, Guerreiro Filho O, Colombo CA. Genetic diversity of cultivated *Coffea arabica* inbred lines assessed by RAPD, AFLP and SSR marker systems. Sci Agric. 2005;62:366-373. DOI: 10.1590/s0103-90162005000400010

[93] Sousa TV, Caixeta ET, Alkimim ER, de Oliveira AC B, Pereira AA, Zambolim L, et al. Molecular markers useful to discriminate Coffea arabica cultivars with high genetic similarity. Euphytica. 2017;213:75. DOI: 10.1007/s10681-017-1865-9

[94] Sánchez E, Solano W, Gatica-Arias A, Chavarría M, Araya-Valverde E. Microsatellite DNA fingerprinting of *Coffea* sp. Germplasm conserved in Costa Rica through singleplex and multiplex PCR. Crop Breed Appl Biotechnol. 2020;20:e27812013. DOI: 10.1590/1984-70332020v20n1a3

[95] Spinoso-Castillo JL, Escamilla-Prado E, Aguilar-Rincón VH, Morales Ramos V, de los Santos GG, Pérez-Rodríguez P, et al. Genetic diversity of coffee (*Coffea* spp.) in Mexico evaluated by using DArTseq and SNP markers. Genet Resour Crop Evol. 2020;67:1795-1806. DOI: 10.1007/s10722-020-00940-5

[96] Geleta M, Herrera I, Monzn A, Bryngelsson T. Genetic diversity of arabica coffee (*Coffea arabica* L.) in Nicaragua as estimated by simple sequence repeat markers. Sci World J. 2012;2012:ID 939820. DOI: 10.1100/2012/939820

[97] Prakash, S.N.; Combes, M.C.; Naveen, S.K.; Graziosi, G; Lashermes P. Application of DNA marker technologies in characterizing genome diversity of selected coffee varieties and accessions from India. In: ASIC 19th. Trieste: Association Scientifique Internationale du Cafe; 2001.

[98] Mishra MK, Sandhyarani N, Suresh N, Satheesh Kumar S, Soumya PR, Yashodha MH, et al. Genetic Diversity Among Indian Coffee Cultivars Determined via Molecular Markers. J Crop Improv. 2012;26:727-50. DOI: 10.1080/15427528.2012.696085

[99] Jingade P, Huded AK, Kosaraju B, Mishra MK. Diversity genotyping of Indian coffee (*Coffea arabica* L.) germplasm accessions by using SRAP markers. J Crop Improv. 2019;33:327-45. DOI: 10.1080/15427528.2019.1592050

[100] Yunita R, Oktavioni M, Chaniago I, Syukriani L, Setiawan MA, Jamsari J, et al. Analysis of genetic diversity of Arabica coffee [*Coffea arabica* L.] in Solok Regency by SRAP molecular markers. In: IOP Conference Series: Earth and Environmental Science.

IOP Publishing; 2020. p. 497. DOI: 10.1088/1755-1315/497/1/012018

[101] Yan L, Ogutu C, Huang L, Wang X, Zhou H, Lv Y, et al. Genetic Diversity and Population Structure of Coffee Germplasm Collections in China Revealed by ISSR Markers. Plant Mol Biol Report. 2019;37:204-213. DOI: 10.1007/s11105-019-01148-3

[102] Kathurima, C.W.; Kenji, G.M.; Muhoho, S.M.; Boulanger, R.; Gichimu, B.M.; Gichuru EK. Genetic diversity among commercial coffee varieties, advanced selections and museum collections in Kenya using molecular markers. Int J Biodivers Conserv. 2012;4:39-46. DOI: 10.5897/ijbc11.231

[103] Aga E, Bryngelsson T, Bekele E, Salomon B. Genetic diversity of forest arabica coffee (*Coffea arabica* L.) in Ethiopia as revealed by random amplified polymorphic DNA (RAPD) analysis. Hereditas. 2003;138:36-46. DOI: 10.1034/j.1601-5223.2003.01636.x

[104] Geletu, K.T.; Govers, K.; Bekele, E.; Borsch T. Genetic Diversity of Wild *Coffea arabica* in Ethiopia: Analyses based on plastid, ISSR and Microsatellite markers. In: 21st International Scientific Colloquium on Coffee. Montpellier: ASIC; 2006.

[105] Aerts R, Berecha G, Gijbels P, Hundera K, Van Glabeke S, Vandepitte K, et al. Genetic variation and risks of introgression in the wild *Coffea arabica* gene pool in south-western Ethiopian montane rainforests. Evol Appl. 2013;6:243-252. DOI: 10.1111/j.1752-4571.2012.00285.x

[106] Legesse A. Climate Change Effect on Coffee Yield and Quality: A Review. Int J For Hortic. 2019;5:1-9. DOI: 10.20431/2454-9487.0504001

[107] Pruvot-Woehl S, Krishnan S, Solano W, Schilling T, Toniutti L, Bertrand B, et al. Authentication of *Coffea arabica* Varieties through DNA Fingerprinting and its Significance for the Coffee Sector. J AOAC Int. 2020;103:325-334. DOI: 10.1093/jaocint/qsz003

[108] Sant'Ana GC, Pereira LFP, Pot D, Ivamoto ST, Domingues DS, Ferreira R V., et al. Genome-wide association study reveals candidate genes influencing lipids and diterpenes contents in *Coffea arabica* L. Sci Rep. 2018;8:465. DOI: 10.1038/s41598-017-18800-1

Section 2

Molecular Markers in the Detection of Genetic Polymorphism

Chapter 6

Potential of Mutation Breeding to Sustain Food Security

Arain Saima Mir, Meer Maria, Sajjad Muhammad and Sial Mahboob Ali

Abstract

Mutation is a sudden heritable change in the genetic material of living organism. Spontaneous mutation, the natural process that develops new allele copies of a gene was the only source of genetic diversity until the 20th century. Besides, mutations can also be induced artificially using physical or chemical mutagens. Chemical mutations received popularity due to its efficiency in creating gene mutations contrary to chromosomal changes. Mutation has played a vital role in the improvement of crop productivity and quality, resultantly > 3,000 varieties of 175 plant species have been developed either through direct or indirect induced mutation breeding approaches worldwide. The advances in plant breeding also achieved through molecular marker technology. The *in vitro* mutagenesis, heavy-ion beam, and space mutation breeding are being efficiently used to create genetic variability to improve various complicated traits in crop plants. In mutation breeding, TILLING (Targeting Induced Local Lesions in Genomes), a more advanced molecular technique is being used to identify specific sequential genomic changes in mutant plants. Therefore, the mutation breeding in combination with molecular techniques could be an efficient tool in plant breeding programs. This chapter will discuss and review the mutation breeding application for the improvement of crop productivity and environmental stresses.

Keywords: biotic and abiotic stresses, climatic changes, food security, ion beam, space mutagenesis, TILLING

1. Introduction

Mutation breeding also called "variation breeding," is the procedure of exposing seeds to chemicals or radiation to produce mutants with desirable traits. The mutants created are called mutagenic plants or mutagenic seeds and can be used directly as a commercial cultivar or used as parent to breed new commercial cultivars. Although mutation breeding in the past fifty years was mainly focused on improving the yield specifically height reduction in wheat and rice, in contrast, today's challenges are environmental stresses and its related effects and to motivating for climate-smart agriculture for food security [1]. In this chapter, the historical background of mutation breeding has been discussed chronologically. The types of mutants and mutations reported in worldwide literature are described. The natural and spontaneous mutations are elaborated with practical examples. All types of physical and chemical mutants and their success stories are discussed citing examples from all over the world. Lastly, the impact of mutation breeding on food security is explained with practical examples and achievements so far. The future prospectus of mutation breeding has also been discussed to highlight the significance of this important plant breeding process. This chapter provides a comprehensive understanding of the process with successful commercial examples of mutation breeding and the potential of this technique to meet future food security challenges. This chapter includes (i) introduction, (ii) historical background/development of mutation breeding (iii) mutation concept and its importance (iv) mutant crop varieties and their impact on food security (v) new breeding techniques (vi) prospects (vii)conclusions

2. Historical background/development of mutation breeding

The story of mutation and development of mutants in crop plants was first described in the book *Lula*, in 300 BC in China. The first natural mutant plant in cereals was found about 2317 years ago in China [2, 3]. Later, many aberrant plants with diverse variations were identified known as the first phase of mutation (1590-1868). The 2nd phase of the mutation was commenced in 1895 with the discovery of X-rays by W.K. Rontgen and the use of mutagens for the first time in 1897 to 1920 with the "Law of homologous series of variations" by N.I. Vavilov [2]. The chapter of mutation breeding was opened with the pioneered use of irradiation to create genetic variation by Lewis John Stadler in the 1920s.In almost the same time period Muller did his mutation experiments on fruit flies, Stadler was working on barley, maize and wheat manifested that radiation has power to create genetic variability in crop plants, although he was more interested in mutation breeding for fruit trees. Many geneticists believe the induction of mutation as a breakthrough in the history of genetics. American researchers were not so optimistic in their findings of the agricultural crops [3–5]. Chromosomal aberrations in Nicotiana were reported by Goodspeed and co-workers [6, 7]. The first-ever mutant variety "Vorsteland" of Tobacco with improved quality traits was released in Indonesia in 1934. Russian scientists, Delaunay and Sapehin reported the first wheat mutants with practical importance. German researchers started using mutation induction very early, but it was only theoretical in nature. The lecture on polymorphic factors in barley delivered by NILSSON-EHLE in Halle opened a new era in the use of induced mutation in Germany during the year 1939. In mid-thirties, he produced mutants at least one of those mutants [8] was equal to the mother genotype in yield performance. The experiments on induced mutation were extended after the cost for all work done till the 1940s was collectively funded by the A.B. Salts Jiiqvarn, Stockholm and Miirten Pehrsons Valsqvarn, Kristianstad, through Professor A. Akerman (head Swedish seed Association). This hard work led to the development of promising mutants in wheat, barley, oats, flax, soybeans, oleiferous and sweet lupine [8]. Stadler reported the production of solitary mutations and an increase in lethality by X-rays. Plants of different crop species respond differently to radiation doses. Cruciferae Seeds showed most insensitive sometimes tolerated 100.000r, whilst, the pea plants in contrast showed very sensitive with a maximum dose of only 10-20 r [9]. In the year 1942 and 1943, Eisleben Lien introduced comprehensive model experiments for barley. The use of induced mutation through radiation in crop breeding in Latin America was started during the 1960s initially in six countries viz., Colombia, Peru, Brazil, Guatemala, Costa Rica, and Mexico [10]. Rice breeding through use induced mutation started by china in 1960 and working continuously to improve the conventional and hybrid varieties. The first variety was the mutant developed in a series called 12 'Zhefu'. The most widely cultivated mutant variety between 1986 and 1994 was the Chinese variety'Zhefu802', which was evolved from Simei No. 2'. [10] Also reported radiation-induced biological effects in coffee breeding. In 1964 after the establishment of joint FAO/IAEA Division of Nuclear Techniques in Food and Agriculture, mutation breeding was acknowledged as a greater tool by plant breeders globally [11]. Nuclear Institute of Agriculture (NIA) released its first mutant wheat variety Jauhar-78 in 1979 with salinity tolerance and shattering resistance and Kiran-95 released in 1996 endowed with better grain quality, tolerance to salinity and rusts. In 1977 Pakistan released its first rice mutant variety 'Kashmir Basmati from Basmati 370.Over 1000 mutant varieties in various crops have been developed by China in collaboration with IAEA and FAO in the past 6 decades. China started the use of space mutagenesis for crop improvement in 1987, Chinese scientists stated to produce giant sweet peppers and improved quality traits in wheat and rice through rare inheritable genetic mutations using space radiation (satellites and high-altitude balloons) [12]. Using space induced radiation, a number of advantageous mutations to make a breakthrough in most desired crop yield was also achieved [13–15]. The officially released mutant varieties in China accumulate around 741 of 45 crops and ornamental species [16, 17]. Recently, China has announced the launching of a new satellite for experiments in space on a variety of industries including agriculture [18].

3. Mutation concept and its importance

The word 'mutation' was coined by Hugo de Vries (1901) to represent a sudden heritable change occurring in the DNA of an organism caused artificially through irradiation, chemicals, viruses, transposons, or chromosomal aberrations that occur during reproductive processes [19]. These changes can be transferred to the offspring and are e differentiated in three general types namely gene mutation, chromosomal and genomic mutations. Induced mutation became the most frequently employed technique for developing novel improved germplasm in crop plants [20]. Mutation breeding is the application of mutagens to plant cells to accomplish crop breeding. Genetic variation makes the basis for the evolutionary process and breeding. In 1940, mutagenesis was adopted by the breeders as a tool that works faster to create mutations in plants [21]. Induced mutation breeding techniques have become most efficient, fast-tracking and widely exploited tools for crop improvement worldwide (**Figure 1**).

Mutation can be differentiated in three general types namely gene mutation, chromosomal and genomic mutations. However, mutation breeding is the application of mutagens to plant cells to accomplish crop breeding. Mutation provides the fundamental basis for a genetic variation on which genetic advancement and genetic drift depend and a single base mutation can cause devastating or beneficial consequences or no effect at all. Mutation breeding has played a significant role in crop breeding and genetics and genomic studies by generating a large amount of genetic diversity. Concurrently, climatic changes also threatening the food supply chain on the global level, resulting in fast loss of biodiversity for food and agriculture. The ongoing unpredictable climatic changes are the core problem in reducing crop yields worldwide, thus continuous development of new improved varieties for sustainable production is unavoidable. While the rate of natural mutations in the

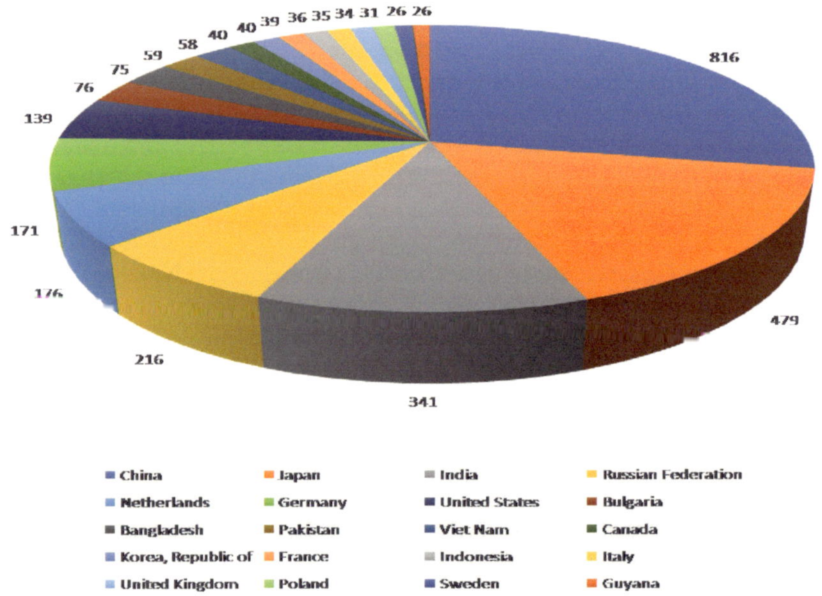

Figure 1.
Number of mutant varieties released in top 20 countries. Source: Mutant varieties database, IAEA accessed on 10th September, 2020.

crop plants is rare, thus use of induced mutation is indispensable to create genetic diversity for the desired traits for use in the breeding programs. Developing a new variety through mutation breeding reduces the time span for varietal development as compared to hybridization (**Figure 2**). Moreover, Mutants with multiple traits can be discriminated through mutation breeding, mutant varieties show a higher survival rate in the face of environmental swings. Mutagenesis is an efficient tool for generating mutations; these mutations can occur naturally or can be induced using mutagens, broadly classified as physical and chemical mutagens [22]. Mutagens offer more chances to acquire desired phenotypic changes and to study the genetic variations in relation to phenotypes and the annotation /deciphering of gene functions [23]. Various genetic resources of crop plants have been developed globally using different mutagenesis sources like EMS, gamma or X-rays and fast neutrons [24]. The crops like tomato have been focused after the availability of whole-genome sequencing data, which led to the identification of millions of single nucleotide polymorphisms (SNPs) and indels in tomato lines and in mutants [23]. In view of the introduction of high throughput next-generations equencing (NGS), several innovative approaches have been introduced for the discrimination of mutations in the mutagenized material. Some remarkable techniques are MutMap (mapping-by-sequencing) and MutChromSeq helpful to identify the basic changes induced through mutagenesis [25]. MutChromSeq helps to assort the desired genes in the shortest time span and has been successfully utilized in wheat and barley. Pakistan Atomic Energy Commission's (PAEC) first agriculture institute, Nuclear Institute of Agriculture (NIA)Tandojam" has exploited mutation breeding techniques since its inception in 1963 and developed 3 mutant varieties of wheat, 7 of rice, 1 of sugarcane, 5 of cotton, one each of lentil, mungbean and rapeseed through mutation breeding techniques. NIA released the first rice variety (Shadab) in 1978 from IR6 using ethyl methane sulphonate (EMS 0.5%) a chemical mutagen, variety had the potential to produce 7 tones/ha with superior grain quality [26]. However, the Nuclear Institute for Agriculture and Biology (NIAB) and other institutes of

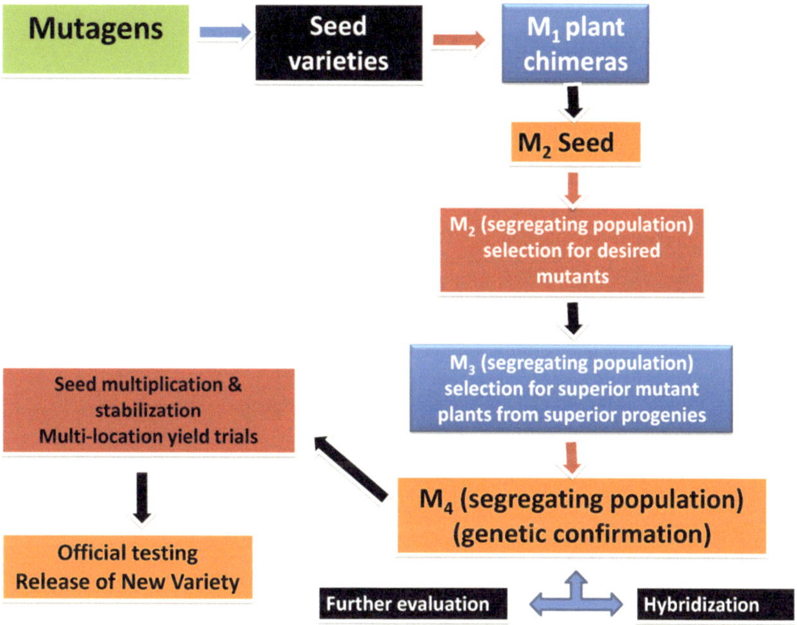

Figure 2.
Scheme of mutation breeding in crop plants.

PAEC have also developed mutant varieties of cotton, castor bean, sesame and mandarin thus helping the farming community by developing these improved varieties and boosting up their socio-economic status.

3.1 Spontaneous mutations

These are the genetic changes that occur due to chromosomal aberrations in the biological processes and serves as raw material for the evolutionary process. These mutations are the alleles of unknown genes which afterward given the name according to the phenotype or other related information viz., super-root (surl-7 to surl-7) [27], maize bronze (bz), carbohydrate accumulation mutant (caml) [28]. In maize spontaneous mutations occur in high frequency in the pollen part of some maize genotypes, but not in others [29]. Recessive mutations (one or two copies of the mutated allele produces the phenotype) are denoted by small letters, whilst dominant (one or two copies of the mutated allele produces the phenotype) and partially-dominant (one mutant allele produces an intermediate phenotype) are denoted by the first letter capital followed by the small letters. Most of the spontaneous mutations are point (single base pair change in the DNA) mutations. Gregor John Mendel was the first to quantitatively evaluate the dominance and recessiveness phenomenon in diploid organisms in 1866 [30].

3.2 Induced mutations

In addition to naturally occurring genetic mutations, novel alleles have been induced in plants by chemical and physical mutagenesis (**Figure 3**). The goal of mutagenesis is to induce genetic variation in cells that give rise to plants while minimizing chimeras, sterility and lethality [31]. Mutagenesis based breeding is primarily used to improve 1 to 2 main traits that effect on productivity or quality traits. More importantly is not under the regulatory restrictions faced by the genetically

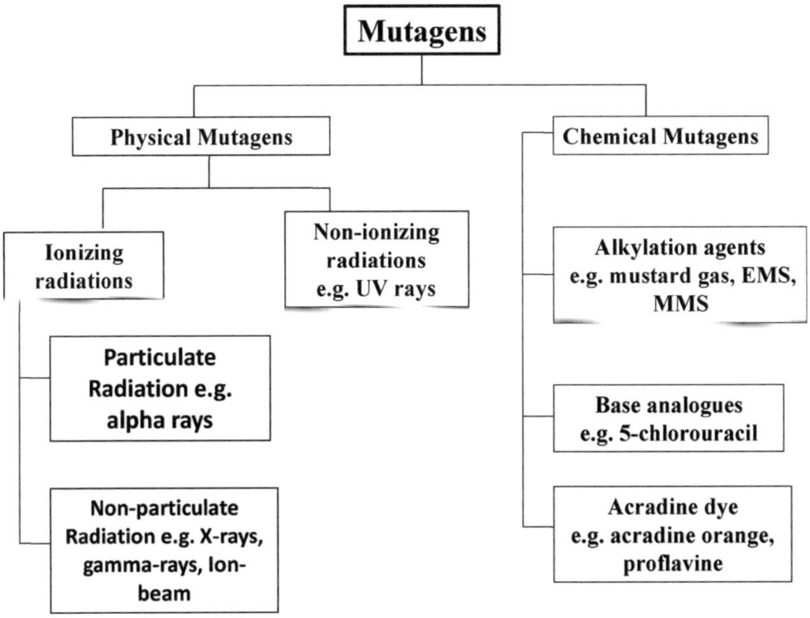

Figure 3.
Common mutagens used in plant mutation breeding. Source: Reproduced from FAO/IAEA, 2018.

modified organisms [32]. In some crops, chemically induced mutagenesis produced the desired phenotype in only several thousand lines. Today's high throughput phenotyping and next-generation sequencing methods have expedited the process to identify the mutants with desired genes (**Figure 4**). The use of engineered nucleases has helped to increase the accuracy of the mutation breeding through gene-specific mutation. Allelomorphic diversity induced in the gene of interest, whether spontaneously or experimentally, can be a great source for breeding programs to inculcate novel agricultural attributes [31]. Wanga et al. [33] used a combination of EMS and gamma radiation in sorghum but results were not recommendable. Although these are two major mutagens used to develop mutations [34, 35].

3.2.1 Physical mutagenesis

Physical mutagens namely X-rays, neutrons-alpha-beta particles, fast and thermal neutron, UV-light, especially gamma rays are used for the induction of mutation [36]. Physical mutagens are more common as compared to chemical mutagens (EMS) for mutagenesis. Physical mutagens like x-rays and gamma rays are preferred by the breeders as compared to the chemical ones. Gamma rays were used more frequently which accounted to improve 1604 mutants than the X-rays which improved 561 mutants [36]. Plant's exposure to X-rays provided the first ever undeniable evidence that phenotypic variability can be induced artificially. Hermann J. Muller was awarded Nobel Prize in 1946 in medicine/physiology for introducing irradiation using X-rays. Gamma-irradiation produces severe genetic mutations due to large chromosomal deletions and the re-enactment of the chromosome. Gamma rays have been used to induce mutations in seeds, cuttings, pollens and calli [37]. Since 1960 gamma irradiation has become the most popular and commonly used mutagens. This radiation-based mutagenesis was broadly used to improve mutant varieties directly as compared to other methods (acclimatization, selection, hybridization), comparatively, time-consuming, laborious and with

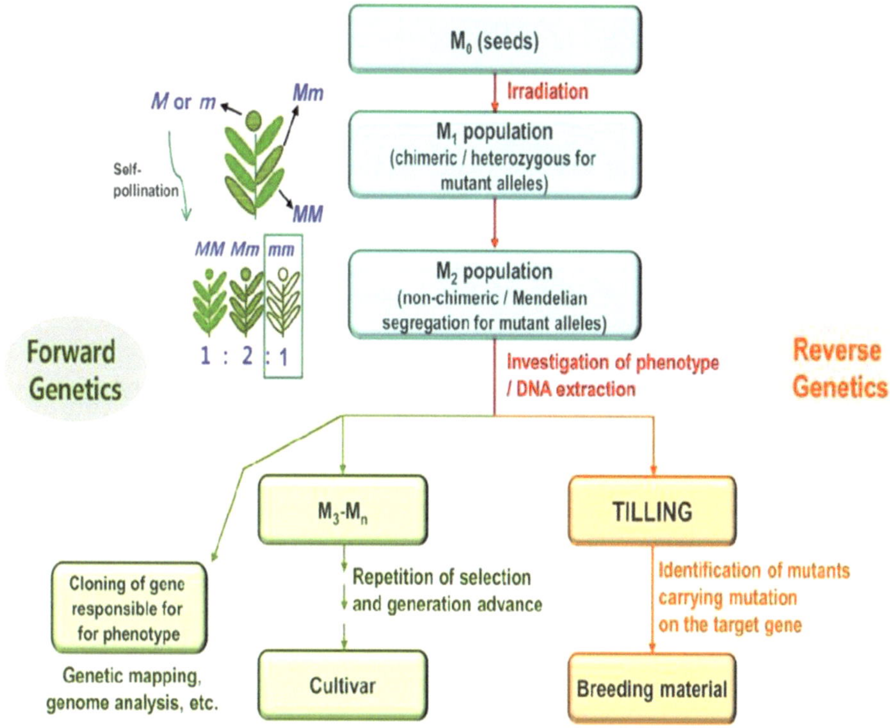

Figure 4.
Mutation breeding integrated use with modern techniques. Source: Directly taken from Jo and Kim, 2019.

lower genetic variation [38]. Fast neutron-induced mutagenesis is an exceptional technique among the other mutagenesis tools being employed in crop science in relation to higher impact. Fast neutrons normally cause deletions from a small number of bases to million bases [39]. Although, previously fast neutron was not as popular as other physical mutagens in plant mutagenesis [40].

3.2.2 Space mutagenesis

Space-induced mutation breeding uses cosmic rays to induce seeds in the space, for this experiment it is carried out in the satellites, space shuttles, and high altitude balloons and are considered beneficial over gamma radiation because of its lower damage to plants as compared to gamma rays on earth. Using space induced radiation, several advantageous mutations to make a breakthrough in yield were also achieved [13, 15, 41]. China has produced 41 varieties developed through space–induced mutation breeding of various crop species viz., rice, wheat, cotton, sesame, pepper, tomato, and alfalfa [42].

3.2.3 Ion beam mutagenesis

Heavy-ion bean is an important tool in mutation breeding since lower radiation doses are found to induce high mutation rates [43]. Due to its dense localized effect on DNA to effectively alter a single trait of the irradiated cultivar without damaging the rest of the characteristics, this technique is effectively being used in China and Japan to produce a large number of mutant varieties [44, 45]. In Japan, several ornamental plant varieties have been developed using high-energy ion beam irradiation while China is using low-energy Ion bean to create improved crop varieties. The initial plant varieties produced using Ion bean mutagenesis included carnation (*Dianthus caryophyllus*), Chrysanthemum (*Dendranthema Grandiflora*), and plants of Verbena sp. Afterward, several color and shape variations of petunia, Dahlia, and Torenia were also developed using this mutagenesis technique. Furthermore, the varieties developed using Ion beam mutagenesis include not only ornamental plants of high commercial demand [46], but also crops like salt-tolerant rice, citrus fruits, coniferous trees, mutant blast-resistant rice [47], mutant muskmelon, and rice varieties with lower fertilizer requirements [48].

3.3 Chemical mutagenesis

Chemical mutagenesis is the most efficient and expedient tool used for a large number of plant species. Ethyl methane sulfonate and sodium azide are the most widely used chemical mutagens to induce mutations in various crop plants like a tomato. The chemical mutagens used in mutation breeding are ethyl methanesulphonate (EMS), hydroxylamine, methyl methanesulphonate (MMS), sodium azide hydrogen fluoride (HF), and N-methyl-N-nitrosourea (MNU) [32]. Although, EMS is the most extensively used mutagen in plants due to its high efficiency at inducing point mutation (changes in a single nucleotide) and deletions (loss of chromosomal segment) in the chromosomal fragments. Mutant populations in various cereal crops using chemical mutagens for seeds or pollens have been developed comprising maize [49], barley [50, 51], rice [52], sorghum [53], and both hexaploid bread wheat [54] and durum wheat [55]. The EMS was exploited for potyvirus resistance in tomato [23].

4. Mutant crop varieties and their impact on food security

Mutation breeding techniques especially gamma and other physical mutagens have helped in generating a large number of mutants and generated a massive quantity of genetic variability that is significantly employed in the studies from plant breeding and genetics and in modern studies (genomics) (**Figure 4**). The mutants are released directly as varieties or furnish as a basic resource in the breeding programs to create genetic variation. The released mutant varieties offer higher yields, disease-resistant, improved quality, and resilient to environmental swings. A huge number of these mutant cultivars have been released in developing regions boosting up the economic status of these countries. These varieties are covering hundreds of millions of ha of agricultural land, whilst the impact on national economies of these countries is measured based on billions of dollars. The technique of mutation breeding is highly successful and its widespread implementation for crop improvement has led to the release of 3333 mutant varieties from 228 plant species (rice, wheat, and fruits like grapefruit, lettuce and others) in over 73 countries globally [56]. More than 1000 mutant varieties of major food crops covering millions of hectares, improving the rural economy, nutrition and helping in sustainable food security. Food insecurity is increasing worldwide and about 2 billion people especially in low and middle-income countries are undernourished. Concurrently, climatic changes also threatening the food supply chain on the global level, resulting in fast loss of biodiversity for food and agriculture. The ongoing unpredictable climatic changes are the core problem in reducing crop yields worldwide, thus continuous development of new improved varieties for sustainable production is unavoidable. While the rate of natural mutations in the crop plants is rare, thus use of induced mutation is indispensable to create genetic

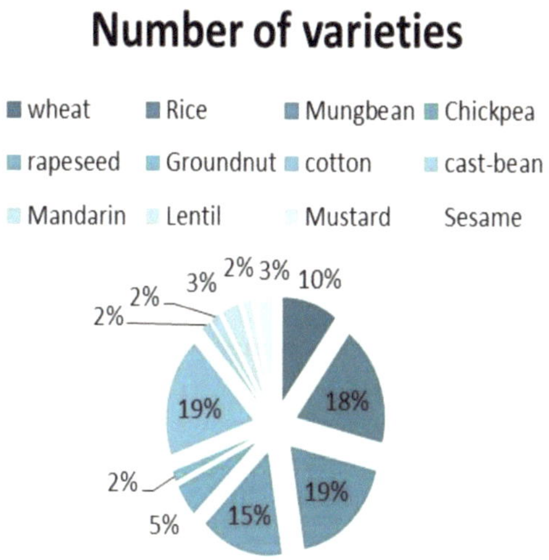

Figure 5.
Mutant varieties released in Pakistan. Data source: MVD/IAEA. Accessed on 13th September, 2020.

diversity for the desired traits for use in the breeding programs. The widespread use of mutation techniques in plant breeding programs throughout the world has led to the official release of more than 3200 mutant varieties from more than 200 different plant species, in more than 70 countries. In Pakistan, more than 59 varieties of different crop species (wheat, rice, cotton, sugarcane, mungbean, lentil, sesame, castor bean, mandarin, rapeseed, mustard, chickpea and groundnut) have been released through the use of mutation breeding (chemical and physical) techniques (**Figure 5**).

In China, only three mutant varieties are covering over 30 million ha and earn US$ 4.9 billion US dollars to uplift socio-economic status. In India, they have developed a huge number of mutant varieties and getting a large amount in return. In Bangladesh, Mutant rice varieties can be harvested a month earlier than the other varieties of rice-producing almost the same yield with superior quality. This variety is planted in three crop rotations and about 10,000 farmers cultivate this variety that is covering almost 80% of the area under rice cultivation. However, in Indonesia, an approximate amount of US$ 2 billion has been received from a single top rice variety. Many farmers' and millions of citizens getting benefits from the mutant varieties released by Indonesia. In Peru, improved barley and amaranth mutant varieties helping farmers to earn 7 million Andean and providing food and economic benefits thus improving their life status. In Vietnam, mutant varieties of rice and soybean helping poor farmers to improve their livelihood and a top rive mutant cultivar earning US$ 3.3 billion with an increase of US$ 537.6 million over old varieties. Whilst soybean mutant varieties bring about US$ 3 billion with 3.5 million farmers get a 30% increase in the economy. In Pakistan, 43 mutant varieties developed by NIAB showed an economic impact with earnings of US dollars 6 billion during 2018.

5. New breeding techniques (NBTs)

New breeding techniques or NBTs are a list of seven plant breeding techniques for incorporating genetic diversity into crop plants using site-specific targeted mutagenesis in the genome with greater accuracy and less off-targeted mutations [57]. The use of these NBT mutations is described as precision breeding. These techniques are zinc finger nuclease (ZFN) technology, oligonucleotide-directed mutagenesis (ODM), cisgenesis and intragenesis, grafting on GM-rootstock, RNA-dependent DNA methylation, agro-infiltration "sensustricto," and reverse breeding. The ZFN tool one of the site-directed nuclease (SDN) can be implemented to create a site-specific mutation in the plant genome. In addition, a number of new SDN techniques have been introduced viz. TALEN and CRISPR/Cas, and the latter is now extensively being used [57]. Recently, IAEA and FAO jointly launched a program known as Plant Mutation Breeding Network (PMBN) on the basis of a large number of crop varieties (2000) in the Asia Pacific region [58]. Out of these, 826 rice varieties to date have been released using mutation breeding, of these 699 were from the Asia-pacific region, with 290 from china. This program will be beneficial to farmers and researchers by developing new improved varieties with a higher yield, stability, and quality traits, disease resistance and resilience to changing climates through mutagenesis. The PMBN will work to further expand these great achievements jointly among the member countries. The conjoint use of classical mutation breeding method through screening of TILLING populations NBT mutations can be employed implicitly in the modification of plant attributes. The main advantage of NBT over the classical mutation technique is its precision and specificity that could be utilized to find robust mutation sites without the unwanted genetic changes that are the main problem in the classical mutation breeding. Resultantly, desired mutations could be retrieved through traditional mutation techniques. This is a lengthy process but of high applicability because of efficient tools to create mutant populations and to screen these mutations for targeted genes [59].

6. Prospects

With the rising food demands, the development of new crop varieties with improved yield potential and better resistance to biotic and abiotic stresses is vital. Modern techniques, molecular, and omics are the tools in hand to speed up the breeding route in integration with conventional (mutation/hybridization) methods. The integrated approach of using genomic and omics data with genetic and phenotypic data helps to unfold the genes/pathways connected with desired traits [60]. The conventional breeding methods have been employed extensively in combination with transformation, gene editing and marker-assisted selection (MAS). The selection of suitable parental materials endowed with desired traits in different crop species is fundamental for any successful breeding program. The highly favored markers known as Single Nucleotide Polymorphisms (SNPs) are helpful to analyze genetic variability and population configuration, in constructing genetic maps and to present genotypes for GWAS (genome-wide association analysis) [61]. Single-nucleotide polymorphisms (SNPs) are markers of choice to detect genetic diversity in crop plants [62]. Genotyping by sequencing (GBS) technique is based on next-generation-sequencing also done with SNP markers to incorporate high throughput genotyping [63]. These molecular techniques in combination with NBTs can do a miraculous job in the future to develop environment resilient cultivars to help fighting food security.

7. Conclusions

Mutation breeding has substantially contributed to crop improvements worldwide. Thousands of mutant crop varieties released in different countries have significantly improved yield potential, nutritional quality, biotic and abiotic stress tolerance. Several mutants with one or few desirable traits in different crops or vegetables are widely used as parents for breeding new commercial cultivars. Besides developing thousands of crop varieties, mutation breeding has created tremendous genetic resources for all major crops and vegetables worldwide. The integration of the latest mutation breeding tools with robust selection and speed breeding tools increases its scope in meeting food security challenges with exponentially increasing human population and climate change scenarios.

Conflict of interest

The authors declare no conflict of interest.

Author details

Arain Saima Mir[1*], Meer Maria[2], Sajjad Muhammad[2] and Sial Mahboob Ali[1]

1 Nuclear Institute of Agriculture (NIA), Tando Jam, PO Box 70060, Pakistan

2 Department of Biosciences, COMSATS University Islamabad (CUI), Park Road, Islamabad, Pakistan

*Address all correspondence to: saimamir_nia@yahoo.com

References

[1] FAO/IAEA. Manual on Mutation Breeding-Third edition. Spencer-Lopes, M.M., Forster, B.P. and Jankuloski, L. (eds.), Food and Agriculture Organization of the United Nations. Rome, Italy. 2018.301 pp.

[2] Harten AM. Mutation breeding: theory and practical applications. Cambridge University Press; 1998 Jun 25.

[3] Solanki RK, Gill RK, Verma P, Singh S. Mutation breeding in pulses: an overview. Breeding of pulse crops. Kalyani Publishers, Ludhiana. 2011:85-103.

[4] Lindstrom EW. Hereditary radium-induced variations in the tomato. Journal of Heredity. 1933 Apr 1;24(4):129-37.

[5] Macarthur JW. X-ray mutations in the tomato. Journal of Heredity. 1934 Feb 1;25(2):75-8.

[6] Goodspeed TH. Cytological and other features of variant plants produced from X-rayed sex cells of *Nicotianatabacum*. Botanical Gazette. 1929 Jun 1;87(5):563-82.

[7] Goodspeed TH, Avery P. The cytogenetics of fourteen types derived from a single X-rayed sex cell of*Nicotianatabacum*. Journal of Genetics. 1934 Oct 1;29(3):327-53.

[8] Gustafsson Å. Mutations in agricultural plants. Hereditas. 1947 Jan;33(1-2):1-00.

[9] Gustafsson Ä. The X-ray resistance of dormant seeds in some agricultural plants. Hereditas. 1944 Jan;30(1-2):165-78.

[10] Moh CC. The use of radiation-induced mutations in crop breeding in Latin America and some biological effects of radiation in coffee. Intern. J. Appl. Radiation Isotopes. 1962 Jul 1;13.

[11] Shu QY, Forster BP, Nakagawa H. Book: plant mutation breeding and biotechnology. Plant breeding and genetics section, Joint FAO/IAEA, Division of Nuclear Techniques in Food and Agriculture International Atomic Energy Agency, Vienna, Austria. 2011.

[12] Cyranoski D, Nature, 410, 19 APRIL 2001. (© 2001 Macmillan Magazines Ltd). www.nature.com

[13] Qiu F, Li JG, Weng ML, Jin DM, Gao HY, Wang PS, Jiang XJ, Wang B. Molecular analysis of long-pod mutant line of mung bean gene rated by space mutagenesis. ScientiaAgriculturaSinica. 1998;31(6):1-5.

[14] Shi J, Fan Q, Wang L, Hu P, Sun G, Li G. Induction of large grain mutation in adzuki bean (*Phaseolusangularis* Wight) by space environmental condition. ActaAgriculturaeNucleataeSinica. 2000;14(2):93-8.

[15] Fuxia L, Moju C, Tingzhao R. Screening a RAPD marker related to the maize male sterility gene obtained by space flight. Sichuan NongyeDaxueXuebao (China). 2005.

[16] Liu LX, Guo HJ, Zhao LS, Zhao S. Advances in induced mutations for crop improvement in China. InProceeding of China-Korea Joint Symposium on Nuclear Technique Application in Agriculture and Life Science 2007 Apr (pp. 22-25).

[17] Liu L, Guo H, Zhao L, Gu J, Zhao S. Achievements in the past twenty years and perspective outlook of crop space breeding in China. ActaAgriculturaeNucleataeSinica. 2007;21(6):589-92.

[18] Anonymous. https://www.technologytimes.pk/2019/07/16/ new-recoverable-satellite-launched-2020.

[19] de Vries H. Die Mutationstheorie: bd. Die Entstehung der Artendurchutation. Veit& comp.; 1901.

[20] Penna S, Vitthal SB, Yadav PV. In vitro mutagenesis and selection in plant tissue cultures and their prospects for crop improvement. Bioremediation, Biodiversity, Bioavailability. 2012;6:6-14.

[21] Wieczorek AM, Wright MG. History of agricultural biotechnology: how crop development has evolved. Nature Education Knowledge. 2012 Jul;3(10):9.

[22] Mishra D, Bhoi L, Dash M, Tripathy SK, Mishra TK, Behera MP, Pradhan B. Mutagenic effectiveness and efficiency of EMS and gamma rays on rice bean (*Vignaumbellate* (Thunb)Ohwi and Ohashi): An underutilized legume crop. IJCS. 2019; 7(3): 2060-4.

[23] Chaudhary J, Alisha A, Bhatt V, Chandanshive S, Kumar N, Mir Z, Kumar A, Yadav SK, Shivaraj SM, Sonah H, Deshmukh R. Mutation breeding in tomato: advances, applicability and challenges. Plants. 2019 May;8(5):128.

[24] Shu QY, Forster BP, Nakagawa H. Plant mutation breeding and biotechnology, CABI. Plant Breeding and Genetics Section. Joint FAO/IAEA Division of Nuclear Techniques in Food and Agriculture. International Atomic Energy Agency, Vienna, Austria. ISBN. 2012:978-2.

[25] Sánchez-Martín J, Steuernagel B, Ghosh S, Herren G, Hurni S, Adamski N, Vrána J, Kubaláková M, Krattinger SG, Wicker T, Doležel J. Rapid gene isolation in barley and wheat by mutant chromosome sequencing. Genome biology. 2016 Dec 1;17(1):221.

[26] Bughio HR, Asad MA, Odhano IA, Bughio MS, Khan MA, Mastoi NN. Sustainable rice production through the use of mutation breeding. Pak. J. Bot. 2007 Dec 1;39(7):2457-61.

[27] Boerjan W, Cervera MT, Delarue M, Beeckman T, Dewitte W, Bellini C, Caboche M, Van Onckelen H, Van Montagu M, Inzé D. Superroot, a recessive mutation in Arabidopsis, confers auxin overproduction. The Plant Cell. 1995 Sep 1;7(9):1405-19.

[28] Eimers K, Wang SM, Lue WI, Chen J. Monogenic recessive mutations causing both late floral initiation and excess starch accumulation in Arabidopsis. The Plant Cell. 1995 Oct 1;7(10):1703-12.

[29] Donner HK, Wang Q, Huang JT, Li Y, He L, Xiong W, Du C. Spontaneous mutations in maize pollen are frequent in some lines and arise mainly from retrotranspositions and deletions. Proceedings of the National Academy of Sciences. 2019 May 28;116(22):10734-43.

[30] Mendel G. Experiments in plant hybridization. Harvard University Press; 1965.

[31] Wilde HD. Induced Mutations in Plant Breeding. InAdvances in Plant Breeding Strategies: Breeding, Biotechnology and Molecular Tools 2015 (pp. 329-344). Springer, Cham.https://doi.org/10.1007/978-3-319-22521-0_11

[32] Parry MA, Madgwick PJ, Bayon C, Tearall K, Hernandez-Lopez A, Baudo M, Rakszegi M, Hamada W, Al-Yassin A, Ouabbou H, Labhilili M. Mutation discovery for crop improvement. Journal of Experimental Botany. 2009 Jul 1;60(10):2817-25.

[33] Wanga MA, Shimelis H, Horn LN, Sarsu F. The Effect of Single and Combined Use of Gamma Radiation and EthylmethaneSulfonate on Early

Growth Parameters in Sorghum. Plants. 2020 Jul;9(7):827.

[34] Leitao JM, Shu QY, Forster BP, Nakagawa H. Plant mutation breeding and biotechnology. Chemical mutagenesis. 2011:135-58.

[35] Mba C, Afza R, Shu QY. Mutagenic radiations: X-rays, ionizing particles and ultraviolet. Plant mutation breeding and biotechnology. 2012:83-90.

[36] Beyaz R, Yildiz M. The use of gamma irradiation in plant mutation breeding. Plant Engineering. INTECH. 2017 Nov 17:33-46.

[37] Oladosu Y, Rafii MY, Abdullah N, Hussin G, Ramli A, Rahim HA, Miah G, Usman M. Principle and application of plant mutagenesis in crop improvement: a review. Biotechnology & Biotechnological Equipment. 2016 Jan 2;30(1):1-6.

[38] Jain SM. Mutagenesis in crop improvement under the climate change. Romanian biotechnological letters. 2010 Mar 1;15(2):88-106.

[39] Kumawat S, Rana N, Bansal R, Vishwakarma G, Mehetre ST, Das BK, Kumar M, Yadav SK, Sonah H, Sharma TR, Deshmukh R. Expanding Avenue of Fast Neutron Mediated Mutagenesis for Crop Improvement. Plants. 2019 Jun;8(6):164.

[40] Gilchrist E, Haughn G. Reverse genetics techniques: engineering loss and gain of gene function in plants. Briefings in functional genomics. 2010 Mar 1;9(2):103-10.

[41] Shi J, Fan Q, Wang L, Hu P, Sun G, Li G. Induction of large grain mutation in adzuki bean (*Phaseolus angularis* Wight) by space environmental condition. ActaAgriculturaeNucleataeSinica. 2000;14(2):93-8.

[42] Liu LX, Guo HJ, Zhao L, Gu J, Zhao S. Advances in crop improvement by space mutagenesis in China. InICSC 2008 (Vol. 4, p. 274)..

[43] Hirano T, Kazama Y, Ishii K, Ohbu S, Shirakawa Y, Abe T. Comprehensive identification of mutations induced by heavy-ion beam irradiation in A rabidopsis thaliana. The Plant Journal. 2015 Apr;82(1):93-104.

[44] Nakagawa H. Induced mutations in plant breeding and biological researches in Japan. Crops. 2009 Jul 1;242(188):48.

[45] Wu YJ, Zhang Y, Yu W, Song M, Yu ZL. The progress of the research and application of ion implantation biotechnology in China. InSolid state phenomena 2005 (Vol. 107, pp. 37-42). Trans Tech Publications Ltd.

[46] Tanaka A, Shikazono N, Hase Y. Studies on biological effects of ion beams on lethality, molecular nature of mutation, mutation rate, and spectrum of mutation phenotype for mutation breeding in higher plants. Journal of radiation research. 2010 May;51(3):223-33.

[47] Nakai H, Asai T, Imada T, Watanabe H, Kitayama S, Takahashi T, Tanaka A, Kobayashi Y. Studies on induced mutations by ion beam in plants. 1996.

[48] Kitamura H, Mori M, Sato D, Nakagawa J, Yoshida T, Yoshizawa K, Kawai T, Hase Y, Tanaka A. Carbon ion beam breeding of rice suitable for low nitrogen input. 2006.

[49] Till BJ, Reynolds SH, Weil C, Springer N, Burtner C, Young K, Bowers E, Codomo CA, Enns LC, Odden AR, Greene EA. Discovery of induced point mutations in maize genes by TILLING. BMC plant biology. 2004 Dec 1;4(1):12.

[50] Caldwell DG, McCallum N, Shaw P, Muehlbauer GJ, Marshall DF, Waugh R. A structured mutant population for forward and reverse genetics in Barley (*Hordeumvulgare* L.). The Plant Journal. 2004 Oct;40(1):143-50.

[51] Talamè V, Bovina R, Sanguineti MC, Tuberosa R, Lundqvist U, Salvi S. TILLMore, a resource for the discovery of chemically induced mutants in barley. Plant biotechnology journal. 2008 Jun;6 (5):477-85.

[52] Suzuki T, Eiguchi M, Kumamaru T, Satoh H, Matsusaka H, Moriguchi K, Nagato Y, Kurata N. MNU-induced mutant pools and high performance TILLING enable finding of any gene mutation in rice. Molecular Genetics and Genomics. 2008 Mar 1;279(3):213-23.

[53] Xin Z, Wang ML, Barkley NA, Burow G, Franks C, Pederson G, Burke J. Applying genotyping (TILLING) and phenotyping analyses to elucidate gene function in a chemically induced sorghum mutant population. BMC Plant Biology. 2008 Dec 1;8(1):103.

[54] Slade AJ, Fuerstenberg SI, Loeffler D, Steine MN, Facciotti D. A reverse genetic, nontransgenic approach to wheat crop improvement by TILLING. Nature biotechnology. 2005 Jan;23(1):75-81.

[55] Weil CF. TILLING in grass species. Plant physiology. 2009 Jan 1;149(1):158-64.

[56] FAO/IAEA. 2018. Manual on Mutation Breeding - Third edition. Spencer-Lopes,M.M., Forster, B.P. and Jankuloski, L. (eds.), Food and Agriculture Organization ofthe United Nations. Rome, Italy. 301 pp.

[57] Holme IB, Gregersen PL, Brinch-Pedersen H. Induced genetic variation in crop plants by random or targeted mutagenesis: convergence and differences. Frontiers in Plant Science. 2019;10.

[58] Gil 2019. https://www.iaea.org/newscenter/news/accelerating-growth-iaea-launches-plant-mutation-breeding-network-for-asia-and-the-pacific

[59] Jankowicz-Cieslak J, Tai TH, Kumlehn J, Till BJ. Biotechnologies for plant mutation breeding: protocols. Springer Nature; 2017.

[60] Langridge P, Fleury D. Making the most of 'omics' for crop breeding. Trends in biotechnology. 2011 Jan 1;29(1):33-40.

[61] Xia W, Luo T, Zhang W, Mason AS, Huang D, Huang X, Tang W, Dou Y, Zhang C, Xiao Y. Development of high-density SNP markers and their application in evaluating genetic diversity and population structure in Elaeisguineensis. Frontiers in plant science. 2019 Feb 12;10:130.

[62] Ren J, Sun D, Chen L, You FM, Wang J, Peng Y, Nevo E, Sun D, Luo MC, Peng J. Genetic diversity revealed by single nucleotide polymorphism markers in a worldwide germplasm collection of durum wheat. International journal of molecular sciences. 2013 Apr;14(4):7061-88.

[63] Tang W, Wu T, Ye J, Sun J, Jiang Y, Yu J, Tang J, Chen G, Wang C, Wan J. SNP-based analysis of genetic diversity reveals important alleles associated with seed size in rice. BMC plant biology. 2016 Dec;16(1):1-1.

Chapter 7

Exploring Plant Genetic Variations with Morphometric and Molecular Markers

Bushra Sadia, Faisal Saeed Awan, Fozia Saleem, Javaria Altaf, Abdullah Bin Umar, Muhammad Nadeem, Samra Hameed, Farwa Ashraf and Mariam Nasir

Abstract

For centuries, crop improvement has served as the basis of food security of ever increasing human population. Though vast germplasm collections are available; their exploitation for crop improvement still depends upon efficient assessment of genetic diversity. Genetic variability is the key element in adaptation of plants to varying climates. While crops with narrow genetic diversity are vulnerable to stresses. The estimation of extent and pattern of genetic variability is a prerequisite for generating superior varieties. Genetic diversity analysis generates key information to dissect genetic variations in crop germplasm with the help of morphometrical, biochemical and molecular tools. Among these, DNA markers provide a reliable and detailed insight into the similarities and differences among crops. In this chapter, we discuss the applications of phenotypic and molecular markers to probe genetic divergence in crops and present case studies that describe the significance of these tools to characterize sorghum germplasm. Furthermore, we spotlight sorghum biodiversity exploration efforts worldwide and propose future directions.

Keywords: molecular markers, *Sorghum bicolor*, PCR, RAPDs, SSRs, SNPs, GWAS, association mapping, UPGMA, dendrogram

1. Introduction

The term "variability" refers to variation in one or more than one characters of living organisms. The cumulative influence of environment and the genetic factors brings about variations in a specific trait. Genetic variation refers to variation in sequences of genes between individuals in a population. Allelic variation is the building block of hereditary variation that is expressed in the form of different phenotypes. Processes like mutation, random mating and fertilization and gene duplication may introduce new genes and alleles thereby increasing genetic variation. Random mutations are the source of genetic variations. Mutations are either heritable or non-heritable; the beneficial heritable mutations exert a great influence on the genetic variations of living organisms. Likewise, gene flow is also a means of introducing new alleles to a population and thereby broadening the genetic diversity of living organisms. Genetic variability provides baseline for genetic diversity;

a broader term that reflects the degree/amount of variation existing within a population. Without genetic variability, populations fail to adapt to varying climatic conditions and are prone to extinction. Genetic variability is a source of natural selection, that is the key driver of evolution of living organisms.

Agriculture is directly influenced by environmental degradation and biodiversity loss leading to compromised quantity and quality of diverse and nutritious foods. Globally, people are relying on three major cereal crops wheat, maize and rice to fulfill their dietary needs and in turn are adopting similar dietary plans. Due to selective exploitation of few crops and large scale cultivation of genetically homogeneous cultivars, other wild and more nutritious crops are wiped out of global atlas. Not only we have compromised our health due to poor nutrition, the resilience of our food system is also at stake due to loss of crop diversity. Such lack of biodiversity was the root cause of Irish potato famine in the nineteenth-century. Presence of genetic diversity is the vital element of all variety development programs. Existence of genetic diversity in crop germplasm aides in the efficient selection of high yielding, better adapted crop plants with possible uses of direct introduction as a variety or one of the parents in crossing scheme of breeders for variety development programs. Since genetically diverse germplasm offers wider tolerance to biotic and abiotic stresses; such programs extensively involve exploring and exploiting diverse crop germplasm.

There is a continuous shift in the focus of agriculture from time to time. Agriculture in ancient times was focused on meeting subsistence food requirement. While, present day agriculture is focused to maximize yields for growing populations. That's why breeders are utilizing crop genetic resources for targeted and sustainable development of new high yielding and nutritious crop varieties in order to address malnutrition and balanced diet of human population. Under prevailing conditions of scarce water resources and escalating temperatures, development of climate resilient crop varieties is gaining momentum. Climate smart agriculture relies on cultivars with novel biotic/abiotic stress tolerance traits. However, depletion of natural variability persists in existing crop germplasm. Targeted breeding to improve specific traits and repeated use of few breeding parents has narrowed the genetic base of existing major crop varieties, raising serious concerns about genetic vulnerability of modern crops and making breeder's task even harder. In this context, new sources of desirable alleles are exploited from wild as well as closely related crop species and mutants. Hence, for ever changing breeding goals, it is imperative to conserve genetic diversity as germplasm resource. Crop genetic diversity is the core element of climate smart agro ecosystem to promise sustainable food availability and thereby to alleviate hunger and poverty.

A dire need exists to brought back underutilized and forgotten crops of every region to the canvas of agriculture for enhancing sustainable food production under anticipated harsh climates of the planet. A huge resource of alternative crops like sorghum, can replace the monoculture of three dominating cereal crops. Sorghum is a grass of multiple uses including food, feed, fiber, sugar, ethanol etc. Exploiting this and other nutritious and hardy crops is the best way to diversify present cropping system and enhance its resilience towards climate change. We need concerted collective efforts to increase awareness of farmers, policy makers and consumers towards benefits of diversification in agricultural systems.

2. Crop genetic diversity assessment methods

The assessment, extent and distribution of genetic divergence is the base line of preservation and exploitation of genetic variability within and between crop species. Initially, morphometric, cytological and biochemical markers were frequently used to evaluate the extent of similarities and differences among crop germplasm. Genetic and molecular markers were developed in the genomics and post genomics era and now are the widely used method for crop genetic divergence estimation.

2.1 Morphological markers

Evaluation of phenotypic traits in glasshouse or field- grown plants has long been used for selection of diverse crop plants. Effective morphometric characterization involves field plantation of large number of plants following specific lay out design. The morphological traits are recorded at vegetative growth (germination percentage, number of leaves, nodes, leaf area index, leaf color, stem thickness etc.), reproductive growth (Days to flowering, days to maturity, flower color, morphology, brix value etc.) and maturity stage (Plant height, yield, dry biomass and grain weight etc.). Plants express physiological and morphological changes under biotic and a biotics stresses. Hence, phenotypic characterization is vital in the selection of tolerant plants under stress environment.

This approach is easy, simple, inexpensive and directly measurable. However, experienced staff is required for effective selection of promising plants. Such field-based evaluation is directly influenced by environmental factors. Moreover, labor and field requirements pose extra work. Morphological evaluations must be detailed involving all growth stages of plants. Presently high throughput phenomics approaches have refined the morphological data recording of large number of entries in the field with precision. The growth-stage dependent physio-morphological characterization provides a base line for breeders to develop diverse genotypes having stress tolerant attributes. Furthermore, good quality phenotypic data is the foundation of new genomics and molecular approaches to successfully dissect the molecular basis of complex quantitative traits such as yield, disease resistance etc. Morphological markers have limitation of delayed expression till the specific developmental stage of the plant. Moreover, genotype x environmental interactions render the morphological markers less reliable than other marker types.

2.2 Cytological markers

These markers are related to morphological variations in chromosome size, shape, number, length, arm ratio, volume, behavior in cell divisions and DNA content etc. These chromosomal features can be identified through microscopy and expressed by chromosome karyotype and bands. The G, Q, R and C banding patterns of chromosomes indicate regions of chromatin that are stained with the help of different fluorescent dyes, viz.; Quinacrine hydrochloride (Q bands) and Giemsa stain (G bands) [1]. The presence or absence of a chromosome band is associated with the specific traits. A thinnest chromosome band hosts over hundreds of genes. These are used to detect cytological mutations and track evolutionary chromosomal rearrangements. The fusion of chromosomal and molecular biology protocols in 1990 introduced fluorescence *in situ* hybridization (FISH) method. It is capable of physical mapping of nuclear content directly on the chromosomes and identifying protein content of a cell. A more advanced variant of *in situ* hybridization, "genomic *in situ* hybridization (GISH)" technique utilizes total genomic DNA of plant as a probe. Both GISH and FISH are powerful tools to characterize alien introgressions in crop species and dissect genetic makeup of natural and artificial hybrids [2]. However, cytological markers have limited use in genetic diversity estimation due to their small number and discrete detection.

2.3 Biochemical markers

Biochemical markers have been among the most widely used markers for assessing variations among and within crop species before the advent of molecular/DNA markers. The alternative forms of protein (isozymes) exhibit specific banding patterns on gel electrophoresis, owing to variations in charge- based protein mobility. Isozymes are the products of different alleles, their position can be mapped on to chromosomes and hence are used to map other genes. Protein/isozyme analysis is still among the simple, rapid and cheap methods and fits well in the projects where low level of genetic diversity estimation is desired. Though protein markers are more reliable than morphological markers, their expression is plant growth stage dependent and is readily influenced by the environment [3, 4].

2.4 Molecular markers

Molecular markers are based on DNA sequence polymorphism and bypass the limitations encountered in the use of morphological, cytological and biochemical markers. These have become the preferred method for evaluating crop genetic variations due to their simple inheritance, high reproducibility, widespread distribution in plant genome and being stable, highly polymorphic with minimum pleiotropic effects [5]. Molecular markers are not plant stage dependent and are least affected by environment. Large number of markers have been mapped on chromosomes of crop plants and livestock. Molecular markers show either dominant or co dominant inheritance mode. The codominant markers are preferred over dominant ones being more reliable and informative [6]. These have been extensively exploited for variety of applications like genetic fingerprinting, hybrid identification, functional genomics etc. In crop breeding, molecular markers help in early identification/selection of desired genotypes thereby shortening variety development time. These markers enhance breeders' capability of targeted breeding. The earlier version of hybridization- or PCR- based markers has now been upgraded to newer types based on sequencing or array platforms. Following are the groups of molecular markers based on principle techniques:

1. Nucleic acid hybridization- based markers: Restriction fragment length polymorphisms (RFLPs).

2. PCR- based markers: Randomly amplified polymorphic DNA (RAPDs), Amplified fragment length polymorphisms (AFLP), Microsatellites, or simple sequence repeats (SSRs), Randomly amplified microsatellite polymorphisms (RAMP), Sequence-related amplified polymorphism (SRAP), Inter simple sequence repeat (ISSR), Target region amplification polymorphism (TRAP)

3. PCR–RFLP markers: Cleaved amplified polymorphic sequences (CAPS)

4. Retrotransposons- based markers: Inter-retrotransposon amplified polymorphism (IRAP), Retrotransposon microsatellite amplification polymorphisms (REMAP), Retrotransposon-based insertion polymorphism (RBIP), Inter-primer binding site (iPBS).

5. Sequence-based markers: Single-nucleotide polymorphism (SNP)

6. Array-based platforms like Diversity Arrays Technique (DArT), restriction site-associated DNA (RAD), single feature polymorphism (SFP), etc. [7, 8]

7. Functional molecular markers (FMM): The term "Functional markers" was proposed by Andersen and Lübberstedt [9] for DNA markers that arise from sequence polymorphism among functional genes that are linked with variations in the desired phenotypic traits. Hence, these are more reliable and informative than all previous PCR- based markers.

Each marker system has its own benefits and disadvantages and variations exist on the basis of development cost, efficiency and reproducibility.

3. Need for genetic diversity assessment of sorghum germplasm

3.1 Sorghum origin

The word sorghum originated from "*Syricum*" in Latin, meaning "Grain of Syria" [10]. Sorghum (*Sorghum bicolor*) belongs to class *Liliopsida*, family *Graminea*, genus *Sorghum Moench* and has five groups named as: *Hetrosorghum*, *Chaetosorghum*, *Spitosorghum*, *Parasorghum* and *Eusorghum*. It is an ancient grain that has been cultivated for thousands of years. It originated mainly from Sudanese and Ethiopian grasslands more than 6000 years ago.

3.2 Global sorghum distribution and production

About 100 countries grow sorghum worldwide (**Figure 1**). USA is the top sorghum producer with five countries viz.; Nigeria, Ethiopia, Mexico, India and China follow in the order of production (**Figure 2**). The countries of Japan, Mexico, and Philippines are the major importers of North American sorghum, while China is the world's largest sorghum importer.

3.3 Sorghum in Pakistan

In Pakistan, sorghum is grown for fodder and forage of livestock. It is grown as kharif fodder in irrigated and rain fed areas of Punjab and Sindh provinces. Production of sorghum (*Sorghum bicolor*) in Pakistan is 1.45 million metric tons in 2020 (www.indexmundi.com). Sorghum is the second largest fodder crop after berseem (Pakistan Bureau of Statistics, 2016). Scarce record exists on use and adoption of

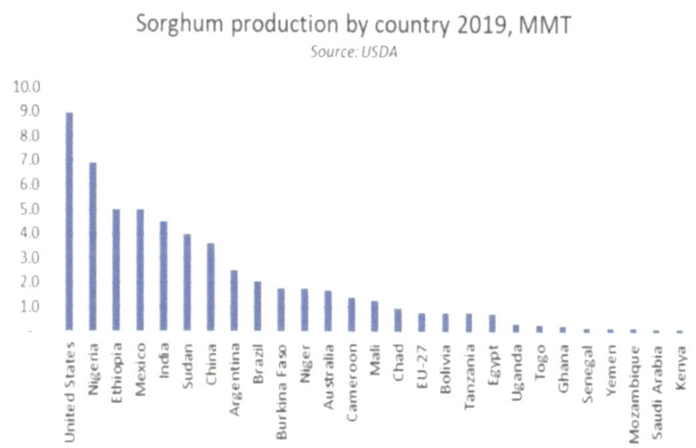

Figure 1.
Country-wise production of sorghum in the world.

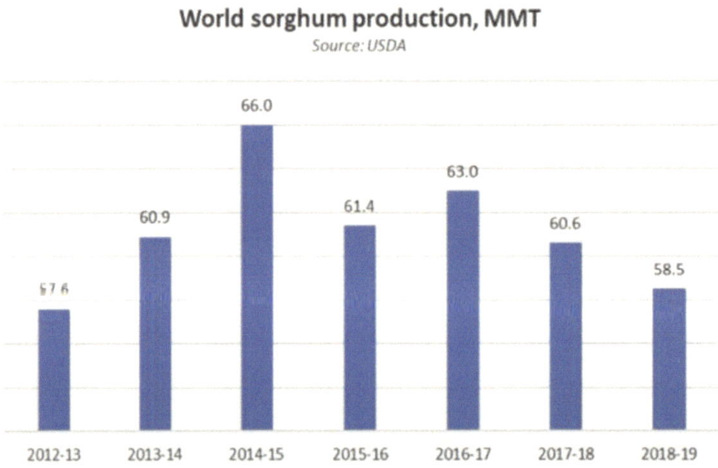

Figure 2.
Worldwide sorghum production statistics from 2012 to 2019.

grain and sweet sorghum types as silage, fodder and bioethanol source in Pakistan. Exploring diversity of different sorghum types is vital to develop better sorghums.

3.4 Multiple uses of sorghum

Sorghum is ranked as 5th most widely grown cereal crop of the world. It has C4 photosynthetic pathway which is useful for global food production. It is a staple food with significant nutritional qualities for about 500 million people around the globe. With growing world population, the demand for reliable food and feed sources has also escalated. In the context of possible limited water supplies and high temperatures, sorghum's role to feed the world will increase in importance owing to its higher adaptability. Sorghum has amazing range of multiple uses:

3.4.1 Sorghum grain as food

Sorghum grain is used for food and biofuels. Grain has an edible hull and retains the majority of its nutrients. It contains 86% total digestible nutrients, up to 15.6% protein and 3772 kcal/kg energy. Sorghum grain has higher levels of magnesium that help in higher absorption of calcium and thereby contribute to bone health. It is abundant in phenolic compounds and antioxidants that safeguard against age-onset degenerative diseases [11]. Sorghum grain is reported to reduce the risk of many important diseases like cancer, cardiac infarction and some neurological disorders [12]. The grain is consumed as whole or ground to nutritious flour for baking. Most importantly, sorghum food products are gluten free, have wide range of color, neutral flavor and low allergenicity.

3.4.2 Sorghum grain as feed

Sorghum grain is second to maize in consumption as feed in the USA. It is a significant component of animal feed in South America, Australia and China, and poultry feed in India. The low-tannin high digestible sorghum (HDS) varieties are quickly replacing corn in poultry feed.

3.4.3 Sorghum as feedstock for biofuels

Sorghum starch, sugar, and biomass are used as feedstocks for biofuel. High biomass sorghums developed by selective breeding are used as biofuel feedstock. Moreover, sweet sorghum has emerged as a promising contender of bioenergy. Its stalk, seeds and syrup are used for biomass and ethanol production [13].

3.4.4 Sorghum as fodder

For livestock feed, sorghum may be utilized in a number of ways like as green chop, grazed and made into hay or silage [14]. By adopting a combination of these systems, sorghum sufficiently meets the year round needs of stock farmers.

3.4.5 Sorghum as a climate smart crop

Worldwide climate change forecasts suggest incidences of low rainfall with variable distribution, flooding, extended droughts and elevated temperatures. Sorghum thrives exceptionally well under low water availability, heat, salinity and low inputs and thus is named as "the camel of crops". It is anticipated to perform high for food security of large number of masses with scant resources in arid zones of the world. According to climate predictions for 2050, sorghum will remain world's top crop to survive coming harsh weathers across the globe [15]. The crop is set to enjoy a relatively healthy future.

3.4.6 Sorghum as a diverse crop

Sorghum exhibits promising diversity in yield and quality traits as well as resilience to different environmental conditions in dry arid, semi-arid, temperate and tropical areas. In order to harness immense benefits of sorghum and for long term maintenance, there is a dire need to preserve this variability in the form of germplasm collections. Once this biodiversity in these collections is lost, it cannot be brought back. A crop with narrow genetic base cannot cope with drastic climatic stresses. Estimation of diversity among and within the species of any crop helps identify the germplasm with maximum variability that can be exploited in developing varieties of wide genetic background to withstand biotic and abiotic stresses.

4. Case studies on morphometric and molecular characterization of sorghum

We report morphological characterization of ten sweet sorghum genotypes from National Agriculture Research Center, Islamabad, Pakistan [16]. Data for Plant height (PH), Days of 50% flowering (DF), Brix value(BV), Number of leaves per plant (NL), Leaf length (LL), Leaf width (LW), Leaf area index (LAI), Stem girth (SG), Flag leaf width (FLW), Flag leaf length (FLL), Flag leaf area index (FLAI), Fresh weight (FW) and Dry weight (DW) were recorded. The means and standard error of means for each trait were calculated [17] and presented in **Table 1**.

Correlation for observed 14 morphological traits is presented in **Table 2**. Number of leaves per plant (NL) indicated positive strong correlation with BV, LL, LAI, DW and PH. Whereas, NL showed moderate to low correlation with DTF, FLL, FLW, FLAI and DTM. The morphological trait DW showed positive higher correlation with NL, BV, SG, LL, LW, LAI and FW. Significant ($p = 0.01$) strong positive correlation was obtained for Plant height (PH) with BV.

Variables	Range		Mean	Std. deviation
	Minimum	Maximum		
NL	8.55	11.89	10.00	1.17
DTF	58.33	77.56	71.45	6.02
BV	6.81	9.87	8.22	0.97
SG	1.60	5.67	3.71	1.12
LL	34.71	76.90	53.70	12.86
LW	3.33	7.23	4.97	1.15
LAI	130.71	518.36	277.51	126.90
FW	56.70	100.80	82.25	13.03
DW	32.55	52.85	41.90	6.19
FLL	21.84	33.75	27.66	4.03
FLW	2.36	3.36	2.71	0.35
FLAI	57.36	113.35	75.78	20.20
PH	158.71	230.02	191.40	22.50
DTM	106.33	124.78	117.45	5.84

Table 1.
Cumulative response of sorghum genotypes for fourteen phenotypic traits.

In PCA, three PCs were selected out of nine because their Eigen value is more than one. Selected PCs cover the character variability (**Tables 3** and **4**).

Bi-Plot (**Figure 3**) showed allocation of genotypes on the basis of performance. The characters which were far away from origin showed more variability.

Our group previously reported RAPD- based genetic diversity evaluation of sorghum germplasm of Pakistan [10]. We also performed molecular diversity analysis of twelve sweet sorghum genotypes with 17 RAPD primers viz.; GLA03, GLB10, GLC01, GLC 02, GLI06, GLL02, GLL05, GLL07, GLL09, GLL10, GLL12, GLL14, GLL15, GLL16, GLL17, GLL18 and GLL19 [18]. These markers yielded 77 fragments of different sizes and 6.41 bands per primer were produced on average (**Figure 4**). RAPD primers identified 83.33% polymorphism among sweet sorghum genotypes (**Figure 5**).

Genetic similarity was assessed among sorghum genotypes via Nei's similarity indices with popgen 1.32. The genotypes MN 2363 and Dobbs showed minimum similarity (76.92%). Whereas, Masaka and Dobbs exhibited the lowest similarity (44.87%) and hence the maximum divergence (**Table 5**).

The genetic relationship among sorghum genotypes was assessed by Popgen 1.32. All twelve sorghum genotypes were clustered in two groups with the help of Cluster analysis. Two genotypes (Malnal and Maska) were present in one group. While the rest of the genotypes constituted the second group. A close similarity was present among Masaka and Malnal that were clustered in Group A. Group B comprised of three genotypes, among these Dobbs and MN 2363 were clustered together and MN 2109 resided separately in this group. The genotypes Chedomba, Kamandri, Dura Huria and Juar were placed in Group C and IS12833, Juar 49 and Early Folger constituted Group D. The highest similarity was observed among Malnal and Masaka. On the other hand, the highest divergence was recorded between Malnal and Early Folger exhibited (**Figure 6**).

In a separate study, we explored genetic divergence of 24 sorghum genotypes with RAPD markers (OPL-7, OPL-8, OPA-13 and OPA-3) [19]. These markers produced

Variables	NL	DTF	BV	SG	LL	LW	LAI	FW	DW	FLL	FLW	FLAI	PH	DTM
NL	**1**	0.347	0.722	0.686	0.770	0.529	0.717	0.643	0.787	0.321	0.403	0.403	0.722	0.466
DTF	0.347	**1**	0.753	0.755	0.640	0.699	0.642	0.123	0.466	0.550	0.340	0.479	0.753	0.958
BV	0.722	0.753	**1**	0.883	0.903	0.937	0.957	0.404	0.739	0.572	0.543	0.614	1.000	0.764
SG	0.686	0.755	0.883	**1**	0.872	0.853	0.898	0.594	0.823	0.487	0.519	0.551	0.883	0.741
LL	0.770	0.640	0.903	0.872	**1**	0.810	0.947	0.488	0.799	0.504	0.367	0.482	0.903	0.662
LW	0.529	0.699	0.937	0.853	0.810	**1**	0.946	0.245	0.569	0.579	0.638	0.670	0.937	0.678
LAI	0.717	0.642	0.957	0.898	0.947	0.946	**1**	0.406	0.719	0.542	0.541	0.599	0.957	0.654
FW	0.643	0.123	0.404	0.594	0.488	0.245	0.406	**1**	0.888	0.004	0.109	0.064	0.404	0.137
DW	0.787	0.466	0.739	0.823	0.799	0.569	0.719	0.888	**1**	0.279	0.299	0.319	0.739	0.470
FLL	0.321	0.550	0.572	0.487	0.504	0.579	0.542	0.004	0.279	**1**	0.683	0.914	0.572	0.631
FLW	0.403	0.340	0.543	0.519	0.367	0.638	0.541	0.109	0.299	0.683	**1**	0.920	0.543	0.396
FLAI	0.403	0.479	0.614	0.551	0.482	0.670	0.599	0.064	0.319	0.914	0.920	**1**	0.614	0.552
PH	0.722	0.753	1.000	0.883	0.903	0.937	0.957	0.404	0.739	0.572	0.543	0.614	**1**	0.764
DTM	0.466	0.958	0.764	0.741	0.662	0.678	0.654	0.137	0.470	0.631	0.396	0.552	0.764	**1**

*,**: *Significant at 5% and 1% probability level.*

Table 2.
Correlation matrix for different traits in sweet sorghum genotypes.

	F1	F2	F3	F4	F5	F6	F7	F8	F9
Eigen value	9.292	2.103	1.154	0.604	0.381	0.281	0.114	0.054	0.017
Variability (%)	66.373	15.024	8.241	4.316	2.722	2.006	0.812	0.385	0.122
Cumulative %	66.373	81.397	89.638	93.953	96.675	98.681	99.493	99.878	100.000

Table 3.
Principle component analysis.

Variables	PC1	PC2	PC3
NL	0.750	0.394	−0.248
DTF	0.768	−0.206	0.528
BV	0.970	0.032	0.104
SG	0.939	0.170	0.066
LL	0.913	0.218	0.092
LW	0.911	−0.148	0.070
LAI	0.948	0.068	0.024
FW	0.469	0.747	−0.311
DW	0.781	0.556	−0.139
FLL	0.668	−0.577	−0.189
FLW	0.632	−0.476	−0.527
FLAI	0.712	−0.568	−0.397
PH	0.970	0.032	0.104
DTM	0.796	−0.229	0.433
Eigen value	9.292	2.103	1.154
Variability (%)	66.373	15.024	8.241
Cumulative %	66.373	81.397	89.638

Table 4.
PCA factor loadings for sorghum genotypes.

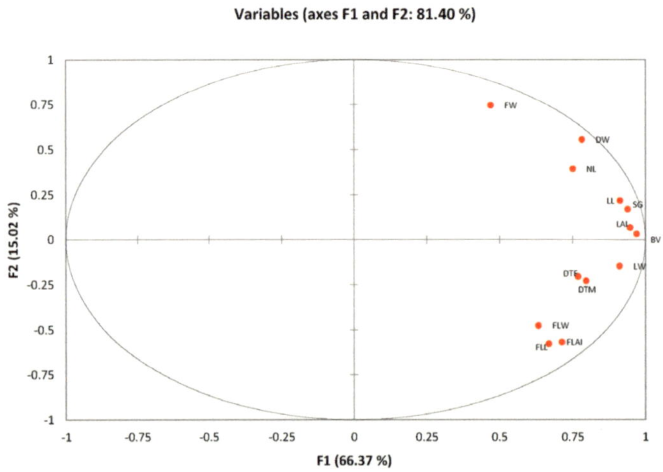

Figure 3.
PCA Biplot.

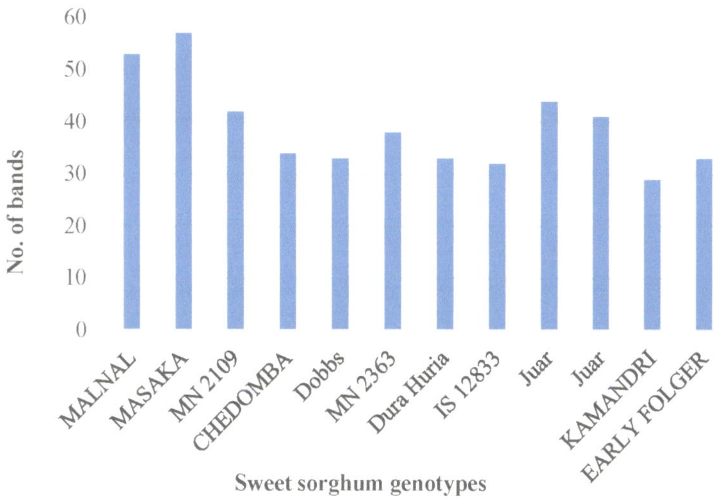

Figure 4.
Number of bands recorded per sorghum genotype.

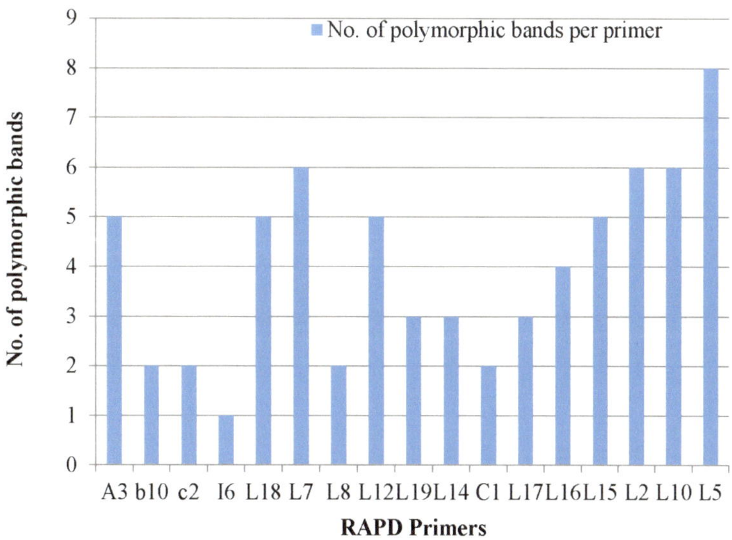

Figure 5.
Number of polymorphic bands per primer in sorghum genotypes.

74 bands of varying sizes/intensities. On average, each primer produced 18.5 bands. RAPD markers revealed 77.13% polymorphism among sorghum genotypes.

While, previous fingerprinting studies showed 58% [20] and 52% polymorphism [21] among various sorghum genotypes. The primer OPL7 produced the maximum number of fragments [22] whereas, the minimum number of fragments were generated by OPA3 (14) (**Figure 7**). The low level of similarity indicated high divergence among the sorghum germplasm under study.

More recently, we exploited sixteen SSR markers for DNA fingerprinting of fifty sorghum genotypes [8]. The molecular analysis indicated significant polymorphism among these genotypes.

The bands varied in size and intensity. The number of bands per primer per genotypes also varied. Some bands showed a high level of polymorphism indicating great variation among the sorghum germplasm (**Figure 8**). Marker diversity among

pop ID	1	2	3	4	5	6	7	8	9	10	11	12
1	****	0.6154	0.6410	0.5769	0.5769	0.5769	0.5000	0.5897	0.6795	0.5256	0.5256	0.6154
2	0.4855	****	0.5385	0.5513	**0.4487**	0.4744	0.5256	0.4872	0.5256	0.5769	0.4231	0.5641
3	0.4447	0.6190	****	0.7308	0.7564	0.7308	0.7308	0.6923	0.6538	0.7051	0.7051	0.5897
4	0.5500	0.5955	0.3137	****	0.6923	0.6410	0.6923	0.6282	0.6154	0.6667	0.7436	0.6795
5	0.5500	0.8014	0.2792	0.3677	****	**0.7692**	0.6154	0.6795	0.6154	0.5641	0.7436	0.6538
6	0.5500	0.7458	0.3137	0.4447	0.2624	****	0.6923	0.7564	0.6154	0.7179	0.6667	0.5513
7	0.6931	0.6431	0.3137	0.3677	0.4855	0.3677	****	0.7051	0.6154	0.6923	0.7436	0.5769
8	0.5281	0.7191	0.3677	0.4649	0.3864	0.2792	0.3494	****	0.7308	0.6538	0.7051	0.5641
9	0.3864	0.6431	0.4249	0.4855	0.4855	0.4855	0.4855	0.3137	****	0.5897	0.6154	0.6795
10	0.6431	0.5500	0.3494	0.4055	0.5725	0.3314	0.3677	0.4249	0.5281	****	0.6154	0.5256
11	0.6431	0.8602	0.3494	0.2963	0.2963	0.4055	0.2963	0.3494	0.4855	0.4855	****	0.5769
12	0.4855	0.5725	0.5281	0.3864	0.4249	0.5955	0.5500	0.5725	0.3864	0.6431	0.5500	****

Nei's genetic identity (above diagonal) and genetic distance (below diagonal). 1: Malnal, 2: Masaka, 3: MN 2109, 4: Chedomba, 5: Dobbs, 6: MN 2363, 7: Dura Huria, 8: IS 12833, 9: Juar 49, 10:Juar 48, 11: Kamandri, 12: Early Folger
***** are symbols just to separate above diagonal and below diagonal values*

Table 5.
Similarity matrix of 12 sweet sorghum genotypes.

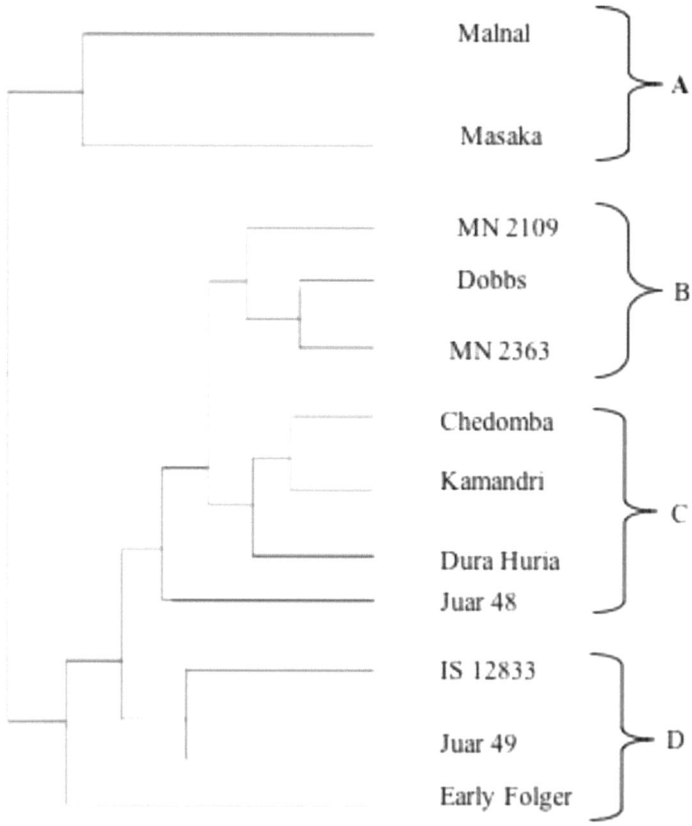

Figure 6.
Dendrogram of 12 sweet sorghum genotypes based on RAPD analysis.

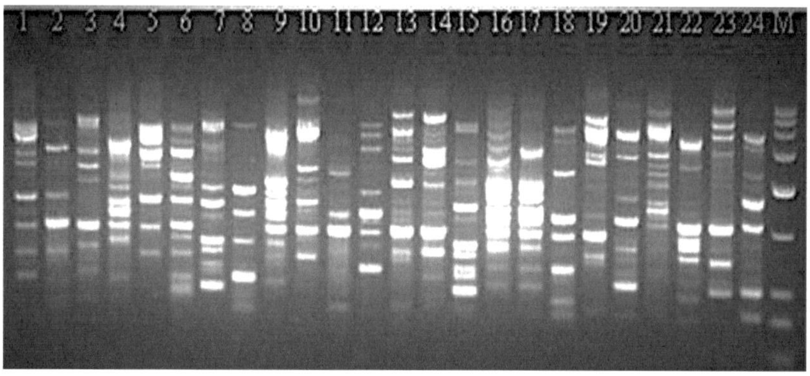

Figure 7.
PCR amplification of 24 sorghum genotypes with RAPD primer L-7. Lanes L: Ladder, 1–24: Sorghum genotypes.

fifty sorghum genotypes was studied using Powermarker software. The number of alleles per locus ranged from 2 to 3 with mean value of 2.875 alleles per locus. Genetic relationship among sorghum genotypes was evaluated by using popgen 1.32. All genotypes were grouped in two major clusters which were further divided into sub-groups. One small group consisted of eight genotypes (15, 39, 16, 35, 20, 22, 24, and 18) and the other large group contained remaining 42 sorghum genotypes. Maximum genetic distance was observed between 1st and 18th genotype.

Figure 8.
No. of bands amplified by SSR primers in fifty sorghum genotypes.

This study revealed positive correlation among the allele number, gene diversity and PIC value. The ease of using these PCR-based markers for diversity evaluation, for allocating genotypes to heterotic groups, and for DNA fingerprinting proved advantageous for selecting biomass- related traits and for sorghum breeding programs.

5. Worldwide sorghum biodiversity exploration efforts

At present, extensive record is available on genetic diversity evaluation of sorghum using molecular markers. A review of global research on sorghum genetic diversity evaluation using morphological and molecular markers is presented in **Tables 6** and 7, respectively. Most of the studies analyzed vegetative and

Sr. #	Sorghum germplasm	Morphological traits	References
1	94 sorghum accessions	Area, Breadth, Circularity, Major axis length, Perimeter length and Rectangularity	Dahlberg et al. [22]
2	45 sorghum accessions including 34 landraces, 6 elite breeding lines and 5 improved cultivars	Ten qualitative (Plant color, Stalk juiciness, Leaf midrib color, Inflorescence exsertion, Panicle compactness and shape, Awns, Glume color, Grain covering, Grain color and Endosperm texture) and 16 quantitative (Days 50% flowering, Leaf number, Leaf length, Leaf width, Leaf area, Internode length, Leaf sheath length, Plant height, Panicle length, Panicle width, Number of primary branches Panicle head weight, Grain yield panicle, 1000-seed weight, Threshing percent and Grain size) traits	Geleta et al. [23]
3	40 sorghum landraces from Tanzania and 2 from Zambia	Five panicles average weight (g), Grain number/panicle, Height (cm), Hundred grain weight (g), Inflorescence length (cm), Inflorescence width (cm), Leaf length (cm), Number of leaves, Leaf width (cm), Leaf senescence, Main stem diameter (cm), Tillers diameter (cm), Number of tillers, Grain yield	Bucheyeki et al. [24]
4	320 sorghum accessions from more than 3500 germplasm collection	Days to heading (DTH), Days to flowering (DTF), Days to maturity (DTM), Culm diameter (CD), Grain weight per panicle (GWP), 100 grain weight (100GW), Culm length (CL), Number of tillers (NoT), Number of panicles (NoP), Panicle length (PL), Leaf	Shehzad et al. [25]

Sr. #	Sorghum germplasm	Morphological traits	References
		length (LL) and Leaf width (LW), Panicle shape (PS), Panicle type (PT), Coleoptile's color (CC), Quantity of lipid white powder on stem and leaves (LWP), Color of midrib (MC), Neck length of panicle (PNL), Awn presence (AP), Glume color (GC), Growth in early stage (GES), Endosperm type (ET), Aphid resistance (AR), Number of regenerated tillers (NRT), Regrowth (RG) and Resistance to insecticides (RI)	
5	124 sorghum from Burkina Faso	28 agro morphological traits (Vigor at emergence 5(Ve), Coleoptile color (Cc), Leaf anthocyanin pigmentation (Lap), Panicle compactness (Pc), Pedicellate spikelet length (Psl) and Persistence (Psp), Glume length (Gl) and opening (Go), Awn (Aw), Kernel shape (Ks), Kernel rotation (Kr), Glume color (Gc), Kernel color (Kc), Anthocyanin spots on kernels (Ask), Glume adherence (Ga), Seed coat or testa (Sc) and Kernel vitreousness (Kv), Plant height (Ph), Leaf number (Ln), Length (Ll) and Width (Lw) of the third leaf under the panicle, Number of effective tillers (Net), Panicle length (Pl), Panicle weight (Pw), Harvested seed weight (Hsw) and 1000-seed weight (1000-Sw)	Barro-Kondombo et al. [26]
6	156 sorghum accessions	Lowering time, Plant height, and panicle type/ inflorescence, Panicle type and glumes coverage, grain color	Sharma et al. [27]
7	25 sorghum genotypes	Seedling vigor, Number of leaves, Leaf area, Stay-green, Peduncle exertion, Panicle length and width, Plant height, Days to flowering and maturity, Grain yield, Biomass and Harvest index under Drought stress	Abraha et al. [28]
8	9 sorghum genotypes from Sudan	Days to flowering (DF), Days to maturity (DM), Plant height (PH) (cm), Panicle length (PL) (cm), Panicle exertion (cm), Head weight (HW) (g), Yield per panicle (YPP) (g), Thousand seed weight (TSW) (g), Biomass (BM) yield (ton/ha) and GY (kg/ha)	Sabiel et al. [29]
9	Recombinant inbred line of *Sorghum bicolor* made by crossing E-Tian, a sweet sorghum accession with Ji2731	Biomass and Biofuel traits	Mocoeur et al. [30]
10	Diallel set of 10 parents and their 90 crosses including reciprocals of sorghum	Days to flowering, Days to maturity, Plant height, Grain yield per plant, Panicle length, Number of tillers per plant, Panicle weight, Panicle exsertion, Thousand seed weight, Grain-filling period	Mohammed et al. [31]
11	40 accessions of sorghum from Tamil Nadu	Days to 50% flowering, Days to maturity, Plant height, Panicle length, Panicle width, Leaf length, Leaf breadth, Number of leaves per plant, Stem girth, Number of primary branches per panicle, Hundred-seed weight, Yield per plant, Panicle weight and Dry matter production	Sinha and Kumaravadivel [32]
12	267 sorghum genotypes from Ethiopia	Leaf rolling, Head compactness, Glume cover, Glume color, Leaf orientation, Midrib color,	Amelework et al. [33]

Sr. #	Sorghum germplasm	Morphological traits	References
		Panicle exsertion, Head shape, Grain color, Stay-green, Leaf color, Head orientation	
13	315 sorghum accessions	Plant height and Seed number	Jing Zhao et al. [34]
14	Two overlapping sets of RILs of sorghum	Grain yield, Flowering time, and Stay-Green traits	Sivakumar Sukumaran et al. [35]
15	100 sweet sorghum accessions	Bioenergy traits, Protein content and Ethanol yield	Da silva et al. [36]
16	Populations of sweet sorghum F4 families made by crosses between 11 tall sweet sorghum cultivars (used as males), and 3 short grain sorghums as females	Relationship between Sugar content and Plant height	Shukla et al. [37]
17	30 sorghum accessions	Days to 50% anthesis, Plant height, Flag leaf area, Brix percentage, Panicle length, Grain weight and Grain yield	Mumtaz et al. [38]
18	54 introgressed sorghum breeding lines	Drought stress imposed at pre-flowering and post-anthesis developmental stages, Panicle area, Width, Percent green leaf, Total above ground, Dry biomass and Dry panicle weight	Emendack et al. [39]
19	75 sorghum lines including 74 indigenous cultivars and 1 exotic cultivar	Glume color, Neck of panicle, Length of flower with pedicel, time of panicle emergence, color of dry anther, panicle length of branches, panicle shape and caryopsis color	Prajapati et al. [40]
20	196 sorghum accessions	Seedling vigor, Days to flowering, Days to maturity, Days to grain filling period, Plant height, Panicle exertion, Number of green leaf at physiological maturity, Panicle length, Panicle weight, Thousand seed weight, Panicle yield, Grain yield, Above ground dry matter, Harvest index	Derese et al. [41]
21	453 diverse photo-period sensitive sorghum lines	Moisture, Plant height	Fernandes et al. [42]
22	194 *Sorghum bicolor* and *S. bicolor sudanese* genotypes	Root system architecture	Parra-Londono et al. [4]
23	93 sweet grain sorghum accessions	Sweet grains in pasty stage	Sawadogo et al. [43]
24	329 accessions of sorghum	Seed morphology	Sakamoto et al. [44]
25	200 Sweet sorghum accessions from Serbia	Plant height, Plant biomass, Stem leaves, Panicle length and Yield of crude biomass	Bojović et al. [45]
26	98 accessions of South African sorghum	Genetic variability, Plant height, Panicle length, Width and exsertion, Rachis number, Panicle weight, Seed weight, Grain yield Per panicle	Mofokeng et al. [46]
27	12 *Sorghum bicolor* genotypes (5 sweet, 4 grain and 3 forage sorghums)	Green leaf area (cm2), Plant height (cm), Leaf number, Fresh biomass yield (t/ha), Cane yield (t/ha), Bagasse yield (t/ha), Brix degree and Juice yield (kl/ha)	Kanbar et al. [47]

Sr. #	Sorghum germplasm	Morphological traits	References
28	Seven groups of 44 parental lines of sorghum	Mid-season drought tolerance, Mid-season drought susceptibility, Stay green lines, Terminal drought tolerance, Saline-tolerance, Saline-susceptibility, High Fe–Zn lines	Pandian et al. [48]
29	Recombinant inbred line derived from a cross between an elite U.S. common parent RTx430 and 10 diverse founders	Inflorescence morphology	Olatoye et al. [49]
30	3 recombinant inbred line mapping populations of sweet sorghum	Stem lodging resistance, Mechanical stability analysis	Gomez et al. [50]
31	210 Ethiopian genotypes of grain sorghum	Days to flowering, Days to maturity, Plant height, Grain yield per plant, Panicle length, Number of tillers per plant, Panicle weight, Panicle exsertion, Thousand seed weight, Grain-filling period	Birhan et al. [51]
32	55 sorghum accessions comprising 11 Bicolor accessions, 15 Caudatum, 10 Durra, 9 Guinea and 10 Kafir	Inflorescence architecture	Li et al. [52]
33	21 diverse sorghum accessions	Transpiration efficiency, the ratio of plant carbon produced to water transpired and carbon isotope discrimination of leaf dry matter	Henderson et al. [53]

Table 6.
Studies on assessment of genetic variations using morphological markers in sorghum.

Sr. No	Sorghum germplasm	Molecular markers	Reference
1	25 accessions of sorghum	Microsatellites	Djè Y et al. [54]
2	415 sorghum accessions consisting of 391 landraces, 8 standard varieties and 16 introduced elite breeding lines	Allozymes and RAPD markers	Ayana, [55]
3	94 sorghum accessions	RAPDs	Dahlberg et al. [22]
4	100 accessions from a core collection of 293 sorghum	SSR markers	Folkertsma et al. [56]
5	45 sorghum accessions	SSRs and AFLP	Geleta et al. [23]
6	1 sorghum accession	SSRs	Wu et al. [57]
7	46 sorghum lines	AFLP and SSRs	Perumal et al. [58]
8	42 grain sorghum landraces	SSRs	Bucheyeki et al. [24]
9	40 sorghum genotypes	SSRs	Assar et al. [59]
10	320 sorghum accessions	SSR markers	Shehzad et al. [3]
11	124 sorghum landraces from Burkina Faso	Microsatellite markers	Barro-Kondombo et al. [26]
12	156 sorghum germplasm accessions	SSRs	Sharma et al. [27]

Sr. No	Sorghum germplasm	Molecular markers	Reference
13	Three populations of backcross-derived lines of sorghum	EST SSRs	Mohamed et al. [60]
14	160 plants of sorghum	SSR markers	Adugna et al. [61]
15	Recombinant Inbred Line of *sorghum bicolor* made by crossing E-Tian x Ji2731	PAV markers and SSRs	Mocoeur et al. [30]
16	Sorghum population derived from a cross between two sorghum landraces, Red Kafir and Takakibi	SSRs	Shehzad et al. [62]
17	Recombinant sorghum line (hugurtay x N-13 (resistance donor)	SSRs	Yohannes et al. [63]
18	Set of 1108 sorghum diverse collections	Microsatellite markers	Salih et al. [64]
19	22 sorghum accessions (landraces)	Microsatellites	Motlhaodi et al. [65]
20	267 genotypes from Ethiopia	SSRs	Amelework et al. [33]
21	Two overlapping sets of RILs of sorghum	SNPs	Sukumaran et al. [35]
22	A random collection of 44 genotypes of sorghum	SPAR - (ISSR, RAPD, DAMD)	Satish et al. [66]
23	315 sorghum accessions	SNP, SQNM	Zhao et al. [34]
24	100 sweet sorghum accessions	SNPs	Da silva et al. [36]
25	80 sorghum accessions	Microsatellites	Sifau et al. [67]
26	300 diverse accessions of sorghum	SNPs	Chopra et al. [68]
27	93 sweet grain sorghum accessions from Burkina Faso	Microsatellites	Sawadogo et al. [43]
28	194 *Sorghum bicolor* and *S. bicolor sudanese* genotypes	SNPs	Parra-Londono et al. [69]
29	41 sorghum accessions	22 SSRs	Danquah et al. [70]
	329 accessions of sorghum germplasm collection	SNPs analysis	Sakamoto et al. [44]
30	Seven groups of 44 parental lines of sorghum	ISSRs, RAPDs, DAMD	Pandian et al. [48]
31	46 accessions of *Sorghum bicolor*	RAPD	Ruiz-Chutan et al. [71]
32	214 sorghum accessions	SNPs	Afolayan et al. [72]
33	12 *Sorghum bicolor* genotypes (5 sweet, 4 grain and 3 forage sorghums)	RAPDs, ISSRs	Kanbar et al. [47]
34	150 accessions of *Broomcorn Sorghum*	SSRs	Zhu et al. [73]
35	20 sorghum accessions	SSRs	Joshi Akansha et al. [74]
36	10 *sorghum bicolor* genotypes collected from USA (Texas)	SSRs	Jessup et al. [75]
37	3 RIL mapping populations of sweet sorghum genotypes	SNPs	Gomez et al. [50]
38	Recombinant Inbred Line derived from a cross between an elite U.S. common parent RTx430 and 10 diverse founders	SNPs	Olatoye et al. [49]
39	21 diverse Sorghum accessions	SNPs	Henderson et al. [53]

Table 7.
Studies on assessment of genetic variations using molecular markers in sorghum.

morphological traits for characterizing sorghum followed by maturity characters. Plant height is the most researched trait in these studies. Majority of efforts related to DNA fingerprinting of sorghum employed SSRs, followed by SNP and RAPD markers.

6. Conclusions

Most of the modern cultivated crops exhibit narrow genetic base due to domestication, selection of few desired traits and repeated use of genetically similar varieties as breeding parents. Climate change poses a serious threat to agricultural communities with possible forecast of high temperature, water scarcity and altered pattern of showers round the globe. Climate variations and shift will be a key driver of crop production especially in arid and semi-arid rain fed areas of the world. Such effects will vary among crops depending upon their physiology and climate resilience traits of particular crop. Sorghum is a C4 grass cultivated in diverse regions of the world for variety of uses. It stands tall among other cereal crops owing to inherent biotic/abiotic stress tolerance and wider adaptability. It is among few climate smart crops with potential to withstand future harsh environmental conditions. Hence, development of high yielding sorghum varieties will contribute towards ensuring global food security. Breeders are exploiting high throughout phenotyping platforms as well as omics- assisted variability evaluation of sorghum germplasm to identify/select highly diverse types that will serve as a base line for breeding of broad genetic base sorghum varieties. So, dissecting phenotypic and molecular diversity of sorghum germplasm is strongly justified.

Conflict of interest

The authors declare no conflict of interest.

Author details

Bushra Sadia[1*], Faisal Saeed Awan[1], Fozia Saleem[1], Javaria Altaf[2], Abdullah Bin Umar[1], Muhammad Nadeem[1], Samra Hameed[1], Farwa Ashraf[1] and Mariam Nasir[1]

1 Centre of Agricultural Biochemistry and Biotechnology (CABB), University of Agriculture Faisalabad (UAF), Pakistan

2 Department of Zoology, Government College University Faisalabad, Pakistan

*Address all correspondence to: bushra.sadia@uaf.edu.pk

References

[1] Nadeem MA, Nawaz MA, Shahid MQ, Doğan Y, Comertpay G, Yıldız M, Hatipoğlu R, Ahmad F, Alsaleh A, Labhane N, Özkan H. DNA molecular markers in plant breeding: current status and recent advancements in genomic selection and genome editing. Biotechnology & Biotechnological Equipment. 2018 Mar 4; 32(2):261–85.

[2] Jiang J, Gill BS. Current status and the future of fluorescence in situ hybridization (FISH) in plant genome research. Genome. 2006 Sep;49(9): 1057–68.

[3] Mateu-Andres I, De Paco L. Allozymic differentiation of the *Antirrhinum majus* and A. siculum species groups. Annals of botany. 2005 Feb 1;95(3):465–73.

[4] Hamouda M. Molecular analysis of genetic diversity in population of *Silybum marianum* (L.) Gaertn in Egypt. Journal of Genetic Engineering and Biotechnology. 2019 Dec 1;17(1):12.

[5] Platten JD, Cobb JN, Zantua RE. Criteria for evaluating molecular markers: comprehensive quality metrics to improve marker-assisted selection. PloS one. 2019 Jan 15;14(1):e0210529.

[6] Jiang GL. Molecular markers and marker-assisted breeding in plants. Plant breeding from laboratories to fields, Sven Bode Andersen. 2013 May 22:45–83. IntechOpen, DOI: 10.5772/52583

[7] RAK Rakshit S, Swapna M. DNA Markers in Diversity Analysis. InSorghum Molecular Breeding 2015 (pp. 23–46). Springer, New Delhi.

[8] Ashraf F. Genetic diversity analysis of exotic sorghum (*Sorghum bicolor* L. Moench) germplasm using single nucleotide polymorphisms. 2018. MPhil Dissertation. University of Agriculture, Faisalabad.

[9] Andersen JR, Lübberstedt T. Functional markers in plants. Trends in plant science. 2003 Nov 1;8(11):554–60.

[10] Iqbal A, Sadia B, Khan AI, Awan FS, Kainth RA, Sadaqat HA. Biodiversity in the sorghum (Sorghum bicolor L. Moench) germplasm of Pakistan. Genetics and Molecular Research. 2010; 9(2):756–64.

[11] Arias S, Bhatia SK. Sorghum. InMedical Applications for Biomaterials in Bolivia 2015 (pp. 33–39). Springer, Cham.

[12] Mrid RB, Bouargalne Y, El Omari R, Nhiri M. New insights into the therapeutic effects of phenolic acids from sorghum seeds. Journal of Reports in Pharmaceutical Sciences. 2019 Jan 1;8 (1):91.

[13] Dar RA, Dar EA, Kaur A, Phutela UG. Sweet sorghum-a promising alternative feedstock for biofuel production. Renewable and Sustainable Energy Reviews. 2018 Feb 1; 82:4070–90.Chris Duran, Nikki Appleby, David Edwards and Jacqueline Batley, " Molecular Genetic Markers: Discovery, Applications, Data Storage and Visualisation", Current Bioinformatics (2009) 4: 16. https://doi.org/10.2174/157489309787158198

[14] Janhi K, Matshaya Z, Chiduza C, Muzangwa L. Clipping Forage Sorghum Twice and Nitrogen Topdressing Offer an Option for Dual-Purpose Use for Cover Cropping and Fodder in Mixed Crop/Livestock Farming Systems. Agronomy. 2020 Jan;10(1):17.

[15] Nanaiah GK, Rakshit S. Genomic Designing for Climate Smart Sorghum. InGenomic Designing of Climate-Smart Cereal Crops 2020 (pp. 171–219). Springer, Cham.

[16] Nadeem M, Biodiversity evaluation of sweet sorghum (*Sorghum bicolor* L. Moench) for biomass potential. 2015. MPhil Dissertation. University of Agriculture, Faisalabad.

[17] Steel RG. Pinciples and procedures of statistics a biometrical approach. 1997.

[18] Hameed S. Assessment of genetic diversity of sweet sorghum (*Sorghum bicolor* L. Moench), an efficient biofuel crop. 2015. MPhil Dissertation. University of Agriculture, Faisalabad.

[19] Irshad B. Chemical and genetic diversity analyses of high biomass USDA sorghum (*Sorghum bicolor* M.) collections. 2017. MPhil Dissertation. University of Agriculture, Faisalabad.

[20] Agrama HA, Tuinstra MR. Phylogenetic diversity and relationships among sorghum accessions using SSRs and RAPDs. African journal of biotechnology. 2003;2(10):334–40.

[21] Nkongolo KK, Nsapato L. Genetic diversity in Sorghum bicolor (L.) Moench accessions from different ecogeographical regions in Malawi assessed with RAPDs. Genetic Resources and Crop Evolution. 2003 Mar 1;50(2): 149–56.

[22] Dahlberg JA, Zhang X, Hart GE, Mullet JE. Comparative assessment of variation among sorghum germplasm accessions using seed morphology and RAPD measurements. Crop Science. 2002 Jan;42(1):291–6.

[23] Geleta N, Labuschagne MT, Viljoen CD. Genetic diversity analysis in sorghum germplasm as estimated by AFLP, SSR and morpho-agronomical markers. Biodiversity & Conservation. 2006 Sep 1;15(10):3251–65.

[24] Bucheyeki TL, Gwanama C, Mgonja M, Chisi M, Folkertsma R, Mutegi R. Genetic variability characterisation of Tanzania sorghum landraces based on simple sequence repeats (SSRs) molecular and morphological markers. African Crop Science Journal. 2009;17(2).

[25] Shehzad T, Okuizumi H, Kawase M, Okuno K. Development of SSR-based sorghum (Sorghum bicolor (L.) Moench) diversity research set of germplasm and its evaluation by morphological traits. Genetic Resources and Crop Evolution. 2009 Sep 1;56(6):809–27.

[26] Barro-Kondombo C, Sagnard F, Chantereau J, Deu M, Vom Brocke K, Durand P, Gozé E, Zongo JD. Genetic structure among sorghum landraces as revealed by morphological variation and microsatellite markers in three agroclimatic regions of Burkina Faso. Theoretical and Applied Genetics. 2010 May 1;120(8):1511–23.

[27] Sharma R, Deshpande SP, Senthilvel S, Rao VP, Rajaram V, Hash CT, Thakur RP. SSR allelic diversity in relation to morphological traits and resistance to grain mould in sorghum. Crop and Pasture Science. 2010 Mar 30;61(3):230–40.

[28] Abraha T, Githiri SM, Kasili R, Araia W, Nyende AB. Genetic variation among Sorghum (Sorghum bicolor L. Moench) landraces from Eritrea under post-flowering drought stress conditions. American Journal of Plant Sciences. 2015;6(09):1410.

[29] Sabiel SA, Noureldin I, Baloch SK, Baloch SU, Bashir W. Genetic variability and estimates of heritability in sorghum (Sorghum bicolor L.) genotypes grown in a semiarid zone of Sudan. Archives of Agronomy and Soil Science. 2016 Jan 2; 62(1):139–45.

[30] Mocoeur A, Zhang YM, Liu ZQ, Shen X, Zhang LM, Rasmussen SK, Jing HC. Stability and genetic control of morphological, biomass and biofuel traits under temperate maritime and continental

conditions in sweet sorghum (Sorghum bicolour). Theoretical and applied genetics. 2015 Sep 1;128(9):1685–701.

[31] Mohammed R, Are AK, Bhavanasi R, Munghate RS, Kavi Kishor PB, Sharma HC. Quantitative genetic analysis of agronomic and morphological traits in sorghum, Sorghum bicolor. Frontiers in plant science. 2015 Nov 3;6:945.

[32] Sinha S, Kumaravadivel N. Understanding genetic diversity of sorghum using quantitative traits. Scientifica. 2016 Jan 1;2016.

[33] Amelework B, Shimelis H, Tongoona P, Laing M, Mengistu F. Genetic diversity of lowland sorghum landraces assessed by morphological and microsatellite markers. Australian Journal of Crop Science. 2016 Mar;10(3):291.

[34] Zhao J, Mantilla Perez MB, Hu J, Salas Fernandez MG. Genome-wide association study for nine plant architecture traits in sorghum. The Plant Genome. 2016 Jul;9(2):1–4.

[35] Sukumaran S, Li X, Li X, Zhu C, Bai G, Perumal R, Tuinstra MR, Prasad PV, Mitchell SE, Tesso TT, Yu J. QTL Mapping for Grain Yield, Flowering Time, and Stay-Green Traits in Sorghum with Genotyping-by-Sequencing Markers. crop science. 2016 Jul;56(4):1429–42.

[36] Da Silva MJ, Pastina MM, de Souza VF, Schaffert RE, Carneiro PC, Noda RW, Carneiro JE, Damasceno CM, Parrella RA. Phenotypic and molecular characterization of sweet sorghum accessions for bioenergy production. PloS one. 2017 Aug 17;12(8):e0183504.

[37] Shukla S, Felderhoff TJ, Saballos A, Vermerris W. The relationship between plant height and sugar accumulation in the stems of sweet sorghum (Sorghum bicolor (L.) Moench). Field Crops Research. 2017 Mar 1;203:181–91.

[38] Mumtaz A, Hussain D, Saeed M, Arshad M, Yousaf MI. Estimation of genetic diversity in sorghum genotypes of Pakistan. Journal of the National Science Foundation of Sri Lanka. 2018 Sep 30;46(3).

[39] Emendack Y, Burke J, Sanchez J, Laza HE, Hayes C. Agro-morphological characterization of diverse sorghum lines for pre-and post-flowering drought tolerance. Australian Journal of Crop Science. 2018 Jan;12(1):135.

[40] Prajapati DK, Pahuja SK, Verma NK, Chaudhary S. Morphological characterization of sorghum [Sorghum bicolor (L.) Moench] germplasm for DUS traits. International Journal of Current Microbiology and Applied Sciences. 2018;7(2):2058–71.

[41] Derese SA, Shimelis H, Mwadzingeni L, Laing M. Agro-morphological characterisation and selection of sorghum landraces. Acta Agriculturae Scandinavica, Section B—Soil & Plant Science. 2018 Oct 3;68(7): 585–95.

[42] Fernandes SB, Dias KO, Ferreira DF, Brown PJ. Efficiency of multi-trait, indirect, and trait-assisted genomic selection for improvement of biomass sorghum. Theoretical and applied genetics. 2018 Mar 1;131(3): 747–55.

[43] Sawadogon N, Batieno TB, Kiebre Z, Ouedraogo MH, Zida WP, Nanema KR, Nebie B, Bationo-Kando P, Traore RE, Sawadogo M, Zongo JD. Assessment of genetic diversity of Burkina Faso sweet grain sorghum using microsatellite markers. African Journal of Biotechnology. 2018 Mar 21;17(12): 389–95.

[44] Sakamoto L, Kajiya-Kanegae H, Noshita K, Takanashi H, Kobayashi M, Kudo T, Yano K, Tokunaga T, Tsutsumi N, Iwata H. Comparison of shape quantification methods for genomic

prediction, and genome-wide association study of sorghum seed morphology. PloS one. 2019 Nov 21;14(11):e0224695.

[45] Bojović R, Popović VM, Ikanović J, Živanović L, Rakaščan N, Popović S, Ugrenović V, Simić D. Morphological characterization of sweet sorghum genotypes across environments. JAPS: Journal of Animal & Plant Sciences. 2019 Jun 1;29(3).

[46] Mofokeng MA, Shimelis H, Laing M, Shargie N. Genetic variability, heritability and genetic gain for quantitative traits in South African sorghum genotypes. Australian Journal of Crop Science. 2019 Jan;13(1):1.

[47] Kanbar A, Shakeri E, Alhajturki D, Horn T, Emam Y, Tabatabaei SA, Nick P. Morphological and molecular characterization of sweet, grain and forage sorghum (Sorghum bicolor L.) genotypes grown under temperate climatic conditions. Plant Biosystems-An International Journal Dealing with all Aspects of Plant Biology. 2020 Jan 2; 154(1):49–58.

[48] Pandian S, Satish L, Shilpha J, Ramesh M. Genetic Diversity Analysis Reveals Strong Population Structure in Sorghum Germplasm Collection. Proceedings of the National Academy of Sciences, India Section B: Biological Sciences. 2020 Mar;90(1):179–90.

[49] Olatoye MO, Marla SR, Hu Z, Bouchet S, Perumal R, Morris GP. Dissecting adaptive traits with nested association mapping: Genetic architecture of inflorescence morphology in sorghum. G3: Genes, Genomes, Genetics. 2020 May 1;10(5):1785–96.

[50] Gomez FE, Mullet JE, Muliana AH, Niklas KJ, Rooney WL. The genetic architecture of biomechanical traits in sorghum. Crop Science. 2020 Jan;60(1): 82–99.

[51] Birhan T, Bantte K, Paterson A, Getenet M, Gabizew A. Evaluation and Genetic Analysis of a Segregating Sorghum Population under Moisture Stress Conditions. Journal of Crop Science and Biotechnology. 2020 Jan;23 (1):29–38.

[52] Li M, Shao MR, Zeng D, Ju T, Kellogg EA, Topp CN. Comprehensive 3D phenotyping reveals continuous morphological variation across genetically diverse sorghum inflorescences. New Phytologist. 2020 Jun;226(6):1873–85.

[53] Henderson AN. The morphological, physiological, and genetic underpinnings of intraspecific salinity tolerance in Sorghum bicolor.

[54] Djè Y, Heuertz M, Lefebvre C, Vekemans X. Assessment of genetic diversity within and among germplasm accessions in cultivated sorghum using microsatellite markers. Theoretical and Applied Genetics. 2000 Apr 1;100(6): 918–25.

[55] Ayana A. Genetic diversity in sorghum (Sorghum bicolor (L.) Moench) germplasm from Ethiopia and Eritrea (Doctoral dissertation, Addis Ababa University).

[56] Folkertsma RT, Rattunde HF, Chandra S, Raju GS, Hash CT. The pattern of genetic diversity of Guinea-race Sorghum bicolor (L.) Moench landraces as revealed with SSR markers. Theoretical and Applied Genetics. 2005 Aug 1;111(3):399–409.

[57] Wu YQ, Huang Y, Porter DR, Tauer CG, Hollaway L. Identification of a major quantitative trait locus conditioning resistance to greenbug biotype E in sorghum PI 550610 using simple sequence repeat markers. Journal of economic entomology. 2014 Sep 26; 100(5):1672–8.

[58] Perumal R, Krishnaramanujam R, Menz MA, Katilé S, Dahlberg J, Magill CW, Rooney WL. Genetic

diversity among sorghum races and working groups based on AFLPs and SSRs. Crop Science. 2007 Jul;47(4): 1375–83.

[59] Assar AH, Uptmoor R, Abdelmula AA, Wagner C, Salih M, Ali AM, Ordon F, Friedt W. Assessment of sorghum genetic resources for genetic diversity and drought tolerance using molecular markers and agro-morphological traits. University of Khartoum Journal of Agricultural Sciences. 2020 Feb 25;17(1).

[60] Mohamed A, Ali R, Elhassan O, Suliman E, Mugoya C, Masiga CW, Elhusien A, Hash CT. First products of DNA marker-assisted selection in sorghum released for cultivation by farmers in sub-saharan Africa. Journal of Plant Science & Molecular Breeding. 2014;3(3):1–0.

[61] Adugna A. Analysis of in situ diversity and population structure in Ethiopian cultivated Sorghum bicolor (L.) landraces using phenotypic traits and SSR markers. SpringerPlus. 2014 Dec 1;3(1):212.

[62] Shehzad T, Okuno K. QTL mapping for yield and yield-contributing traits in sorghum (Sorghum bicolor (L.) Moench) with genome-based SSR markers. Euphytica. 2015 May 1;203(1): 17–31.

[63] Yohannes T, Abraha T, Kiambi D, Folkertsma R, Hash CT, Ngugi K, Mutitu E, Abraha N, Weldetsion M, Mugoya C, Masiga CW. Marker-assisted introgression improves Striga resistance in an eritrean farmer-preferred sorghum variety. Field Crops Research. 2015 Mar 1;173:22–9.

[64] Salih SA, Herslman L, Labuschange MT, Mohammed AH. Assessment of genetic diversity of sorghum [sorghum bicolor (l.) Moench] germplasm in East and Central Africa. World Journal of Biotechnology. 2016;1 (03):113–20.

[65] Motlhaodi T, Geleta M, Chite S, Fatih M, Ortiz R, Bryngelsson T. Genetic diversity in sorghum [Sorghum bicolor (L.) Moench] germplasm from Southern Africa as revealed by microsatellite markers and agro-morphological traits. Genetic resources and crop evolution. 2017 Mar 1;64(3): 599–610.

[66] Satish L, Shilpha J, Pandian S, Rency AS, Rathinapriya P, Ceasar SA, Largia MJ, Kumar AA, Ramesh M. Analysis of genetic variation in sorghum (Sorghum bicolor (L.) Moench) genotypes with various agronomical traits using SPAR methods. Gene. 2016 Jan 15;576(1):581–5.

[67] Sifau MO, Oduoye OT, Oluwasanya OA, Aladele SE. Assessment of Genetic Variability in Sorghum Accessions (Sorghum bicolor L. Moench) at the National Centre for Genetic Resources and Biotechnology, Ibadan, Nigeria. Journal of Applied Sciences and Environmental Management. 2017;21(6):1143–7.

[68] Chopra R, Burow G, Burke JJ, Gladman N, Xin Z. Genome-wide association analysis of seedling traits in diverse Sorghum germplasm under thermal stress. BMC Plant Biology. 2017 Dec 1;17(1):12.

[69] Parra-Londono S, Kavka M, Samans B, Snowdon R, Wieckhorst S, Uptmoor R. Sorghum root-system classification in contrasting P environments reveals three main rooting types and root-architecture-related marker–trait associations. Annals of botany. 2018 Jan 23;121(2): 267–80.

[70] Danquah A, Galyuon IK, Otwe EP, Asante DK. Genetic diversity in some Ghanaian and Malian sorghum [Sorghum bicolor (L) Moench] accessions using SSR markers. African Journal of Biotechnology. 2019 Jul 31;18 (27):591–602.

[71] Ruiz-Chután JA, Salava J, Janovská D, Žiarovská J, Kalousová M, Fernández E. Assessment of genetic diversity in sorghum bicolor using RAPD markers. Genetika. 2019;51(3): 789–803.

[72] Afolayan G, Deshpande SP, Aladele SE, Kolawole AO, Angarawai I, Nwosu DJ, Michael C, Blay ET, Danquah EY. Genetic diversity assessment of sorghum (Sorghum bicolor (L.) Moench) accessions using single nucleotide polymorphism markers. Plant Genetic Resources. 2019 Oct;17(5):412–20.

[73] Zhu M, Chen J, Yuyama N, Luo L, Xiao X, Lv Y, Liu Y, Cai H. Genetic Diversity and Population Structure of Broomcorn Sorghum Investigated with Simple Sequence Repeat Markers. Tropical Plant Biology. 2020 Mar;13(1): 62–72.

[74] Joshi Akansha R, Kale Sonam S, Chavan Narendra R. Genetic diversity among elite sorghum (Sorghum bicolor L.) accessions genotyped with SSR markers to enhance use of global genetic resources. IJCS. 2020;8(2):1691–7.

[75] Jessup RW, Abed ZA, Najeep HF, Al-Azawi NM. Genetic analysis of sorghum cultivars from USA using SSR markers. Plant Archives. 2020;20(1): 1121–5.

Chapter 8

Genetic Diversity of Fusarium Wilt Disease of Banana

Gilberto Manzo-Sánchez, Marco Tulio Buenrostro-Nava, Carlos L. Leopardi, Mario Orozco-Santos and Mauricio Guzman-Quesada

Abstract

Bananas and plantains (*Musa* spp.) represent the fourth most important crop in the world. In 2017, an area of 5,637,508 hectares and a production of 153 million tons were reported. Fusarium wilt caused by the fungus *Fusarium oxysporum* f. sp. *cubense* (Foc), is considered one of the most destructive diseases of bananas and plantains worldwide. The pathogen *Foc* causes a typical wilt syndrome on infected plants, it has a saprophytic and parasitic phase in its life cycle. Fusarium wilt is a "polycyclic" disease. This pathogen shows a relatively diverse population genetic structure for a fungus apparently of asexual reproduction and is composed of different evolutionary lineages, which has 24 groups of vegetative compatibility (VCGs), two clades and nine clonal linage. Foc is a genetically diverse pathogen, although the available evidence so far indicates that it does not use the mechanisms of sexual reproduction, such as recombination, to increase its genetic diversity. Furthermore, the population of this fungus in Southeast Asia shows a high degree of variation, suggesting that Foc lineages evolved together with their hosts in Southeast Asia. Alternatively, it has been suggested that Foc has multiple independent evolutionary origins, both within and outside of the Musaceae origin center.

Keywords: genetics, diversity, fusarium wilt disease, banana

1. Introduction

Bananas and plantains (*Musa* spp.) represent the fourth most important crop in the world, since only rice, wheat and corn surpass it [1]. The fruit has a high content of carbohydrates, potassium, phosphorus, magnesium, vitamins A and C, folic acid and tannins [2]. This fruit is produced throughout the year. Therefore, we can always consume bananas, regardless of the month we are in.

These crops are produced in 135 countries in the tropical and subtropical regions. India contributes 31% of the total, followed by China with 10% and the Philippines with 9% of world production. In 2017, an area of 5,637,508 hectares and a production of 153 million tons were reported, with the main exporting countries being Ecuador, Costa Rica, the Philippines, Guatemala and Colombia, who ship their products to the United States, Canada, Europe, Russia, and the Asian Pacific region. The commercialization of this fruit represents an important source of income for the Latin American region. Most of the producers are farmers who grow it for domestic consumption or for local markets and only 15 percent of production is for export [3].

The production of bananas and plantains is seriously affected by various phytopathogenic agents, such as fungi, nematodes, viruses, bacteria, and insects. Some of the pathogens spread during the distribution of Musaceae germplasm native to Southeast Asian occurred in the 20th century to new agricultural areas (Latin America and the Caribbean), since by nature their spread occurs on a smaller scale and hardly at long distances [1].

Fusarium wilt caused by the fungus *Fusarium oxysporum* f. sp. *cubense* (Foc), is considered one of the most destructive diseases of bananas and plantains worldwide [4, 5]. The disease greatly hinders the production mainly of the genotypes of *Musa acuminata*, *M. balbisiana*, *M. schizocarpa* and *M. textilis* and their hybrids [4, 6].

Once Foc enters the fields, it is difficult to control; this is due to the fact that the pathogen persists in the soil for long periods. This is the reason because the use of plants derived from tissue culture have been considered as one of the disease management strategies, this in order to avoid the introduction of Foc in pathogen-free fields; as well as the implementation of safety practices to avoid its dispersion [5]. However, the most effective means of controlling the disease is the replacement of susceptible cultivars by those who are resistant, although today the main markets demand the 'Giant Dwarf' clone from the Cavendish subgroup.

History indicates that the pathogen probably originated in Southeast Asia; however, the first report was in Australia in 1876 affected by the cultivar 'Silk', also known as 'Manzano' (AAB) [4] and in 1890 it occurred in plantations in Costa Rica and Panama. About 30,000 hectares were lost in this country between 1940 and 1960 [4]. In total, it was estimated that more than 40,000 hectares of bananas were lost in a 50-year period in Central and South America [4, 7]. Also, epidemics have been reported on other continents. For example, in Bali, banana production decreased from 134,000 to 54,000 tons in 1997, due to the disease [8].

Given the damage caused by Fusarium wilt, there is a probability that the pathogen could be distributed through the planting material (corms or suckers) of 'Gros Michel', since this was used for use in new plantations [4, 9]. At that time, large shipments of suckers and rhizomes may also have been transported between countries by transnational companies to supplement local stocks of commercial cultivars, thereby promoting the spread of disease. The stage was set for a major epidemic to emerge [10].

2. Banana importance

Banana and plantain (*Musa* spp.) are believed to have originated in Southeast Asia and are cultivated in a wide variety of environments in the tropics and sub-tropics regions of the world. *Musa*, including the dessert banana and the cooking types or plantains are produced in 155 countries.

Throughout history *Musa* has provided humans with food, medicine, clothing, tools, shelter, furniture, paper, and handicrafts. *Musa* are rich in vitamin C, B6, minerals (particularly potassium), and dietary fiber. They are also a rich energy source, with carbohydrates accounting for 22% and 32% of fruit weight for banana and plantain, respectively. It is cholesterol free, high in fiber, and low in sodium.

In terms of total fruit crops production, the banana ranks after oranges, grapes, and apples, but when plantain production is added, it becomes the world's number one fruit crop. According to the Food and Agriculture Organization of the United Nations (FAO), in 2018 more than 11.3 million hectares of banana and plantain were harvested worldwide and were produced a total of 155.2 million tonnes:

Rank	Country/territory	Production (tonnes)
1	India	30,808,000
2	China	11,221,700
3	Philippines	9,358,785
4	Colombia	7,287,997
5	Indonesia	7,264,383
6	Ecuador	7,157,603
7	Brazil	6,752,171
8	Cameroon	5,144,258
9	Congo, Democratic Republic of the	5,066,203
10	Uganda	4,337,747
11	Guatemala	4,294,121
12	Ghana	4,264,258
13	Tanzania	4,045,568
14	Angola	3,492,184
15	Nigeria	3,093,872
16	Costa Rica	2,633,788
17	Mexico	2,354,479
18	Peru	2,329,480
19	Cote d'Ivoire	2,280,368
20	Dominican Republic	2,224,403

Source: http://www.promusa.org/Banana-producing+countries+portal

Table 1.
Top 20 of banana-producing countries and overseas territories.

115.7 million tonnes under their bananas crop item (75%) and 39.5 million tonnes under their plantains crop item (25%). However, the estimated production for the same year published by [11] is 139.5 million tonnes: 79.6 million tonnes of Cavendish (57%), 17.5 million tonnes of other dessert bananas (13%), 20.9 million tonnes of Plantain (15%), and 21.4 million tonnes of other cooking bananas (15%).

In the **Table 1**, the list of top 20 of banana-producing countries and overseas territories and the number of tonnes they each produced in 2018 is showed. Production is measured in tonnes and represent the total of the bananas and plantains categories into, according FAO statistics.

3. Fusarium wilt caused by the fungus *Fusarium oxysporum* f. sp. *cubense* (Foc)

Fusarium is a genus comprises several species of filamentous ascomycetes, including pathogenic and non-pathogenic species for agricultural crops. One of the best known is *F. oxysporum*, this causes vascular wilt and root rot in more than 100 plant species [12].

In the *Fusarium* system, Foc belongs to the *Fusarium oxysporum* species complex (FOSC), four clades have been identified from this, using the translational elongation factor 1-alpha (tef1) and the rDNA of the mitochondrial subunit (mtssu), in Foc isolates, which were grouped as baseline lineage [13].

The pathogenic isolates of *F. oxysporum* have been classified in more than 100 special forms. Members of a special form usually cause disease in a particular range of host species, with some special forms capable of colonizing a wider range of plants [14]. A special form can be subdivided into races based on characteristic virulence patterns in differential host cultivars [15].

Taxonomic classification:
Domain: Eukaryota
Kingdom: Fungi
Phylum: Ascomycota
Class: Ascomycetes
Subclass: Sordariomycetidae
Order: Hypocreales

One of the most devastating special forms is responsible for Fusarium wilt of bananas and plantains [9], which is caused by *Fusarium oxysporum* Schlect. f. sp. *cubense* (E.F. Smith) Snyder & Hansen, who lives in the soil. The sexual phase (teleomorphic) of the fungus is unknown and cannot be distinguished morphologically between different strains. This pathogen produces three types of asexual spores, these are macroconidia, microconidia and chlamydospores, which function as mechanisms of dispersal, reproduction and survival [16].

The microconidia (5–16 × 2.4–3.5 μm) are oval in shape and consist of a single cell, generally without septa, may be oval, elliptical to reniform, and develop abundantly on false heads on short monophialides. While macroconidia (27–55 × 3.3–5.5 μm) are abundant, slightly curved, and relatively thin, they have 4–8 cells, with 3–5 septa (generally 3 septa) see **Figure 1A**. The apical cell is attenuated or hook-shaped in some isolates. The basal cells are shaped like a foot. Macroconidia develop into single hyphal fialids (**Figure 1B**). Micro and macroconidia occur on branched or unbranched monophial cuts [17, 18].

Chlamydospores (7–11 μm in diameter) are generally globose and form individually or in pairs, they are abundantly formed in hyphae or conidia, single or in chains, generally in pairs, this type of spores constitutes resistance structures of the fungus, These have thick cell walls, and their production is abundant on infected tissues in advanced stages of the disease [4]. They can be interspersed or in the terminal part of the hyphae [17].

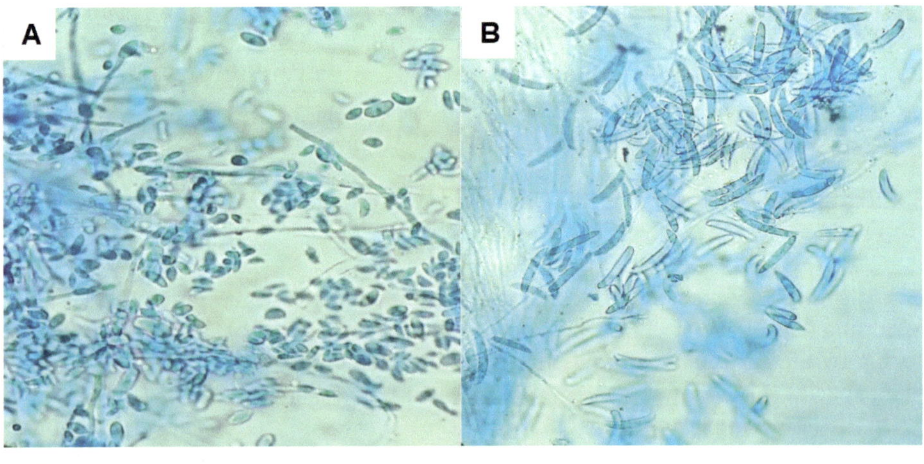

Figure 1.
Reproductive structure of Fusarium oxysporum *f.sp.* cubense *(A) Microconidia y (B) Macroconidia.*

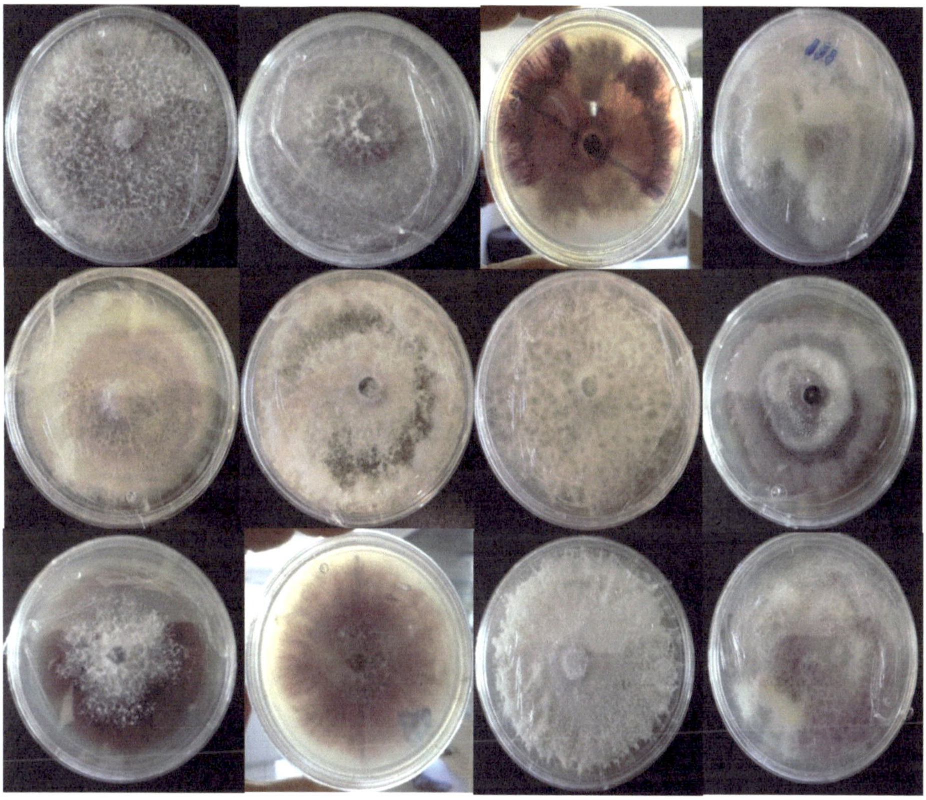

Figure 2.
Colony morphotypes of Fusarium oxysporum *f. sp.* cubense *isolated from bananas and plantains in different states of Mexico.*

In vitro development of the pathogen onto potato dextrose agar (PDA) culture medium, has a variable morphology, its growth is 4 to 7 mm per day at 24°C, forming colonies with abundant aerial mycelium and variable color pigmentation from white, salmon to pale violet. In general, the *F. oxysporum* strains cannot be morphologically distinguished between different races or groups of vegetative compatibility (VCGs). The fungus *F. oxysporum* generally produces black to violet sclerotia, while the pigmentation of the colonies is pale violet to dark red on PDA culture media [9, 19], as shown on **Figure 2**.

Some isolates rapidly mutate from pionnotal (with abundant fatty or shiny aggregates of conidia) to a flat, moist pale yellowish-white to peach mycelium grown on a PDA culture [9, 19].

4. Symptomatology

The pathogen *Fusarium oxysporum* f.sp. *cubense* causes a typical wilt syndrome on infected plants, it has a saprophytic and parasitic phase in its life cycle. It begins as a saprophyte in the soil as chlamydospores, which are dormant and immobile until plant exudates stimulate their germination to spread towards the roots [9]. These germinated chlamydospores develop a thallus that produces conidia after 6–8 hours. The conidia germinate and adhere to the roots of the host plant where they penetrate the epidermal cells and then invade and colonize the vascular system [20, 21].

After successfully infecting the roots, the pathogen grows towards the rhizome and pseudostem, causing a deficiency in the absorption of water and consequently an eventual wilting of the leaves and finally causing the death of the plant [9, 16]. This pathogen has the ability to invade all the organs of the plant with the exception of the fruit [16].

Externally, the first signs of the disease are usually wilting and yellowing of the older leaves around the margins (**Figure 3A**), the older chlorotic leaves collapse (**Figure 3B**), the old leaves hang down and dry forming a skirt (**Figure 3C**), the suckers are shown asymptomatic (**Figure 3D**), while internally the vascular bundles of the pseudostem turn reddish brown (**Figure 3E**), the corm shows an abnormal dark brown discoloration (**Figure 3F**), the base of the pseudostem shows fissures (**Figure 3G**) and the midrib of the leaves shows a dark brown discoloration (**Figure 3H**) [5, 6, 9].

To better understand the process of the Foc-banana interaction, some investigations have emerged using isolates transformed from Foc with the gene for the

Figure 3.
External and internal symptoms caused by Fusarium oxysporum *f.sp. cubense in banana and plantain plants (*Musa *spp.). Chlorosis in older leaves around the margins (A). Older leaves collapsed (B). Hanging and dried leaves forming a skirt (C). Asymptomatic children (D). Reddish-brown vascular bundles of the pseudostem (E). Corm with abnormal dark brown discoloration (F). Fissures at the base of the pseudostem (G). Central rib with dark brown discoloration (H).*

green fluorescent pigment (GFP), with the aim of studying the movement of the pathogen from the soil towards the roots and rhizome [22, 23].

Recently [24], using GFP they demonstrated the movement of the pathogen before the appearance of external symptoms, as well as the presence of inoculum on the external surface of the veins of senescent or decomposing leaves, followed by the substantial production of macroconidia and chlamydospores, these results demonstrate that there may be serious implications regarding the spread of the pathogen. In addition, chlamydospore production occurs inside and outside the veins of the leaves, which increases the risk of spores returning to the ground through leaf removal. Also, it was possible to identify the progress of the pathogen in the pseudostem before the development of external symptoms. The authors suggest that future studies are required on the possible wind-borne spread of inoculum and the potential of the pathogen to infect a healthy plant through aerial inoculation.

5. Epidemiology

Fusarium wilt is a "polycyclic" disease. However, several cycles of infection can occur in affected banana plantations. Losses can eventually develop, even when very small amounts of the pathogen inoculum manage to infest fields and the disease is initially of little concern to growers [6]. For example, the first outbreaks of TR4 reported in China and the Philippines were not taken with great importance; this resulted in devastation and uncontrollable problems in the affected plantations [25].

In addition to prevention, early recognition and rapid containment of a disease outbreak is necessary to prevent epidemic development. A good understanding of the key factors responsible for the development of the disease is required when designing practical protocols for the destruction of infected plants, the treatment of the surrounding infested soil, and the reduction of inoculum in plant residues and soil [26].

Foc was shown to have the ability to survive for decades in infested soil, as "Gros Michel" production was generally impossible in plantations previously affected by Foc [9]. Chlamydospores of Foc in dead host material play a role in their survival, but their persistence for long periods is probably due to their ability to infect weed species [6]. For example, in studies in tropical America and Australia, Foc was isolated from the roots of various weed species (*Chloris inflata, Euphorbia heterophylla Tridax procumbens, Cyanthillium cinereum, Commelina diffusa*, *Ixophorus unisetus*, *Panicum purpurascens*, *Cyperus luzulae*, *Paspalum fasciculatum*), present in banana plantations that were affected by R1 and TR4 [27]; however, these are asymptomatic and their presence in banana fields could be of high risk and therefore it is important to carry out a targeted control to reduce their presence. Foc's ability to survive in the absence of its host is an important factor in the management of this disease [6].

Foc has been shown to spread in various ways, with infected suckers being the most efficient, since they are the most used as vegetative material for new plantations [9]. In many cases, the suckers are washed and treated with fungicides. However, infected suckers were the main material before tissue culture seedlings were available [6], being practically impossible to establish plantations free of the pathogen. However, even after it was possible to produce tissue culture material, secondary contamination of plantations by Foc was common. For example, TR4-affected Cavendish plantations were routinely established with tissue culture seedlings [6].

Foc has the ability to spread in the soil, which indirectly contaminates in and around plantations, but unfortunately it is also used in nurseries for the propagation of seedlings used for field establishment [25]. Surface waters are easily polluted and use for irrigation of polluted river or pond water is highly risky. In addition, Foc is spread by contaminated tools (shovels, machetes, hoes, etc.), agricultural machinery, clothing and footwear [9, 28]. Any or all of these ways can facilitate the spread of Foc in and around a plantation, and may be possible through other means [6, 28].

Studies carried out in Australia detected TR4 spores in the exoskeletons of the banana weevil (*Cosmopolites sordidus*) and suggested that the insect could be a predisposing agent as a vector of the disease [29].

The recent transcontinental disseminations of TR4, suggest that something other than vegetative material (suckers) was responsible for these long-distance disseminations. Although these outbreaks may have been the result of something as simple as workers' boots impregnated with soil contaminated by Foc spores from plantations in Southeast Asia, or some other means could be responsible such as the entry of machinery from affected areas. Better knowledge is needed to understand the long-distance spread of this pathogen [6].

6. Genetic diversity and evolution of Fusarium wilt of bananas

Fusarium oxysporum Species Complex (FOSC) are widely distributed and it is mostly non-pathogenic and it is commonly found in roots and soil associated fungus in asymptomatic crop plants. It has been found to be associated with plants as endophyte, saprophyte or just latent in agro-ecosystems [30]. Both, studies on FOSC isolated from non-cultivated species and form cultivated crops have reported a considerable variability based on the morphology of the asexual reproductive structures [31] and latter at the DNA sequence [32, 33]. Understanding its genetic variability is relevant to implement an earlier detection system and implement a proper disease surveillance program.

Recent studies on molecular genetics of *Fusarium* from cultivated plants have shown a high diversity and this variation relays on environmental conditions and are classified in groups and vegetative compatible groups (VCG) as described latter in this chapter. *Fusarium* has evolved heavily depending on its interaction with plant genotypes, such is the case for both 'tropical' and 'subtropical' race 4, which attacks different cultivars, depending of the geographical region [34] as well as for those FOSC from non-cultivated species [30].

Knowledge of the genetic diversity of populations of phytopathogenic fungi and their mode of reproduction are important for the application of management strategies, this with the aim of reducing the impact of the disease [35]. In the case of Foc, this pathogen shows a relatively diverse population genetic structure for a fungus apparently of asexual reproduction and is composed of different evolutionary lineages [33], which has 24 groups of vegetative compatibility (VCGs, VCG0120 to VCG0126 and VCG0128 to VCG01224) distributed worldwide [34, 36–40].

However, in recent samplings in Latin America it was possible to identify 20 new VCGs (new VCG 1 to new VCG 20), these were distributed over the three main clades (clade 1, clade 2 and clade 3), these results show that the majority of the new VCG are grouped in clade 3 and these originate from Latin America [41], this supports the hypothesis on the evolution of Foc, in which it is mentioned that the local populations of *F. oxysporum* evolved and they became pathogenic in the introduced bananas [36, 42–44].

Studying VCGs has been a useful means of subdividing Foc into genetically isolated groups, but it does not, however, measure the genetic relationship between the isolates. Furthermore, VCG are phenotypic markers that can undergo a selection process. Direct identification of VCG is a relatively objective, but time-consuming test, and the results indicate genetic similarity rather than genetic difference [31]. Therefore, VCGs represent good phenotypic traits for assessing diversity within populations, but the genetic relationships between VCGs must be assessed using other molecular tools.

Fourie and collaborators [39], classified Foc into two clades, clade 1 and clade 2, these based mainly on their evolutionary origins. In the case of clade A, the Foc groups that co-evolved with bananas of genome A belong, while those that belong to clade B evolved with their hosts having genome B or both genome A and genome B.

The teleomorph for Foc has never been reported and the pathogen is likely to manifest mutations or parasexualism, as the main basis for its genetic diversity. Although PCR analysis has shown the presence of both MAT idiomorphs, therefore the pathogen can potentially reproduce by sexual means.

The race concept has been widely used in the *F. oxysporum* classification system by plant pathologists. Based on the published data, it can be inferred that the Foc-TR4 isolates recently evolved from the predecessors in Foc-R1. Foc-R1 showed greater phylogenetic diversity than Foc-TR4. Once established, both races apparently co-evolved in the same region, which means that possible horizontal gene transfer could be involved in the high level of diversity seen in Foc-R1, as well as in the appearance of Foc-TR4.

Three races of Foc are known; but nevertheless, the term race is used in a less formal way in relation to this pathosystem (*Musa*-Foc), since the genetic bases of susceptibility and resistance have not yet been characterized. The Foc races currently described refer to strains of the pathogen, which have been found to be pathogenic to specific cultivars in the field [9, 38]. For example, race (R1) is pathogenic to cultivars of 'Silk', 'Manzano' (AAB) and 'Gros Michel' (AAA). While race 2 (R2) is pathogenic to cooking bananas such as 'Bluggoe' and 'Pear' (ABB) and race 4 (R4) affects all cultivars of the Cavendish subgroup (AAA) and those susceptible to R1 and R2 [4, 5, 45]. Previously, a population of *Fusarium oxysporum* in Central America was considered as race 3 causing wilt in *Heliconia* spp., but is no longer considered to be part of Foc [5].

A Foc race 4 variant was reported in Taiwan affecting Cavendish cultivars in the tropics in 1967 [4, 38]. Therefore, it was necessary to separate the populations that only affected Cavendish cultivars in the subtropics from those populations that affected in the tropics, so two divisions of Foc R4 were generated: race 4 sutropical (STR4) and race 4 tropical (TR4) [38], however, TR4 was pathogenic under tropical and subtropical conditions affecting Cavendish cultivars [25, 39]. In the case of VCGs, they have been associated with STR4 (0120, 01201, 01202, 01209, 01210, 01211, 01215, 0120/15; 0129/11), while only one VCG to TR4 (01213/16) [25, 40].

Visser and collaborators [46], carried out a study on the characterization of tropical Foc race 4 populations affecting 'Cavendish' plantations in South Africa. Only VCG 0120 and idiomorph MAT-2 could be identified, while phylogenetic analysis of the TEF sequence revealed that the isolates from South Africa were pooled with other isolates belonging to VCG 0120 from Australia and Asia. Suggesting, the introduction and dispersal mainly by infected material within the country.

In Latin America and the Caribbean, the composition of the populations has been limitedly studied. For example, the Cuban populations belong to VCG 01210 (mostly race 1), 0124, 0124/0125 and 0128 (mostly race 2); the isolates did not produce lacinias in K2 medium and the production of volatiles was independent of the race, while in Venezuela VCG 01215 and race 1 are reported. A study using AFLP markers grouped VCG 01210 into a subgroup and showed the presence of common alleles with VCG 0124 [47]. On the other hand, the pathogenicity studies with representative isolates of each VCG in Cuba, showed a differentiated aggressiveness on different clones between VCG 0124 and 0128, belonging to race 2, indicating lack of genetic sense in the racial classification. It is required to determine in Latin America and the Caribbean the VCG present in the different countries and the pathogenic relationships between them.

In order to better understand how races 1 and 4 are related, genome and transcriptome analysis of *F. oxysporum* f. sp. *cubense* has shown common sequences of single-copy genes from Race 1 and Race 4, showing that there is a close relationship and suggesting that they share a common ancestor. Furthermore, a comparative genomics study among *F. oxysporum* f. sp. *licoperci*, *F. graminearum* and *F. verticillioides* showed that there is transfer of lineage-specific (LS) genomic regions that have pathogenicity related genes with distinct evolutionary profiles, indicative of horizontal acquisition and suggesting that there is transfer of LS chromosomes between genetically isolated *Fusarium* species. This is of high relevance and of particular concern for agricultural systems, because non-pathogenic *F. oxysporum* strains that are already endophytic to crop plants could suddenly become pathogenic [48] and give origin to new pathogenic lineages of *F. oxysporum*. It is clear that in the last decade a large amount of DNA sequence information has been published on *F. oxysporum*, but there is a lack of consistency in the data and a larger study needs to be conducted in which DNA sequences of isolates from non-cultivated species is included and even from *Fusarium* species that are thought not to be related.

Foc genetic diversity studies were initiated using various molecular methods, including random amplified polymorphic DNA markers (RAPDs) [49]; Restriction Fragment Length Polymorphisms (RFLP) [43]; Amplified fragment length polymorphism (AFLP) [50]; DNA sequence analysis [32, 44]; microsatellites or simple repetitive sequences [51]; simple repetitive inter sequence (ISSR) [52]. These studies showed that the population of this fungus in Southeast Asia shows a high degree of variation, suggesting that the Foc lineages evolved together with their hosts in Southeast Asia.

Alternatively, Foc has been suggested to have multiple independent evolutionary origins, both within and outside the *Musa* genetic center [36]. Using the phylogenetic genealogical approach, [32] identified five Foc-independent genetic lineages in a global population. Using a similar approach and additional data, [44] found three additional lineages. However, none of these studies included Indonesian populations, and therefore there is only limited information available on Foc diversity at the center of origin of bananas.

F. oxysporum f.sp. *cubense* probably coevolved with its host species within its center of origin [32, 36, 44]. For example, various studies that have used deoxyribonucleic acid (DNA) markers have revealed the polyphyletic origin of Foc and the separation of two main clades and eight to ten lineages, as some VCGs are taxonomically closer to other special forms of *F. oxysporum* than some Foc VCGs [32, 36, 42, 44, 50, 53].

Furthermore, strains belonging to various VCGs infect particular banana cultivars and, therefore, were grouped in the same race, suggesting that the pathogenicity towards a specific cultivar evolved in a convergent way [32, 38, 44] or as a result of horizontal gene transfer between members of the *F. oxysporum* complex [48, 54].

High resolution genotyping sequencing analyzes using (DArTseq) validated and expanded these findings [55]. According to the DArTseq markers of 24 Foc strains (representing all the known VCG so far) they were divided into two groups. These results strongly corroborate the clades mentioned in previous studies, except VCG0123, VCG01210, VCG01212 and VCG01214, which were occasionally grouped into opposing clades, VCG 01221 and 01224, which were never classified before but now clearly belong to clade 2 [55].

In the advent of high throughput DNA sequencing technology [56] has allowed scientist to better understand the molecular weaponry used by this pathogen. The pathogen molecular tools include genes involved in root attachment, cell degradation, detoxification of toxins produced by the plant's defense mechanism and signal transduction, among others [16]. In Ref. [57], the authors have reported a predicted genome size for several *F. oxysporum* f. sp. *cubense* with a size of 48.56 Mb for Foc Race 1 and 48.81 for Foc Race 4, comprising and estimated of 15,865 and 14,506 genes, respectively. This genome information was compared and aligned to 11 of the 15 chromosomes contained in *F. oxysporum* f. sp. *licopersici*, including those regions reach in transposable elements; which might explain its high genetic variability and lack of chromosome stability [57].

Recently, in a study samples of musaceae with wilt symptoms were collected in the regions of Indonesia, Java, Sumatra, Kalimantan, Sulawesi, Papua and Nusa Tenggara, this demonstrated by phylogenetic analysis that the Foc lineages were genetically different, and it was achieved to identify 11 new species of *Fusarium* affecting musaceae, these were: *Fusarium cugenangense*, *F. duoseptatum*, *F. gros-michelii*, *F. hexaseptatum*, *F. kalimantanense, F. odoratissimum, F. phialophorum, F. purpurascens, F. sangayamense, F. tardichlamydosporum,* and *F. tardicrescens*, placing them in the Banana Fusarium Complex (FOBC), as well as showing that *F. odoratissimum* II-5 comprises TR4 [58].

7. Conclusions

Fusarium wilt disease of banana caused by soil-born pathogen *Fusarium oxysporum* f.sp. *cubense* (Foc) is considered of the most destructive diseases of bananas and plantains worldwide. Foc produces three types of asexual spores, these are macroconidia, microconidia and chlamydospores, which function as mechanisms of dispersal, reproduction and survival. Foc is a genetically diverse pathogen, although the available evidence so far indicates that it does not use the mechanisms of sexual reproduction, such as recombination, to increase its genetic diversity. Furthermore, the population of this fungus in Southeast Asia shows a high degree of variation, suggesting that Foc lineages evolved together with their hosts in Southeast Asia. Alternatively, it has been suggested that Foc has multiple independent evolutionary origins, both within and outside of the Musaceae origin center. Actually, more than 24 vegetative compatibility groups and three races have been reported. This genetic diversity is accommodated in two large clades and nine clonal lineages.

Author details

Gilberto Manzo-Sánchez[1*], Marco Tulio Buenrostro-Nava[1], Carlos L. Leopardi[1] , Mario Orozco-Santos[2] and Mauricio Guzman-Quesada[3]

1 Facultad de Ciencias Biológicas y Agropecuarias, Universidad de Colima, Colima, México

2 Instituto Nacional de Investigaciones Forestales, Agrícolas y Pecuarias (INIFAP) Campo Experimental Tecomán, Colima, México

3 Departamento de Fitopatología, Coorporación Bananera Nacional, La Rita de Pococí, Limón, Costa Rica

*Address all correspondence to: gmanzo@ucol.mx

References

[1] Jones, D. R. Diseases of Banana, Abaca and Enset. CABI Publishing, 2000.

[2] Wall, M. M. Ascorbic acid, vitamin A, and mineral composition of banana (*Musa* sp.) and papaya (*Carica papaya*) cultivars grown in Hawaii. J. Food Compos. Anal. 2006;19:434-445.

[3] FAOSTAT. Available at: http://www.fao.org/faostat/es/#data/QC. (Accessed: 11th September 2020)

[4] Ploetz, R. C. & Pegg, K. G. Fusarium wilt. in Diseases of Banana, Abaca and Enset (ed. Jones, D. R.) CABI Publishing 2000;143-159.

[5] Dita, M., Barquero, M., Heck, D., Mizubuti, E. S. G. & Staver, C. P. Fusarium wilt of banana: Current knowledge on epidemiology and research needs toward sustainable disease management. Frontiers in Plant Science (2018) 871:1468. doi: 10.3389/fpls.2018.01468

[6] Ploetz, R. C. Fusarium wilt of Banana. Phytopathology® 2015; 105:1512-1521. doi: 10.1094/PHYTO-04-15-0101-RVW

[7] Stover, R. H. Banana, plantain and abaca diseases. Commonwealth Mycological Institute UK, 1972.

[8] Sudarma, I. & Suprapta, D. N. Diversity of soil microorganisms in banana habitat with and without Fusarium wilt symptom. J. Int. Soc. Southeast Asian Agric. Sci. 2011;17:147-159.

[9] Stover, R. H. Fusarial wilt (Panama disease) of bananas and other *Musa* species. Commonwealth Mycological Institute UK, 1962.

[10] Jones, D. R. Disease and pest constraints to banana production. Acta Horticulturae. 2009;828:21-36

[11] Lescot, T. Banana genetic diversity: Estimated world production by type of banana. FruiTrop 2020;269:98-102. doi: 10.17660/ActaHortic.2009.828.1

[12] Agrios, G. N. Plant pathology. LIMUSA, 2016.

[13] O'Donnell, K., Ward, T. J., Geiser, D. M., Corby Kistler, H. & Aoki, T. Genealogical concordance between the mating type locus and seven other nuclear genes supports formal recognition of nine phylogenetically distinct species within the *Fusarium graminearum* clade. Fungal Genet. Biol. 2004; 41: 600-623. doi: 10.1016/j.fgb.2004.03.003

[14] Okubara, P. A. & Paulitz, T. C. Root defense responses to fungal pathogens: A molecular perspective. Springer, Dordrecht, 2005. p. 215-226. doi:10.1007/1-4020-4099-7_11

[15] Gordon, T. R. & Martyn, R. D. The evolutionary biology of *Fusarium oxysporum*. Annu. Rev. Phytopathol. 1997;35:111-128. doi: 10.1146/annurev.phyto.35.1.111

[16] Guo, L. *et al.* Genome and transcriptome analysis of the fungal pathogen *Fusarium oxysporum* f. sp. *cubense* causing banana vascular wilt disease. PLoS One 2014;9:e95543. doi: 10.1371/journal.pone.0095543

[17] Fusarium: disease, biology, and taxonomy. Pennsylvania State University Press. 1981.

[18] Leslie, J. H. & Summerell, B. A. *The Fusarium Laboratory Manual*. Wiley-Blackwell, 2006.

[19] Ploetz, R. C. Population biology of *Fusarium oxysporum* f. sp. *cubense*. in *Population biology of Fusarium oxysporum* f. sp. *cubense* (ed. Ploetz, R. C.) APS Press, 1990;63-76.

[20] Stover, R. H. & Waite, B. H. Studies on Fusarium wilt of bananas: V. Pathogenicity and distribution of *F. oxysporum* f. *cubense* races 1 and 2. Can. J. Bot. 1960;38:51-61.

[21] Beckman, C. H. & Roberts, E. M. On the nature and genetic basis for resistance and tolerance to fungal wilt diseases of plants. *Adv. Bot. Res.* 1995; 21:35-77. doi: 10.1016/S0065-2296(08)60008-7

[22] Li, C. *et al.* The use of GFP-transformed isolates to study infection of banana with *Fusarium oxysporum* f. sp. *cubense* race 4. Eur. J. Plant Pathol. 2011;131:327-340. doi: 10.1007/s10658-011-9811-5

[23] Li, C., Yang, J., Li, W., Sun, J. & Peng, M. Direct root penetration and rhizome vascular colonization by *Fusarium oxysporum* f. sp. *cubense* are the key steps in the successful infection of brazil cavendish. Plant Dis. 2017;101:2073-2078. doi: 10.1094/PDIS-04-17-0467-RE

[24] Warman, N. M. & Aitken, E. A. B. The movement of *Fusarium oxysporum* f.sp. *cubense* (sub-tropical race 4) in susceptible cultivars of banana. Front. Plant Sci. 2018:**9**. doi: 10.3389/fpls.2018.01748

[25] Buddenhagen, I. Understanding strain diversity in *Fusarium oxysporum* f. sp. *cubense* and history of introduction of 'tropical race 4' to better manage banana production. Acta Horticulturae 2009;828:193-204. doi: 10.17660/ActaHortic.2009.828.19

[26] Pegg, K. G., Coates, L. M., O'Neill, W. T. & Turner, D. W. The epidemiology of Fusarium Wilt of banana. *Front. Plant Sci.* 2019;10:1395. doi.org/10.3389/fpls.2019.01395

[27] Hennessy, C., Walduck, G., Daly, A. & Padovan, A. Weed hosts of *Fusarium oxysporum* f. sp. *cubense* tropical race 4 in northern Australia. Australas. Plant Pathol. 2005;34:115. doi: 10.1071/AP04091

[28] Ploetz, R. C. Panama disease: Return of the first banana menace. Int. J. Pest Manag. 1994;**40**:326-336. doi: 10.1080/09670879409371908

[29] Meldrum, R. A., Daly, A. M., Tran-Nguyen, L. T. T. & Aitken, E. A. B. Are banana weevil borers a vector in spreading *Fusarium oxysporum* f. sp. *cubense* tropical race 4 in banana plantations? Australas. Plant Pathol. 2013;42:543-549. doi: 10.1007/s13313-013-0214-2

[30] Laurence, M. H., Burgess, L. W., Summerell, B. A. & Liew, E. C. Y. High levels of diversity in *Fusarium oxysporum* from non-cultivated ecosystems in Australia. Fungal Biol. 2012;116:289-297. doi: 10.1016/j.funbio.2011.11.011

[31] Kistler, H. C. Genetic diversity in the plant-pathogenic fungus *Fusarium oxysporum*. Phytopathology 1997;87:474-479. doi: 10.1094/PHYTO.1997.87.4.474.

[32] O'Donnell, K., Kistler, H. C., Cigelnik, E. & Ploetz, R. C. Multiple evolutionary origins of the fungus causing Panama disease of banana: Concordant evidence from nuclear and mitochondrial gene genealogies. Proc. Natl. Acad. Sci. 1998;95:2044-2049. doi: 10.1073/pnas.95.5.2044

[33] O'Donnell, K. *et al.* A two-locus DNA sequence database for typing plant and human pathogens within the *Fusarium oxysporum* species complex. Fungal Genet. Biol.2009;46:936-948. doi: 10.1016/j.fgb.2009.08.006

[34] Ploetz, R. C. Vegetative compatibility among races of *Fusarium oxysporum* f. sp. *cubense*. Plant Dis. 1988;72:325. doi: 10.1094/PD-72-0325.

[35] McDonald, B. A. & Linde, C. The population genetics of plant pathogens and breeding strategies for durable resistance. Euphytica 2002;124:163-180. doi: 10.1023/A:1015678432355

[36] Bentley, S., Pegg, K. G., Moore, N. Y., Davis, R. D. & Buddenhagen, I. W. Genetic variation among vegetative compatibility groups of *Fusarium oxysporum* f. sp. *cubense* analyzed by DNA fingerprinting. *Phytopathology* 1998;88:1283-1293. doi: 10.1094/ PHYTO.1998.88.12.1283

[37] Kistler, H. C. *et al.* Systematic numbering of vegetative compatibility groups in the plant pathogenic fungus *Fusarium oxysporum* . Phytopathology. 1998;88:30-32. doi: doi.org/10.1094/ PHYTO.1998.88.1.30

[38] Ploetz, R. C. *Fusarium*-induced diseases of tropical perennial crops fusarium wilt of banana is caused by several pathogens referred to as *Fusarium oxysporum* f. sp. *cubense*. 2006;96: 653-656. doi: doi.org/10.1094/ PHYTO-96-0653

[39] Fourie, G., Steenkamp, E. T., Ploetz, R. C., Gordon, T. R. & Viljoen, A. Current status of the taxonomic position of *Fusarium oxysporum* formae specialis cubense within the *Fusarium oxysporum* complex. Infection, Genetics and Evolution 2011;11:533-542. doi: 10.1016/j.meegid.2011.01.012

[40] Mostert, D. *et al.* The distribution and host range of the banana *Fusarium* wilt fungus, *Fusarium oxysporum* f. sp. *cubense*, in Asia. PLoS One 2017;12:e0181630. doi: 10.1371/journal. pone.0181630

[41] Ordoñez, N. A global genetic diversity analysis of *Fusarium oxysporum* f.sp. *cubense*. Wageningen University. 2018. doi:9789463432986

[42] Boehm, E. W. A. Statistical analysis of electrophoretic karyotype variation among vegetative compatibility groups of *Fusarium oxysporum* f. sp. *cubense*. Mol. Plant-Microbe Interact.1994;7:196. doi: 10.1094/MPMI-7-0196

[43] Koenig, R. L., Ploetz, R. C. & Kistler, H. C. *Fusarium oxysporum* f. sp. *cubense* consists of a small number of divergent and globally distributed clonal lineages. Phytopathology 1997;87:915-923. doi: 10.1094/ PHYTO.1997.87.9.915

[44] Fourie, G., Steenkamp, E. T., Gordon, T. R. & Viljoen, A. Evolutionary relationships among the *Fusarium oxysporum* f. sp. *cubense* vegetative compatibility groups. Appl. Environ. Microbiol.2009;75: 4770-4781. doi: 10.1128/AEM.00370-09

[45] Su, H. J., Hwang, S. C. & Ko, W. H. Fusarial wilt of cavendish bananas in Taiwan. Plant Dis.1986;70: 814-818. doi: 10.1094/PD-70-814

[46] Visser, M., Gordon, T., Fourie, G. & Viljoen, A. Characterisation of South African isolates of *Fusarium oxysporum* f.sp. *cubense* from Cavendish bananas. S. Afr. J. Sci. 2010;106. doi: 10.4102/sajs. v106i3/4.154

[47] Batlle Viera Luis Pérez Vicente, A. Variabilidad genética de las poblaciones de *Fusarium oxysporum* f. sp. *cubense* en bananos y plátanos de Cuba. Fitosanidad. 2009;13:169-186.

[48] Ma, L.-J. *et al.* Comparative genomics reveals mobile pathogenicity chromosomes in *Fusarium*. Nature 2010;464:367-373. doi: 10.1038/ nature08850

[49] Bentley, S., Pegg, K. G. & Dale, J. L. Genetic variation among a world-wide collection of isolates of *Fusarium oxysporum* f. sp. *cubense* analysed by RAPD-PCR fingerprinting. Mycol. Res.1995;99:1378-1384. doi: 10.1016/ S0953-7562(09)81225-2

[50] Groenewald, S., Van Den Berg, N., Marasas, W. F. O. & Viljoen, A. The application of high-throughput AFLP's in assessing genetic diversity in *Fusarium oxysporum* f. sp. *cubense*. Mycol. Res. 2006;110:297-305. doi: 10.1016/j.mycres.2005.10.004

[51] Bogale, M., Wingfield, B. D., Wingfield, M. J. & Steenkamp, E. T. Simple sequence repeat markers for species in the *Fusarium oxysporum* complex. Mol. Ecol. Notes. 2005;5:622-624. doi: 10.1111/j.1471-8286.2005.01015.x

[52] Thangavelu, R., Kumar, K. M., Devi, P. G. & Mustaffa, M. M. Genetic diversity of *Fusarium oxysporum* f.sp. *cubense* isolates (Foc) of India by inter simple sequence repeats (ISSR) analysis. Mol. Biotechnol. 2012;51:203-211. doi: 10.1007/s12033-011-9457-8

[53] Baayen, R. P. *et al.* Gene genealogies and AFLP analyses in the *Fusarium oxysporum* complex identify monophyletic and nonmonophyletic formae speciales causing wilt and rot disease. Phytopathology. 2000;90:891-900. doi: 10.1094/PHYTO.2000.90.8.891

[54] Czislowski, E. *et al.* Investigation of the diversity of effector genes in the banana pathogen, *Fusarium oxysporum* f. sp. *cubense*, reveals evidence of horizontal gene transfer. Mol. Plant Pathol. 2018;19: 1155-1171. doi: 10.1111/mpp.12594

[55] Ordonez, N. *et al.* Worse comes to worst: Bananas and Panama disease—when plant and pathogen clones meet. PLoS Pathogens. 2015;11:e1005197. doi: 10.1371/journal.ppat.1005197

[56] Warmington, R. J. *et al.* High-quality draft genome sequence of the causal agent of the current Panama disease epidemic. Microbiol. Resour. Announc. 2019;8: e00904-19. doi: 10.1128/MRA.00904-19

[57] Yun, Y. *et al.* Genome data of *Fusarium oxysporum* f. sp. *cubense* race 1 and tropical race 4 isolates using long-read sequencing. Mol. Plant-Microbe Interact. 2019;32: 1270-1272. doi: 10.1094/MPMI-03-19-0063-A

[58] Maryani, N. *et al.* Phylogeny and genetic diversity of the banana Fusarium wilt pathogen *Fusarium oxysporum* f. sp. *cubense* in the Indonesian centre of origin. Stud. Mycol. 2019;92:155-194. doi: 10.1016/j.simyco.2018.06.003

Section 3

Genetic Diversity and Health

Chapter 9

Adaptation to Mediterranea

Jelena M. Milenković, Dijana R. Stojanović and Jelena G. Najdanović

Abstract

The Mediterranean region encompasses countries that surround Mediterranean Sea. Due to its position at the intersection of Eurasia and Africa it has often been a route of human migrations during history, which contributed to its high biodiversity. People living in this area had been exposed to the episodes of natural selection that led to the establishment of specific genetic variations, for which is thought to carry a certain adaptation. Some recent studies have shown that genetic adaptations are probably related to the immune defense against infectious pathogens. One of the most recognizable disease of the region is familial Mediterranean fever (FMF), a prototype of a monogenic autoinflammatory disease. FMF is predisposed by the mutations in the Mediterranean fever (*MEFV*) gene that encodes inflammasome regulatory protein - pyrin. Specific variations of several other genes have been proposed to confer a protection against *Plasmodium malariae* parasite. Some of these are hemoglobin S (HbS), thalassemia, glucose-6-phosphate dehydrogenase deficiency, ovalocytosis, and mutation in the Duffy antigen (FY). In this chapter we will summarize important genetics and pathogenesis features of diseases commonly encountered in the Mediterranean region with a short discussion of potential adaptations that they may carry.

Keywords: familial Mediterranean fever, thalassemia, malaria, hemoglobin S, Duffy antigen, heterozygote advantage

1. Introduction

The Mediterranean region encompasses the lands surrounding the Mediterranean Sea; on the north there is Southern Europe and Anatolia, on the south North Africa, and on the east the countries of Levant. The Mediterranean region has a specific climate, with mild winters and hot, dry summers, which supports the characteristics of the Mediterranean flora and fauna. The region's location at the intersection of Eurasia and Africa has contributed to the high biodiversity of its inhabitants, including people. Beside climate, this region has historically been the most frequent route of human migrations, as it is today.

Given the specific environment influences, several important genetic variations occurred and persist in people living in this area. Some of the genetic adaptations carry a certain degree of protection against infectious agents, but at the same time, when in an inadequate genotype, they can cause health disorders. This chapter will describe and discuss the most common types of genetic variations in the Mediterranean area related to adaptation and / or susceptibility to disease.

2. Mediterranean fever (*MEFV*) gene mutations

2.1 Protein pyrin

One of the most recognizable diseases of the Mediterranean region is Familial Mediterranean Fever (FMF), a prototype of a monogenic autoinflammatory disease, associated with mutations in the *MEFV* gene that encodes for protein pyrin. Autoinflammatory disorders are characterized by dysregulation of innate immune response, unlike autoimmune diseases that are primarily mediated by adaptive immunity. However, approximately in a third of FMF patients pathogenic *MEFV* mutation is not identified, hence the diagnostic criteria for FMF still rely on clinical manifestations [1, 2]. The *MEFV* gene is composed out of 10 exons and 13 introns, which make 781 amino acids (aa) long, multifunctional, protein pyrin. One of its first described functions is the assembly of an inflammasome. Pyrin acts as a pattern recognition receptor (PRR) that senses intracellular danger signals after which it binds to an adaptor protein and oligomerizes to form a pyrin inflammasome. Subsequently, inflammasome recruits and activates caspase-1, which further cleaves pro-inflammatory molecules, such as interleukin (IL)-1β and IL-18 [1, 3, 4].

As a PRR, pyrin seems to recognize downstream effects of a pathogen-driven modification and/or inactivation of RhoA GTPases - molecules that regulate actin dynamics [3]. By sensing a disturbance in actin signaling, pyrin recognizes common virulence mechanisms and starts an immune response. Several pathogen bacteria employ actin cytoskeleton for their invasion and survival, and by secretion of Rho-inactivating cytotoxins they were shown to activate pyrin inflammasome (e.g. *Clostridium, Vibrio parahaemolyticus, Bordetella pertussis, Yersinia pestis*) [5–7].

Besides, pyrin regulates process of autophagy a highly specific degradation of inflammasomes components. With this process pyrin suppresses IL-1β production, thereby preventing an excessive inflammation. Additionally, autophagy-based secretory pathway enables a group of proteins to exit cytoplasm without entering Golgi apparatus, among which is IL-1β [8, 9]. Hence, *MEFV* mutation-induced alterations affect this pathway and may facilitate interleukins secretion.

The pyrin activation requires at least two independent processes: dephosphorylation and pyrin inflammasome maturation involving microtubule dynamics [2, 7]. In FMF patients with pathogenic *MEFV* variants there is a hyperreactive state of the pyrin inflammasome. It seems that the second control mechanism of pyrin activation is lacking, and that pyrin is maintained inactive only by phosphorylation. Mutations in the exon 10 do not impact pyrin phosphorylation but may affect the control mechanism of microtubule dynamics [4, 7].

2.2 The *MEFV* mutations and their effect

According to the Infevers registry (the registry of hereditary autoinflammatory disorders mutations) there are 377 nucleotide variants identified in the *MEFV* gene so far. Most of them are benign and not involved in pathogenesis of FMF.

The FMF has long been considered an autosomal recessive disease, but with description of cases with heterozygote *MEFV* mutations this definition has changed. The mutations may express their effect in either a recessive or a dominant manner, depending on their location in the gene. Generally, those in the exon 10 are considered recessive, while other manifest their effect in a heterozygous state and are considered dominant (gain-of-function). The most frequently identified FMF-causing *MEFV* variants are in the exon 10 and encompass M694V (c.2080A > G), M680I (c.2040G > C), and V726A (c.2177 T > C) missense mutations, with the carrier frequency of ~10% in the populations of the Mediterranean region [2, 4, 8, 10].

In order to achieve a better classification of pathogenic *MEFV* mutations, and to set a guidelines for genetic diagnostic testing of hereditary recurrent fevers, Shinar et al. [11], adopted the final consensus document that proposed a group of bialelic mutations to be used for definition of FMF. This group comprises 14 mutations, 9 of which are clearly pathogenic (M694V, M694I, M680I, V726A, R761H, A744S, I692del, E167D, and T267I), while 5 mutations are designated as of unknown significance (E148Q, K695R, P369S, F479L, and I591T).

Depending on the type of mutations, permissive environmental factors and genetic background, clinical picture of FMF may vary from typical recurrent inflammatory attacks to mild symptoms or asymptomatic cases. The most frequent symptoms are recurrent episodes of fever with serosal inflammation, arthralgia or arthritis, abdominal pain, and localized erythematous skin rash. Episodes are self-limited and usually resolve within 48-72 h. Heterozygous patients usually have milder symptoms and shorter and less frequent attacks. Some asymptomatic carriers may have elevated inflammatory and oxidative stress biomarkers [4, 10, 12, 13]. The most concerning long-term complication is renal amyloidosis, and renal transplantation is the choice in most end-stage renal disease [14, 15]. Standard treatment is a prolonged use of colchicine, although there are resistant cases. One potential cause of resistance is the vitamin D deficiency in these patients [16–18].

The M694V homozygote mutation is mostly associated with early onset of disease and severe course. It seems that environmental factors have stronger influence on this mutation [19, 20]. It is interesting to note the impact of environmental factors, since patients of the same ethnicity have different phenotype depending on the country they live in, *i.e.* the Eastern or Western Europe. Besides, a set of additional factors influence the phenotype, such as patient's age and sex, micro-RNAs, immune factors (HLA I gene A), and microbiota [11, 20–22].

2.3 Potential heterozygote advantage *of MEFV* mutations

The higher frequency of *MEFV* mutations among multiple populations in the Mediterranean region suggests an existence of a heterozygote advantage. Mostly accepted theories explaining this assumption are those that recognize mutations as an adaptation to yet undetermined endemic infectious agent or, more probably, a group of agents. It seems that *MEFV's* exon 10 had been exposed to the episodic positive selection in primates [2, 4, 23, 24].

Several bacteria secrete invasive factors (toxins) that covalently modify RhoA or its regulators. Other may inhibit pyrin inflammasome assembly by keeping it in the phosphorylated state, such as a protein YopM produced by bacteria *Yersinia enterocolitica*. Pyrin activation occurs when RhoA GTPases are disabled to promote their downstream signaling. In that sense, bacteria *Yersinia pestis* (bubonic plague-bacterium) is proposed as a possible agent that had led to the selection of gain-of-function *MEFV* mutations [25]. Other hypotheses imply that mutated pyrin may confer a protection against tuberculosis or brucellosis, but without direct evidence [2]. Potential association between mutated pyrin and defense against tuberculosis is within the processes of autophagy and inflammasome activation. *Mycobacterium tuberculosis* is capable of arresting phagolysosome biogenesis in macrophages and prevents inflammasome activation by its Zn-metalloprotease, while mutated pyrin mediated stimulation of autophagic pathways may overcome this block [2, 26, 27].

2.4 Diversity of *MEFV* mutations

Higher frequency of pathogenic *MEFV* mutations in the Mediterranean basin is mainly explained by a founder effect and balancing selection. The M694V and

V726A mutations seem to emerge in human genome about 2000 years ago, according to their association with specific microsatellite haplotypes in different populations [21, 28]. Presence of the E148Q (c.442G > C) mutation in different ethnic groups in the region supports the hypothesis of its recurrent nature or founder effect, probably stemming from the Asian countries, such as China and India, where E148Q is also frequent [20, 29]. The M694I (c.2082 G > A) mutation is merely present in North Africa and is estimated that occurred in an indigenous population of Berbers before colonizations in the 7th century BC [4, 20].

The M694V, M694I, M680I, and V726A mutations are most common in the Eastern Mediterranean countries, that is in Turkish, Armenian, Arab, and Jewish populations (**Figure 1**). The carrier rate of the mutations in these populations is estimated to be 1:5 to 1:7. Consequently, FMF mainly affects people of these ethnicities [2, 5, 19, 30–32]. The prevalence of *MEFV* mutations and FMF is much lower in the Western Mediterranean countries (France and Spain). Actually, the ethnic origin of these patients is usually from populations with higher mutation frequencies. Also, higher prevalence of homozygous mutations in the East might reflect the consequence of a local custom of consanguinity marriages [20, 33, 34].

Beside differences in *MEFV* mutations distribution between countries, there are variations within countries as well. For example, in Turkey, 94% of diagnosed FMF patients were from central-western parts of the country. However, more than a half of them had a family origin from the eastern provinces, pointing to the migration routes of mutations and disease [19, 35, 36]. The similarity between *MEFV* mutations present in Turks and Jordanians can also be explained by the local migrations, during the Ottoman Empire [20, 37]. The M694V mutation is the most common in Arab FMF patients. Unlike others, Arabs in North Africa have higher rate of M694I mutation that is probably acquired through the intermarriages with the local autochthonous population [20, 31, 38–43].

There is a dissimilar pattern of *MEFV* mutations among Jewish population, as there is a number of distinct Jewish ethnic groups in Europe. Although the aforementioned mutations are mostly present in Jewish FMF patients, their exact frequencies differ depending on a country and ethnic group [16, 23]. For example, high carrier rate of M694V is identified in Jewish FMF patients living in North Africa (Morocco) (11.1%) and Iraq (2.9%), but is rarely observed among Ashkenazim [20], while the V726A is prevalent among Ashkenazi (7.4%) and Iraqi Jews (12.8%) [44]. In the study of *MEFV* mutation prevalence in the Israeli society, M694V was common mutation among non-Ashkenazi Jews, E148Q was observed in

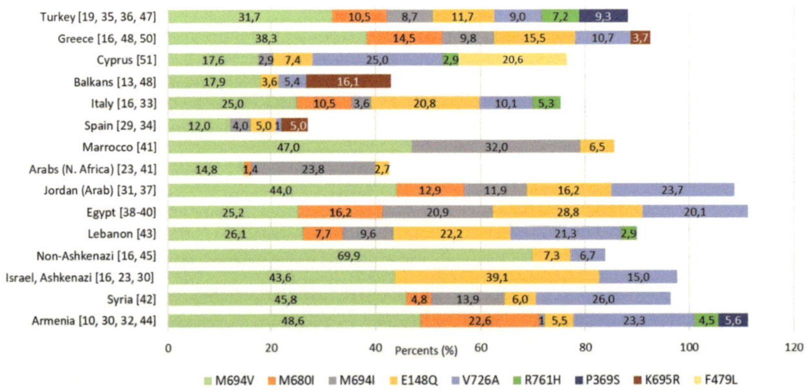

Figure 1.
Allele frequencies of common MEFV mutations in FMF patients (%).

patients of all ethnic groups, while K695R (c.2084A > G) seem to be characteristic mutation present in Jews [45]. In Sephardic Jews, the overall *MEFV* mutation carrier rate is between 1:8 and 1:16 [46], M694V is predominant, while other mutations, such as E148Q, P369S (c.459G > T), K695R and V726A, are rare [16].

The M680I mutation is common in Armenians and is associated with milder phenotype of the disease. Nevertheless, Armenian patients with FMF have common pathogenic mutations as the previous populations, such as M694V (~50%), followed by V726A, M680I, and R761H (c.2282G > A) [10, 30, 32].

The E148Q mutation is the most frequent sequence alteration in the general population, but its clinical significance is still debatable. It is mostly encountered in a heterozygous state in asymptomatic individuals. When in homozygous state it is associated with FMF-like disease, with mild symptoms and later onset of disease. Thus, with regard to the SHARE recommendations heterozygous E148Q does not support the diagnosis of FMF [2, 5, 10, 11].

The P369S and K695R are rare mutations with reduced penetrance, often found in asymptomatic carriers or in complex alleles in FMF patients [23, 30, 47]. They were relatively common in general Ashkenazi Jewish sample in the USA, with the carrier frequency of ~1:5 in FMF patients [16]. The P369S was the most frequent mutation in healthy Armenians and it might ameliorate the phenotypic presentation of the co-existing exon 10 mutations in patients. P369S homozygotes were even observed among the asymptomatic Ashkenazi Jews [16, 30].

Unexpectedly high K695R mutation rate was determined in the countries of Central and South-Eastern Europe. This region is characterized with limited heterogeneity of *MEFV* mutations, with only eight different mutations determined in healthy subjects and FMF patients (K695R, E148Q, V726A, M694V, F756C, I591T, S730F and A744S). The K695R mutation was most common mutation, found in 40% of healthy and 32% of FMF patients, which supports the idea that this is a common mutation for this region [13, 48].

One another *MEFV* variation of unknown significance is the R202Q (c.605G>A), often considered a polymorphism due to its high heterozygous frequency, among healthy individuals. Due to its poor conservation during evolution it is assumed it appeared later compared to other common mutations [49]. The R202Q alteration in homozygous state was associated with FMF-like symptoms in some cases [47, 50]. In our study of *MEFV* distribution in Serbia, 45% of healthy individuals had heterozygous R202Q, while 10% were homozygotes. Although considered healthy, the homozygotes reported self-limited episodes of fever of unknown origin and unspecific abdominal pain [13]. The results indicate a pathogenic role of R202Q homozygosity, perhaps along with other permissive environmental and genetic factors of a patient.

In isolated populations there is a greater chance for arising of specific genotypes. One example of specific *MEFV* mutation distribution is an island of Cyprus, due to its distinct ancestry and relative isolation in the Mediterranean. The *MEFV* mutations carrier rate in the Greek-Cypriot patients suspected for FMF is 1:25, with V726A, M694V, E167D (c.501G > C) and F479L (c.1437C > G) being the most common mutations. F479L is very rare elsewhere but in Greek-Cypriots its frequency is 20.6%. Interestingly, F479L was always co-inherited in *cis* with E167D mutation. It is hypothesized that F479L originated in Cyprus as a founder mutation, from where it spread further [51].

It would be ideal when every population would perform a genetic testing for *MEFV* mutations and accordingly establish a set of the most frequent, which could be used in a screening for suspected FMF patients, as well as other inflammatory conditions, since *MEFV* mutations are found to be a modifying factor in a number of inflammatory and autoimmune diseases [52].

3. Behcet's disease

Behcet's disease (BD) is an autoinflammatory and polygenic disease, more frequent in Mediterranean countries than in rest of Europe. Most cases are identified in countries of the Middle East and along the ancient Silk Route. The highest prevalence among Mediterranean countries is probably in Turkey, with estimated prevalence of 4.2/1000 in Istanbul [53]. This is a rare, sporadic, multi-systemic disease with undetermined cause. The main clinical features are constitutional symptoms and recurrent fever, oral aphthous, genital ulcers, with gastrointestinal, musculoskeletal, neurological, and vascular involvement [54].

Several host genetic factors are implicated in the pathogenesis of BD. The strongest is the association with the major histocompatibility complex HLA–B51 allele, which increases the risk of disease for about 6-fold. Approximately 50% of BD patients possess this HLA variant. Besides, HLA-B51 contributes to the specific clinical features in BD such as less severe disease course, but a higher frequency of ocular manifestations [55, 56].

Behçet's disease can be a comorbidity of FMF, and vice versa, *MEFV* mutations are common finding in BD patients. Some *MEFV* alterations are detected more often in BD patients than healthy subjects, such as P706 polymorphism. In a cohort of Turkish patients, clinical association was found between heterozygous *MEFV* mutation, principally M694V, and vascular involvement [51, 55, 57].

Interestingly, arthritis in BD is self-limiting and nondestructive in nature, pointing to the existence of an inherited protective factor/s. Such a role has been observed for plasminogen activator inhibitor 1 (PAI-1), which levels were higher in synovial fluid of BD patients than healthy. PAI-1 acted protective against destructive arthritis but had promoting effect towards hyperfibrinolysis in BD vasculopathy. However, PAI-1 common polymorphism 4G/5G was not associated with pathogenesis nor development of thrombosis in these patients [58–60].

Besides, several other alterations are described to influence BD occurrence and course, including MHC class 1 polypeptide-related sequence, T cell mediated cytokine dysregulation (of IL-6, IL-8, IL-10), DNA methylation, etc. [55, 61].

4. Pathophysiology of β-thalassemia syndromes

4.1 Introduction

Genetic disorders referred as the ἀ- and β-thalassemias are caused by defective hemoglobin (Hgb) chains (ἀ or β) synthesis and are mostly inherited as a Mendelian recessive [62–64]. The name of the disease" thalassemia" is derived from the Greek words: *thalassa* (sea) and *haima* (blood), implicating the geographical region where the disease was initially described due to its high prevalence. B-thalassemia occurs mostly in people with origins near the Mediterranean Sea, Greece, Italy (Sicily, Calabria and Sardinia), Turkey, Middle East, India, Southern China, Sub Saharan Africa, south America and in the populations of Sephardic Jews and Arabs, with Cyprus (14%) and Sardinia (10,3%) having the highest carrier frequency. However, the other form of the disease, ἀ-thalassemia, is the most common among the people form the Far East, China, Vietnam, Laos, and Cambodia [63, 64].

Although considered as the rare form of the disease, it is confirmed that around 68,000 children annually are born with the various forms of thalassemia syndromes, whereas 1.5-5% of the worldwide population are considered as the carriers of these genetic abnormalities [65–68]. The high frequency of these mutations is considered as an evolutional answer to the malaria infections, providing protection against *Plasmodium falciparum* for the genetic mutation carriers. The aberrant Hgb synthesis reduces the half-life of erythrocytes which disables completion of parasite maturation cycle [63]. Moreover, the same type of genetic aberrance has been confirmed in consanguineous marriages in some countries [64]. However, high rate of the migrations of populations caused that individuals with thalassemia-syndromes may be found in the US, Australia, Canada, South America and North Europe, making it a global health care burden [65–68]. Moreover, the general epidemiological estimation is that the prevalence of thalassemia-syndromes is about to increase, taking into the consideration the fact that infant mortality declines in low-income and middle-income countries [68].

Thalassemias are heterogeneous, inherited, monogenic, Hgb disorders and are initially classified as ἀ or β, depending whether genes that control ἀ- or β-globin chains synthesis are defective. This knowledge implicates that β-thalassemias occur when synthesis of the β-globin chains is reduced (β+) or absent (β-) [62–64]. Moreover, clinical, and hematological manifestations depend on how many of the genes that code β-globin synthesis are defective and whether those defects are homozygous or heterozygous. The phenotype diversity and wide range of disease severity lead to introduction of the concept of β-thalassemia-syndromes.

According, three culprit forms that comprise the β-thalassemia-syndromes are defined and classified by increasing severity of the symptoms: 1) β-thalassemia carrier state, also known as β-thalassemia minor, "heterozygous thalassemia" or "thalassemia trait", 2) β-thalassemia intermedia and 3) β-thalassemia mayor, also referred as "Cooley's anemia" and "Mediterranean anemia", very severe phenotype, that requires blood transfusion for survival (transfusion-dependent anemia) and has a questionable outcome. Besides these forms, there are other identified types of β-thalassemias, that are associated with various Hgb and/or clinical abnormalities or may be autosomal dominant [63, 64]. Persons with most severe forms (major) are homozygotes or compound heterozygotes for β0 or β + genes, intermedia type may be homozygotes or compound heterozygotes, while the mildest form is predominantly heterozygotes [62–64].

In the past two decades, individuals affected with β-thalassemia-syndromes are experiencing tremendous improvement in the quality of life and overall survival, due to the timely diagnosis, adequate therapy, and monitoring of the disease. However, up to date, the only cure for the disease represents allogeneic hemopoietic stem-cell transplantation.

4.2 Molecular basis

The synthesis of β-globin chains in Hgb molecule physiologically is under control of two genes. Any genetic abnormality of the controlling genes, therefore, results in the absence or the reduction of the β-chain. The gene for β-chain is located in the short arm of chromosome 11, sharing the region and being arranged in the order of the development expression, with the functional genes for δ-globin, embryonic ε-globin, the fetal A-γ-globin and G-γ-globin, as well as a pseudogene (ψβ1) [63]. The molecular and clinical diversity of the β-thalassemias emerges from the data that more than 200 genetic mutations have been described up to date [63, 64, 68–70]. Accordingly, clinical, and hematological manifestation and patients' prognosis depend on the basis of imbalance of the ἀ- and β-chains synthesis, therefore from the type and the extent of the genetic disturbance.

The identified and defined genetic aberrations are silent mutations (silent β-globin), mild mutations (relative reduction of β-globin) and severe mutations (complete absence of β-globin, β0) [68]. Nevertheless, these mutations are identified mostly as single-nucleotide substitutions and insertions of single nucleotides or small oligonucleotides causing the frameshifts in genes that code β-chains. The typical genetic abnormalities that were described are promoter mutations, being responsible for the milder phenotypes, whereas nonsense, initiation codon, splicing and frameshift mutations have been documented in more severe forms of thalassemia-syndromes, characterized with the complete absence of β-chains [62, 63]. Deletions of the gene are randomly identified aberrations, where the deletional removal of one or several genes from the chromosome 11 causes very rare forms of thalassemias, designated as δβ-, γδβ- and εγδβ-thalassemia [62].

An autosomal recessive pattern of thalassemia inheritance implicates that both parents have to be heterozygotes, owing a copy of a β-globin gene mutation. Possible outcomes in the affected family may be that every child has: 1) 25% chance of being affected, 2) 50% of being an symptom free and carrier, and 3) 25% of not being affected nor a carrier [63, 64].

4.3 Genetic modifiers

Pathophysiological perception why individuals with beta-thalassemia syndromes may clinically appear very heterogeneous, is based on the perseverance of three group of factors that may modify the disease. These factors are designated as genetic modifiers and are explained as genetic variants that induce differences in disease phenotype [64]. Genetic variants that may impact the imbalance of globin chains are categorized as primary modifiers. The other pathogenetic factors that may alleviate the severity of β-thalassemia major are: coinheritance of an α-thalassemia gene and fetal Hgb production, within the β-globin cluster and are classified as secondary modifiers [62, 68].

Coexistence of α-thalassemia enables decreased α-globin chain synthesis, therefore significantly reduces imbalance between the α/non-α-chain in erythrocytes [63]. Increased γ-chain synthesis, in adult life, encounters the excess of α-chains, therefore enables the survival of the erythrocytes that contain fetal hemoglobin, marked as HbF cells. It may be that deletion mutation or point mutation within the β-globin gene cluster simultaneously trigger a rise in fetal Hgb production [62]. According to some research, the increase of HbF synthesis indicates a single nucleotide polymorphism in one of the γ-globin gene promoters or somewhere in the globin locus, resulting in the overexpression of the related gene [62]. It was reported that HbF, that is highly predominant in individuals with severe forms of thalassemia, may account for their improved survival [71]. Moreover, the inverse correlation of HbF levels and factors that reflect disease morbidity was observed, so as the finding that milder phenotypes present with the increased numbers of HbF cells [62, 72]. In addition to this knowledge, it was suggested that certain therapeutic treatments (hydroxyurea) may induce the production of HbF, hence produce less of a need for blood transfusion [73].

Tertiary modifiers are recognized to be genetic and environmental factors that modulate disease complication rates. The results of the molecular studies revealed genetic polymorphisms as possible pathogenetic factors involved in cardiac iron overload, hyperbilirubinemia, and Gilbert disease, osteoporosis, and infections susceptibility, that occur in patients with β-thalassemia syndromes [63, 68, 74–76].

4.4 Pathophysiology

Essential pathophysiological determinant in β-thalassemia syndromes is the uncoupling of the synthesis of the α- and the β-chain, where β-chain synthesis is reduced or absent, resulting in the accumulation of α-globin tetramers in the erythroid precursors [62–64]. This phenomenon eventually leads to an ineffective erythropoiesis, that is a key feature responsible for various pathophysiological consequences during the course of the disease. Erythrocytes and its precursors (mostly polychromatophilic erythroblasts) are filled with precipitated ἀ-globin tetramers, forming inclusive bodies, causing oxidative membrane damage and subsequent apoptosis [62–64, 77]. Physiologically, biochemical detoxification would be efficient to eliminate harmful proteins from the affected cells. Nevertheless, in the severe forms of β-thalassemias these pathways are inefficient [62].

Premature erythroid cell death in the bone marrow (ineffective erythropoiesis) and in the peripheral blood (hemolysis) cause chronic microcytic-hypochromic hemolytic anemia, that is a persistent finding in persons with thalassemia. Interestingly, hemolysis is less notable in individuals with severe phenotypes of the disease [64]. Chronic hypoxia induces intensive and continuous erythropoietin production, resulting in the great expansion of the bone marrow (25–30 times), subsequent skeletal deformities and the loss of the bone mass [63, 64, 68]. Simultaneously, a compensatory extramedullary hematopoiesis occurs, creating organomegaly, predominantly of spleen and liver [68]. Nevertheless, if the stimulus is extremely potent, all the cell in the body that express hematopoietic potential will be affected, resulting in the formation of the pseudotumors [78]. Hemolysis will trigger the formation of the gall stones and cholelithiasis and also contributes to splenomegaly development. Besides, the thalassemia-syndrome is recognized as a hypercoagulable state, since erythroid precursors, during the ineffective erythropoiesis, may become prothrombotic. Moreover, in association with platelets and coagulation disruption, the condition may result in serious vascular manifestations such as venous thrombosis [63, 64, 68, 79].

Besides ineffective erythropoiesis and anemia, iron overload also represents very important mechanism in the pathogenesis of the thalassemia, contributing to development of complications. Iron deposition within the reticuloendothelial system in the transfusion- dependent forms of β-thalassemia (major and intermedia) represents associated and secondary mechanism in the pathogenesis of iron overload. However, it is well defined that the most important pathogenetic factor in the hemochromatosis development represents increased iron absorption [80], due to the hepcidin downregulation and its deficiency [62, 63].

The apoptosis of the erythroid precursors causes subsequent synthesis and secretion of many factors that most likely inhibit hepcidin synthesis in the liver [62]. Coupled with this, it should be underlined that hepcidin functions as a negative iron regulator, delivering the information between the liver and the red blood cells. Its decreased concentrations result in the increased dietary iron absorption and in release of the iron from its storage (macrophages and hepatocytes). The final result is paradoxically and significant dietary iron absorption, regardless of the iron tissue deposition due to the blood transfusions and eventually hemochromatosis [62].

The identified molecules that are released from the apoptotic erythroid precursors are growth differentiation factor 15, twisted gastrulation 1, and erythroferrone, and all function as hepcidin expression inhibitors [68, 80–83]. The results have been conflicting so far, since some research demonstrated their significant increase in individuals with β-thalassemia [81, 82], while the others confirmed only increase of erythroferrone in animal models [84]. However, their exact function in the pathogenesis is yet to be elucidated. Nevertheless, the substitution of the synthetic hepcidins represents justified therapy option in patients with the severe forms, as already proven experimentally. However, this extensive and progressive iron overload in synergy with anemia may deteriorate already insufficient hematopoiesis. Iron overload, regardless of its pathogenesis, leads to hemochromatosis and organ damage [83, 85].

4.5 Clinical findings

The main clinical features of β-thalassemia syndromes are anemia and iron overload, leading to severe and life threating consequences. The onset and the degree of the symptoms severity depend whether the affected individuals present as a homozygous phenotype (thalassemia major) or as a homozygotes or compound heterozygotes (thalassemia intermedia). Correspondingly, individuals with β-thalassemia minor are usually asymptomatic and may be discovered incidentally, having only the discrete changes in the hematological findings.

The onset of symptoms will appear 12 months after the birth, [67–85], at the moment when HbF production switches to adult and physiological synthesis of HbA is yet to be established [86]. The infants will experience feeding problems, recurrent fevers, diarrhea, enlargement of the abdomen and the growth retardation. If the child has not been diagnosed prenatally, this is the point when the diagnosis of thalassemia is determined, and transfusion indicated [63, 64].

Microcytic-hypochromic hemolytic anemia is an obligatory finding in the affected individuals, predisposing them to progressive paleness and jaundice. Bone marrow expansion secondary to erythroid hyperplasia, lead to significant skeletal changes, creating abnormalities of the face and body. People with severe phenotypes most often experience frontal bossing, depression of the bridge of the nose, mandible and maxilla enlargement with the upper teeth exposure, bone pain, osteopenia and osteochondrosis. If spinal impairment occurs during the childhood, linear growth is delayed, resulting in the discordance in the length of upper and lower limbs [63, 64, 87]. The progressive enlargement of the abdomen is due to the hepatosplenomegaly, whereas the masses of extramedullar hematopoietic tissue may also be found in the chest or spinal column [63, 64].

Iron overload predominates in the most severe clinical phenotypes. Brown pigmentation of the skin, particularly in the areas exposed to the sun, reflects systemic hemochromatosis. Predominant sites for iron deposition tend to be spleen, liver, myocardium, pancreas, and endocrine glands. Although significant liver deposition of iron could be found, its function may be preserved for a long time [62]. Ultimately, liver cirrhosis may develop. Cardiac manifestations stand for the most adverse outcome of iron overload, whereas arrhythmias, dilated cardiomyopathy, and atrial or/and ventricular failure during the course of the disease lead to congestive cardiac failure. Endocrine complications primarily develop due to the insufficiency of the growth hormone (growth retardation) and sex hormones (hypogonadism). Additionally, hormone substitution therapy is commonly required for maintaining normal fertility. Other endocrine disturbances may be very diverse, including diabetes mellitus, hypothyroidism, hypoparathyroidism, hypocorticism, and panhypopituitarism. Pulmonary hypertension may contribute to the complexity of the cardiovascular manifestations by deteriorating left heart function [79, 88].

Other clinical features in β-thalassemia syndromes are osteoporosis, subclinical fractures, nutritional deficiencies, venous thrombosis, chronic B and/or C hepatitis, and infections. The risk of hepatocellular carcinoma in patients who develop liver cirrhosis remains unchanged even if the proper therapy is performed, due to the oxidative DNA damage triggered by chronic iron accumulation [79, 88].

4.6 Laboratory findings and Hgb analysis

Laboratory diagnosis of thalassemia is confirmed based on established red blood cells parameters, qualitative and quantitative Hgb analysis and, when necessary, molecular assessment. Erythrocyte count may be relatively high, whereas Hgb is reduced <7 mg/dL, mean corpuscular volume (MCV) is between 50 and 70 fL and mean corpuscular Hgb (MCH) 12-20 pg. Peripheral blood smear demonstrates microcytosis, hypochromia, anisocytosis, poikilocytosis (dacrocytes and elliptocytes), along with the erythroblasts. The number of reticulocytes may remain normal, without any diagnostic accuracy. In order to differentiate iron deficiency anemia form the thalassemia-syndromes, few formulas are available to calculate a thalassemic index, but should be performed with caution [63, 64, 89]. In biochemical terms, typical β-thalassemia presents with elevated ferritin levels >12 ng/mL, transferrin saturation increased to 75-100% and unconjugated hyperbilirubinemia [62, 88].

The most accurate method for β-thalassemia differentiation is quantitative HbA2 determination. Considering that physiological HbF in adult population is commonly less than 1.5%, the results for HbA2 ranging between 3.6 and 7% are considered as definite thalassemia values. Nonconclusive or borderline cases, with HbA2 ranging between 3.2 and 3.6%, respectively, require further analysis [86, 89]. Additionally, PCR-based procedures or β-globin gene sequence analysis are necessary for diagnosis confirmation. Besides, in couples with increased risk, a prenatal diagnosis of thalassemia may be achieved by chorionic villi sampling (11th gestational week) or DNA analysis from harvested fetal cells (15-18th gestational week) [63, 64, 89].

4.7 Therapy approach

Conventional management of β-thalassemia syndromes includes blood transfusion, iron chelation, splenectomy and hemopoietic stem-cell transplantation. The introduction of blood transfusion in regular management of β-thalassemia has enormously improved quality of life and survival of the affected individuals [62–68]. The mayor indication for its initiation, in previously diagnosed patients, should be low Hgb level (<7 g/dL), that lasts at least two weeks [64], concerning other clinical signs such as growth retardation, skeletal changes and splenomegaly. The therapeutic aim of transfusion is to maintain Hgb level at 9-10 g/dL or 11-12 g/dL in cases of confirmed cardiovascular disease [63, 64, 68, 86]. Although life-saving approach, blood transfusion has several adverse effects, with iron overload and viral infections (hepatitis B, C) being the most common [62–68].

The knowledge that iron cannot be excreted form the human body and that patients requiring constant blood transfusions tend to develop iron overload, lead to the regular assessment of iron body status. Most conventional method is determination of serum ferritin levels, that may be monitored in order to initiate chelation therapy or may be used as a biomarker of iron chelators efficiency. However, more reliable, yet non-invasive method of tissue iron accumulation has been developed. Magnetic resonance imaging has been successfully used for liver and cardiac iron overload, measuring a tissue iron concentration in mg of iron per gr of dry liver/heart weight [63, 90, 91]. Also, iron binders (chelators) enable its elimination through feces and/or urine and should be initiated after approximately 10-20 performed transfusions or with ferritin levels above 1000 mg/gL [64, 68].

Splenectomy is indicated in the following cases: enlarged spleen with the risk of rupture, severe cytopenia and in patients with the significant blood requirements. In patients with splenectomy, infections and subsequent sepsis remain the leading cause of mortality [63].

However, the only curable therapy for the thalassemia represents hematopoietic stem-cell transplantation [63, 64, 68]. Nevertheless, it was documented that a disease-free survival may be achieved in 80% in matched donors and even 65% in unrelated donors and umbilical blood cord stem-cells transplantation. Nevertheless, this therapy option is still associated with risk and complications, even in the high-income countries [92].

Considering the monogenic nature of the disease, the most challenging, yet possible therapy approach, may be an interference in the globin chains imbalance, achieved by gene therapy and genome editing [68]. Alternative pharmaceutical approaches would be use of agents acting as potent stimulators of late stage erythropoiesis and increased hepcidin expression, throughout its substitution or stimulation of its endogenous production. Even though there has been a substantial progress in the development of therapy options for individuals affected with thalassemia, the best approach to the disease management remains prevention of thalassemia births throughout national screening programs [68].

5. Association of HbS, G6PD and FY gene polymorphisms and malaria

5.1 Introduction

Understanding the molecular mechanisms that underlies the adaptation is of crucial importance in evolutionary biology. Among the plethora of genes that causes adaptive variation in fitness-related features in natural populations, very few are identified [93, 94]. The hemoglobins, oxygen-carrying proteins, tightly connect cell metabolic activities with environmental conditions and thus represent convenient system for analyzing adaptive changes [93, 94]. Also, inherited disorders of hemoglobin are the most common human monogenic diseases [95]. Each year, there are between 300,000 and 400,000 newborns with some of the serious hemoglobin disorders and up to 90% of them are born in low- or middle-income countries [96].

Hemoglobin is the oxygen-carrying protein of red blood cells (RBCs), normally formed of two α-globins and two β-globins that constitute adult hemoglobin A (HbA). Without specific medical treatment, the most severe hemoglobinopathies — HbSS homozygosity (sickle-cell disease) and the thalassemias major are not compatible with life after early childhood. People with HbAS, HbAC, HbCC, HbAE, HbEE, and the thalassemias minor have usually normal life expectancy and are rarely directly associated with morbidity [97].

Plasmodium spp. parasites represent vector-borne pathogens which attack the red cells of reptiles, birds, and primates [98]. Out of five Plasmodium species that parasitize and cause malaria in humans, *P. falciparum* and *P. vivax* are the most common in human populations. *P. falciparum* is endemic in tropical areas worldwide, including Mediterranean [99]. Like the other four Plasmodium species, *P. falciparum* is injected into a human skin via female *Anopheles spp*. mosquitoes as a vector. Then, the sporozoites migrate to the liver where they attack hepatocytes and develop within them for 7–10 days. As a consequence, numerous merozoites are formed which, subsequently, enters erythrocytic stage of RBSs life cycle. In that time, typical features of malaria clinical picture develop [97, 98].

As a disease which is a main cause of morbidity and mortality, malaria caused by *P. falciparum* imposed remarkable evolutionary pressure on the human genome.. Also, malaria caused by *P. falciparum* is in relation with numerous genetic polymorphisms that are responsible for protection against this disease [97, 100]. Hemoglobin mutants S, glucose-6-phosphate dehydrogenase (G6PD) deficiency, and Dufy Antigen/Receptor for Chemokines (DARC) gene mutation are mostly distributed in the areas where *P. falciparum* malaria is endemic. These genes expression have high levels of prevalence in malaria endemic areas which is considered to be the consequence of their protective role against *P. falciparum* [101]. In this context, malaria can be defined as an infectious disease that has pronouncedely higher selective pressure on the human genome in comparison with all other infectious diseases [101]. Polymorphysms of the above-mentioned genes are typical examples of Haldane's idea of balanced polymorphism. According to this author, balanced polymorphism exists when certain genes have fixed high frequency in susceptible populations since enhanced fitness encompanied with heterozygotes multiple times overweights morbidity and mortality associated with homozygotes and compound heterozygotes [102].

5.2 HbS gene polymorphisms and malaria

Sickle hemoglobin (HbS) is best characterized genetic polymorphism tightly interconnected with malaria. HbS represents a structural variant of normal adult hemoglobin (HbAA) and results from a single point mutation (Glu → Val) on the sixth codon of the beta globin gene [103]. Homozygotes for hemoglobin S (HbSS) have sickle cell disease that further causes high morbidity and mortality. Also, heterozygous for HbS have 10-fold lower risk of dying from malaria compared to homozygous [97, 104, 105]. Heterozygotes (HbAS) have generally asymptomatic sickle cell disease which does not endanger their lives [106].

It has been found that, in the conditions of selection for fitness against malaria, nearly 45 generations (or 1000 years) were necessary to pass until sickle gene frequency reached a stable equilibrium [107]. People with HbAS have 50–90% lower parasite density [105] in comparison with individuals with normal hemoglobin (HbAA). Sub-Saharan Africa is an area with closely 80% of people born with sickle cell anemia and where most *P. falciparum* malaria cases and deaths occur [108]. Besides sub-Saharan Africa, sickle cell anemia is present, although rarely with frequency higher than 20–25%, in the Mediterranean region, the Middle East, and the Indian subcontinent [95]. There is a strong connection between high HbS allele frequency and high malarial endemicity in the world although this finding is based on the observations made in Africa: HbS allele frequency gradually increases from epidemic areas to endemic areas in Africa which is in accordance with the hypothesis that malaria protection by HbS includes the enhancement of innate and acquired immunity to *P. falciparum* [109].

Knowledge of the existing relationship between malaria infection and extension and prevalence of hemoglobinopathies in Mediterranean region are not new [102, 110]. Sickle-cell homozygous persons have short life expectancy and commonly die before adulthood. However, the gene responsible for sickle cell disease "hidden" within the genotype of heterozygous carrier can achieve high frequency due to resistance to *P. falciparum* [111].

There are lots of described biological mechanisms that are considered to be responsible for protection against malaria. First, there were only two mechanisms described regarding a manner in which the presence of HbS in heterozygotes protects against malaria: sickling of circulating infected RBCs and impaired parasite growth and oxidant damage [101]. It has been found that formation of sickle RBSc shapes under low oxygen presure occured more frequently in RBCs infected with *P. falciparum* compared to uninfected RBCs [112]. When parasite triggers sickling of erythrocytes once, sickled cells are removed by macrophages [113]. This action of may macrophages'possibly occurs due to their ability to produce and release numerous cytokines that further recruit more phagocytic cells [114]. In addition, it has been discovered that enchanced sickling was limited to RBCs infected with small Plasmodium forms [115]. On the other hand, impaired parasyte growth and oxygen damage was discovered thanks to *in vitro* studies [112]. In the conditions of normal oxygen pressure, there were no differences in the invasion, growth, and multiplication of *P. falciparum* in HbAS cells compared to HbAA RBCs. In the opposite, hypoxic consitions caused reduced fraction of *P. falciparum* in HbAS cells and a block in the maturation of ring forms to trophozoites and schizonts.

In addition, sickling and destruction of parasites in HbAS and HbSS RBCs at lower oxygen tensions (1–5%) more closely mimiced the micro-aerophilic environment of post-capillary venules *in vivo* [112].

The guiding hypothesis regarding the protective effect against malaria in people with HbAS suggests that decreased *P. falciparum* Erythrocyte Membrane Protein 1 (PfEMP1) [116] expression on infected HbAS RBCs results in lower binding of infected cells to the endothelium [117]. As a consequence, only approximately one-half the cytoadherence was seen in infected HbAS RBCs. Archer and associates have proposed that oxygen-dependent HbS polymerization is a key factor for HbAS malaria resistance [118]. They found that intraerythrocytic *P. falciparum* parasites in HbAS RBCs at low oxygen concentrations arrest in cell cycle before DNA replication and that HbS polymerization is responsible for this growth arrest.

Among the genetic factors responsible for the protection from malaria is one of the complement regulatory proteins – complement receptor 1(CR1). The frequency of CR1 polymorphisms is high in a numerous of malaria endemic areas [100]. A major receptor for RBCs infected with *P. falciparum* is human protein CD36 [119]. CD36 can be involved in malaria by sequestering infected RBCs thus disabling the immune response to this parasite [120]. Some African populations have extremely high frequency of CD36 mutation and this CD36 deficiency causes susceptibility to severe form of malaria [121]. Important genetic factors involved in resistance to malaria are erythrocyte-binding antigens. Special attention was given to erythrocyte binding antigen-175 (EBA-175), a protein that binds to glycophorin A, thus enabling merozoite entry into erythrocytes [122].

An interesting study regarding host genetic factors responsible for malaria resistance was conducted in Senegal, in the population of children and young adults that were 2 to 18 years old. Thanks to the results of this study, three candidate regions in the genome of these children were detected and one of them contains a gene related with the malaria infection in the 5q31q33 region [123].

One of the newest studies revealed that unfavorable microRNA (miRNA) composition in heterozygous HbAS or homozygous HbSS erythrocytes, leads to resistance versus *P. falciparum*. When erythrocytes are infected with *P. falciparum*, a part of erythrocyte miRNAs can translocate into the parasite. LaMonte et al. found that HbAS and HbSS erythrocytes had high number of miR-451 and let-7i integrated into essential parasite messenger RNAs, as well as that these miRNAs, together with miR-223 are negative regulators of parasite growth [124]. miR-451 fuse with transcripts of the regulatory subunit of the parasite's cAMP-dependent protein kinase (PKA-R) and reduce its translation. Therefore, it up-regulates the activity of its substrate PKA and disrupts multiple parasite developmental pathways [124].

Piel and associates created extensive geodatabase of HbS allele frequency and investigated geographical distribution of malaria [125]. Their HbS allele frequency map has shown that throughout majority of the African continent and in localized areas in Mediterranean, this allele is present with the frequency of >0.5%. According to this geodatabase research, in the Chalkidiki region of Greece, southeastern Turkey and in Central Sudan, frequency of this allele was even above 6% [125].

One of the models of how hemoglobiopathies protect from malaria is proposed by Killian associates [126] and reveals association between reduced cytoadherence phenotype and parasitized hemoglobinopathic erythrocytes. This team used conditional protein export system and tightly synchronized cultures of *P. falciparum*. They have showed that exportation of proteins encoded by parasites across the parasitophorous vacuolar membrane is more advanced, faster and increased in amount in parasitized wild type erythrocytes in comparison with hemoglobinopathic erythrocytes.

Severe malaria is in relation with intraerythrocytic life cycle of *P. falciparum* and the pathological cytoadhesive behavior of parasitized erythrocytes [127, 128]. When parasite adheres to the endothelial cells of venular capillaries, it avoids clearance mechanisms of spleen. As a consequence, pathological sequelae form within the affected blood vessel [127, 128].

Pathological consequences of *P. falciparum* malaria can possibly be mediated by adhesion of infected cells to vascular endothelium either to other uninfected red cells (rosetting) or to platelets (clumping). It has also been found that variant of erythrocytes infected with *P. falciparum* do not have noticeable differences regarding their adhesive phenotypes in comparison with erythrocytes of normal individuals infected with this parasite [129].

There are two main phenotypes of parasite-infected RBCs (iRBCs) and both express PfEMP1 [130, 131]. First type of iRBCs mediate iRBCs binding to the endothelial receptors ("cytoadherence") [132] and the second mediate iRBCs binding to uninfected RBCs ("rosetting") [133, 134]. Different iRBCs phenotypes differ in various PfEMP1 that are responsible for binding of iRBCs to microvascular endothelial cells, placental syncytiotrophoblasts or uninfected RBCs [135–137].

Usually, hemoglobin S does not increase IgG responses to various *P. falciparum* proteins [138], but it can potentially enhance IgG responses to PfEMP1, which is the main cytoadherence ligand and virulence factor [139]. In an *in vitro* study, HbAS affected the trafficking system that directs PfEMP1 to the surface of infected erythrocytes. Using cryo-electron tomography, it has been shown that within the cytoplasm of normal RBCs, the parasite proteins are transported to the surface via a parasite-generated host-derived actin cytoskeleton. In addition, hemoglobin oxidation products disrupted this process in HbAS red cells [140].

Exact pathogenic mechanisms of malaria caused by *P. falciparum* are still unknown due to numerous parasite virulence factors, host susceptibility traits, and innate and adaptive immune responses that modify the occurrence of various malaria syndromes [141, 142].

The most important reason for the high frequency of hemoglobin disorders in tropical countries is natural selection through protection of heterozygotes against severe malaria. Protection observe in HbAS is reflected in protection against severe form of malaria and probably, to some extent, against mild malaria [129]. Natural selection is not the only mechanism responsible for high HbS gene frequency [102, 106, 107]. The others are high frequency of consanguineous marriages and epidemiological transition [98]. In addition, different distribution of some hemoglobin disorders in different populations is an example of founder effects by their original inhabitants [143].

Thanks to the studies conducted *in vitro*, various researches united on general hypothesis that protection from malaria is the result of impairment in the invasion and growth of *P. falciparum* parasites into HbAS red cells under conditions of low oxygen tension that were physiologically representative of *in vivo* conditions [112, 144]. Afterwards, a lot of alternative hypothesis have been developed including the one that refers to the enhanced removal of parasite infected HbAS RBCs. This mechanism could be related with sickling of these cells under low oxygen tension [112, 115, 145] causes their premature destruction in the spleen [112]. Specifically, Shear and associates have observed that protective effect of HbS can be lost on the model of transgenic mice that were subjected to splenectomy [146].

There are few researches that suggest that protection against malaria can be achieved not only through innate immunity, but also via acquired immunity. For example, in populations naturally exposed to *P. falciparum*, protective effect of HbAS increases with age [147, 148]. This is in accordance with recent studies on a mouse model which proposes an immuno-modulatory mechanism mediated throughout hemoxygenase-1 [149]. The problem with these findings when translating into human populations is metabolic difference between sickling disorders of mice and human sickle-cell traits [95].

5.3 Glucose-6-phosphate dehydrogenase (G6PD) gene polymorphisms and malaria

The Glucose-6-phosphate dehydrogenase (G6PD) is a "housekeeping" gene located on long (q) arm of the X chromosome at position 28 – Xq28 [150]. This gene encodes an enzyme named glucose-6-phosphate dehydrogenase that acts in almost all types of cells thus providing normal carbohydrates processing [151]. The most important role of G6PD is in RBCs, where this enzyme is involved in protection of RBCs from damage and early destruction [152].

Glucose-6-phosphate dehydrogenase deficiency is a genetic disorder that mainly affects RBCs, thus causing premature destruction of these cells called hemolysis [153]. Besides causing hemolytic anemia, G6PD has an evolutive advantage regarding the protection against malaria. A consequence of the reduced amount of functional G6PD makes difficult pathway for parasites to invade RBSc [154]. G6PD gene insufficiency is the most frequent in malaria endemic areas. When it comes to Mediterranean, the highest noted frequency of this gene is in Mediterranean parts of Africa, southern Europe and in the Middle East [153].

Interestingly, G6PD deficient patients in Africa, where this type of deficiency is endemic, have milder consequences as well as relatively higher enzyme activity in comparison with patients from Mediterranean and Asia [155].

G6PD deficiency gives especially high protection from *falciparum* malaria infection [156, 157]. Among more than 400 variants of G6PD that differs in biochemical characteristics, enzyme kinetics, physicochemical characteristics, and other parameters [158] is G6PD B+ which is the most common variant of this enzyme. G6PD B+ is used as standard for normal enzyme activity and electrophoretic mobility and, therefore, for identification of other variants. In the area of Mediterranean, special place belongs to G6PD Mediterranean variant [159] which has less than 10% of the enzyme activity of G6PD B+ while its electrophoretic mobility is similar to G6PD B+ [160]. Two-point mutations in gene for this enzyme were identified. One mutation is cytosine to thymine mutation at nucleotide number 563, which causes substitution of serine with phenylalanine [161]. At nucleotide number 1311, change of cytosine with thymine represents a silent mutation [162].

A research conducted by Barišić et al. in the Dalmatinian region of Croatia resulted in discovering a new variant of G6PD named G6PD Split [163]. Change of cytosine to guanine at nucleotide 1442 caused substitution of proline with arginine which led to moderate enzyme deficiency. Besides this novel variant of G6PD discovered in one patient, other 23 unrelated patients with low G6PD activity had five other well-known variants and three patients had uncharacterized forms of G6PD mutations. The most represented form found in nine patients was G6PD Cosenza. G6PD Cosenza was first found in Calabria region of southern Italy and represents the consequence of change of guanine into cytosine at nucleotide 1376. This substitution changes Arginine to Proline [164]. G6PD Cosenza mutation is severe G6PD deficiency frequently jointed with hemolysis.

Around 400 million people from all over the World carry at least one deficient variant of G6PD gene. The frequency of those mutations varies in different populations [165]. In Africans and Afro-Americans G6PD A- is the most common mutation which has a gene frequency of 11%. G6PD B (Mediterranean) is a more severe deficiency usually found in Mediterranean area. Since Mediterranean represents a large region, the prevalence of this mutation varies from 2 to 20% in Greece, Turkey, and Italy, up to the 70% which is the prevalence characteristic for Kurdish Jews [165, 166].

5.4 FY gene polymorphisms and malaria

In addition to the role it plays in transfusion incompatibility and hemolytic disease of newborns, Duffy Blood Group System is important in medicine due to its association with the invasion of RBCs by the parasite *P. vivax*. Outside Africa, *P. vivax* is the most widespread malaria parasite species, with 40% of cases in the Eastern Mediterranean [167]. Without Duffy antigens on their surface, RBCs are relatively resistant to *P. vivax* [168]. There are six types of Duffy antigens (Fy^a, Fy^b, Fy3, Fy4, Fy5, and Fy6), out of which only Fy3 has a clinical significance. Duffy antigens are also receptors for chemicals secreted by blood cells during inflammation [169].

Duffy-Antigen Chemokine Receptor (DARC) is a glycosylated transmembrane protein receptor which, among other roles, serves as a receptor for *P. vivax*. DARK crosses the membrane seven times and has an extracellular epitope, N-terminal domain responsible for RBC invasion by *P. vivax* merozoites [170, 171]. Two exons (FyA and FyB) of FY gene are encoded by the co dominant FyA and FyB alleles located on human chromosome 1 [172]. The difference between these two alleles is a non-synonymous mutation, specifically substitution of guanine to adenine at nucleotide 125, which was enough to determine the two antithetical antigens [173]. Based on this variation, four phenotypes within Duffy Blood Group System were identified: Fy (a + b-), Fy (a-b+), Fy (a-b-) and Fy (a + b+) [174]. The non-functional allele Fy*O is the consequence of a mutation in the gene promoter at −33 nucleotide that changed thymine to cytosine which abolish its expression in the erythrocyte cell lineage [175, 176].

Individuals with Fy (a-b-) phenotype are resistant to *P. vivax* invasion [177]. This was shown in the study which included 11 volunteers. The individuals affected with malaria were Fy (a+) or Fy (b+). In the countries of West Africa, frequency of the Fy (a-b-) phenotype is a high while the incidence of *P. vivax* malaria is low [178]. Virtual absence of *P. vivax* malaria in populations with widespread DARC negativity is the proof of the substantial importance of the Duffy binding protein (DBP)–DARC interaction [179]. It is important to emphasize that Fy (a − b−) does not protect from *P. falciparum* which therefore can infect RBCs of any Duffy phenotype [169].

While *P. falciparum* can enter human RBCs through series of receptors on their surface, RBCs invasion by *P. vivax* depends on an interaction with the Fy^a or Fy^b antigens [169, 180]. Therefore, in the regions of Africa where Fy (a-b-) phenotype is stable within various ethnic groups, the transmission of *P. vivax* is not usual [181]. On the other hand, individuals with Fy (a-b+) or Fy (a + b-) genotypes that express half the level of Duffy antigens on RBCs compared to Fy (a-b-) homozygotes are less sensitive to blood stage infection by *P. vivax*. Therefore, parasitemia by *P. vivax* might be inhibited by total or partial restriction access of *P. vivax* to Duffy antigen [182, 183].

Phenotypic differences in susceptibility to malaria are the results of Fy gene polymorphism. Individuals that carry Duffy antigen-negative allele hidden within heterozygous genotype have significantly reduced adherence of the DBP ligand domain (DBPII) to erythrocytes [184]. On the other hand, people with Fy^a phenotype have 30–80% lower risk of clinical *vivax* malaria, but not for falciparum malaria [185]. In the countries of Southeast Asia that are the source of *P. vivax*, the Fy^a allele is fixed [186], while Fy^b is represented in the populations in North and Northern-central Europe. This kind of distribution of Fy alleles indicates a selective advantage against *P. vivax* malaria [185].

6. Conclusions

Genetic specificities of Mediterranean region described in this chapter are a good example of how important role genetic diversity plays in adaptability to local environment. Recognition of these specific genetic traits, of the region, is of considerable clinical importance. Carrier identification, genetic counseling and prenatal diagnosis still represent a corner stone in the management of thalassemia-syndromes and various hemoglobiopathies. Due to variable frequencies of different *MEFV* mutations, their genotypes and prevalence should be determined in every population, in order to make reference frameworks for mutation screening when needed.

Acknowledgements

The work is supported by the Grants No. III 41017, III 41018, and 43012 of Ministry of Education, Science, and Technological Development of the Republic of Serbia. The research was supported by the Project of Serbian Academy of Science O-06-17 and by an Internal project of University of Nis, Faculty of Medicine of Nis, 2020, conducted by Dijana Stojanovic MD, PhD.

Conflict of interest

The authors declare no conflict of interest.

Author details

Jelena M. Milenković[1*], Dijana R. Stojanović[1] and Jelena G. Najdanović[2,3]

1 Department of Pathophysiology, Faculty of Medicine, University of Niš, Niš, Serbia

2 Department of Biology and Human Genetics, Faculty of Medicine, University of Niš, Niš, Serbia

3 Department for Cell and Tissue Engineering, Faculty of Medicine, University of Niš, Niš, Serbia

*Address all correspondence to: jelenaradovic982@gmail.com; jelena.milenkovic@medfak.ni.ac.rs

References

[1] Martinon F, Hofmann K, Tschopp J. The pyrin domain: a possible member of the death domain-fold family implicated in apoptosis and inflammation. Current Biology 2001;11:R118–R120. DOI: 10.1016/S0960-9822(01)00056-2

[2] Manukyan G, Aminov R. Update on Pyrin Functions and Mechanisms of Familial Mediterranean Fever. Front. Microbiol. 2016;7:456. DOI: 10.3389/fmicb.2016.00456

[3] Xu H, Yang J, Gao W, Li L, Li P, Zhang L, Gong YN, Peng X, Xi JJ, Chen S, Wang F, Shao F. Innate immune sensing of bacterial modifications of Rho GTPases by the Pyrin inflammasome. Nature. 2014;513(7517):237-241. DOI: 10.1038/nature13449

[4] Schnappauf O, Chae JJ, Kastner DL, Aksentijevich I. The Pyrin Inflammasome in Health and Disease. Front. Immunol. 2019;10:1745. DOI: 10.3389/fimmu.2019.01745

[5] Ozen S, Batu ED, Demir S. Familial Mediterranean Fever: Recent Developments in Pathogenesis and New Recommendations for Management. Front. Immunol. 2017;8:253. DOI: 10.3389/fimmu.2017.00253

[6] de Zoete MR, Flavell RA. Detecting "different": Pyrin senses modified GTPases. Cell Research 2014;24:1286-1287. DOI: 10.1038/cr.2014.101

[7] Magnotti F, Lefeuvre L, Benezech S, Malsot T, Waeckel L, Martin A, Kerever S, Chirita D, et al. Pyrin dephosphorylation is sufficient to trigger inflammasome activation in familial Mediterranean fever patients. EMBO Molecular Medicine 2019;11:e10547. DOI: 10.15252/emmm.201910547

[8] Kimura T, Jain A, Choi SW, Mandell MA, Schroder K, Johansen T, Deretic V. TRIM-mediated precision autophagy targets cytoplasmic regulators of innate immunity. The Journal of Cell Biology 2015;210:973-989. DOI: 10.1083/jcb.201503023

[9] Jiang S, Dupont N, Castillo EF, Deretic V. Secretory versus Degradative Autophagy: Unconventional Secretion of Inflammatory Mediators. Journal of Innate Immunity 2013;5:471-479. DOI: 10.1159/000346707

[10] Moradian MM, Sarkisian T, Ajrapetyan H, Avanesian N. Genotype-phenotype studies in a large cohort of Armenian patients with familial Mediterranean fever suggest clinical disease with heterozygous MEFV mutations. Journal of Human Genetics 2010;55:389-393. DOI: 10.1038/jhg.2010.52

[11] Shinar Y, Obici L, Aksentijevich I, Bennetts B, Austrup F, Ceccherini I, Costa JM, et al. Guidelines for the genetic diagnosis of hereditary recurrent fevers. Annals of the Rheumatic Diseases 2012;71:1599-1605. DOI: 10.1136/annrheumdis-2011-201271

[12] Radović J, Vojinović J, Bojanić V, Jevtović-Stoimenov T, Kocić G, Milojković M, Veljković A, Marković I, Stojanović S, Pavlović D. Lipid peroxidation and oxidative protein products in children with episodic fever of unknown origin. J. Med. Biochem. 2014;33:197-202. DOI: 10.2478/jomb-2013-0023

[13] Milenković J, Vojinović J, Debeljak M, Toplak N, Lazarević D, Avčin T, Jevtović-Stoimenov T, Pavlović D, et al. Distribution of MEFV gene mutations and R202Q polymorphism in the Serbian population and their influence on oxidative stress and clinical manifestations of inflammation. Pediatr. Rheum 2016;14:39. DOI: 10.1186/s12969-016-0097-1

[14] Erdem E, Karatas A, Kaya C, Dilek M, Yakupoglu YK, Arık N, Akpolat T. Renal transplantation in patients with familial Mediterranean fever. Clin. Rheumatol. 2012;31(8):1183-1186. DOI: 10.1007/s10067-012-1992-6

[15] Stojanovic D, Cvetkovic TP, Stojanovic MM, Bojanic VV, Stefanovic NZ, Radenkovic SB, Ljubisavljevic S, Pavlovic DD. Crosstalk of inflammatory mediators and lipid parameters as early markers of renal dysfunction in stable renal transplant recipients with regard to immunosuppression. Ann. Transplant 2013;18:414-423. DOI: 10.12659/aot.889239

[16] Dodé C, Pêcheux C, Cazeneuve C, Cattan D, Dervichian M, Goossens M, Delpech M, Amselem S, Grateau G. Mutations in the MEFV gene in a large series of patients with a clinical diagnosis of familial Mediterranean fever. Am. J. Med. Genet 2000;92:241-246. DOI:10.1002/(SICI)1096-8628(20000605)92:4<241::AID-AJMG3>3.0.CO;2-G

[17] Kisacik B, Kaya SU, Pehlivan Y, Tasliyurt T, Sayarlioglu M, Onat AM. Decreased vitamin D levels in patients with familial mediterranean fever. Rheumatol. Int 2013;33:1355-1357. DOI: 10.1007/s00296-011-2278-z

[18] Radovic J, Markovic D, Velickov A, Djordjevic B, Stojnev S. Vitamin D immunomodulatory effect. Acta Medica Medianae 2012; 51(4): 58-64. DOI:10.5633/amm.2012.0409

[19] Tunca M, Akar S, Onen F, Ozdogan H, Kasapcopur O, Yalcinkaya F, Tutar E, Ozen S, Topaloglu R, et al. Turkish FMF Study Group. Familial Mediterranean fever (FMF) in Turkey: results of a nationwide multicenter study. Medicine. (Baltimore). 2005;84(1):1-1)11. DOI: 10.1097/01.md.0000152370.84628.0c

[20] Yepiskoposyan L, Harutyunyan A. Population genetics of familial Mediterranean fever: a review. Eur. J. Hum. Genet. 2007;15:911-916. DOI:10.1038/sj.ejhg.5201869

[21] Yasunami M, Nakamura H, Agematsu K, Nakamura A, Yazaki M, Kishida D, et al. Identification of disease-promoting HLA class I and protective class II modifiers in Japanese patients with familial Mediterranean fever. PLoS ONE 2015;10:e0125938. DOI: 10.1371/journal.pone.0125938

[22] Turkcapar N, Tuncali T, Kutlay S, Burhan BY, Kinikli G, et al. The contribution of genotypes at the MICA gene triplet repeat polymorphisms and MEFV mutations to amyloidosis and course of the disease in the patients with familial Mediterranean fever. Rheumatol. Int. 2007;27:545-551. DOI: 10.1007/s00296-006-0255-8

[23] Aksentijevich I, Torosyan Y, Samuels J, Centola M, Pras E, Chae JJ, Oddoux C, Wood G, Azzaro MP, Palumbo G, Giustolisi R, Pras M, Ostrer H, Kastner DL. Mutation and haplotype studies of familial Mediterranean fever reveal new ancestral relationships and evidence for a high carrier frequency with reduced penetrance in the Ashkenazi Jewish population. Am. J. Hum. Genet. 1999;64:949-962. DOI:10.1086/302327

[24] Fumagalli M, Cagliani R, Pozzoli U, Riva S, Comi GP, Menozzi G, et al. A population genetics study of the familial Mediterranean fever gene: evidence of balancing selection under an overdominance regime. Genes. Immun. 2009;10:678-686. DOI: 10.1038/gene.2009.59

[25] Loeven NA, Medici NP, Bliska JB. The pyrin inflammasome in host-microbe interactions. Curr. Opin. Microbiol. 2020;54:77-86. DOI: 10.1016/j.mib.2020.01.005

[26] Gutierrez MG, Master SS, Singh SB, Taylor GA, Colombo MI, Deretic V. Autophagy is a defense mechanism inhibiting BCG and Mycobacterium tuberculosis survival in infected macrophages. Cell. 2004;119:753-766. DOI: 10.1016/j.cell.2004.11.038

[27] Master SS, Rampini SK, Davis AS, Keller C, Ehlers S, Springer B, Timmins GS, Sander P, Deretic V. Mycobacterium tuberculosis prevents inflammasome activation. Cell. Host. Microbe 2008;3:224-232. DOI: 10.1016/j.chom.2008.03.003

[28] The International FMF Consortium. Ancient missense mutations in a new member of the RoRet gene family are likely to cause familial Mediterranean fever. Cell. 1997;90:797-807. DOI: 10.1016/s0092-8674(00)80539-5

[29] Booth DR, Lachmann HJ, Gillmore JD, Booth SE, Hawkins PN. Prevalence and significance of the familial Mediterranean fever gene mutation encoding pyrin Q148. Q. J. Med. 2001;94:527-531. DOI: 10.1093/qjmed/94.10.527

[30] Sarkisian T, Ajrapetian H, Beglarian A, Shahsuvarian G, Egiazarian A. Familial Mediterranean Fever in Armenian population. Georgian. Med. News. 2008;(156):105-111. DOI: 10.5144/0256-4947.2019.382

[31] Majeed HA, El-Khateeb M, El-Shanti H, Rabaiha ZA, Tayeh M, Najib D. The spectrum of familial Mediterranean fever gene mutations in Arabs: report of a large series. Semin. Arthritis. Rheum. 2005;34(6):813-818. DOI: 10.1016/j.semarthrit.2005.01.010

[32] Schwabe AD, Peters RS. Familial Mediterranean fever in Armenians. Analysis of 100 cases. Medicine. (Baltimore).1974;69:453-462. DOI: 10.1097/00005792-197411000-00005

[33] La Regina M, Nucera G, Diaco M, Procopio A, Gasbarrini G, Notarnicola C, Kone-Paut I, Touitou I, Manna R. Familial Mediterranean fever is no longer a rare disease in Italy. Eur. J. Hum. Genet. 2003;11(1):50-56. DOI: 10.1038/sj.ejhg.5200916

[34] Aldea A, Calafell F, Aróstegui JI, Lao O, Rius J, Plaza S, Masó M, Vives J, Buades J, Yagüe J.The West Side Story: MEFV haplotype in Spanish FMF patients and controls, and evidence of high LD and a recombination "hot-spot" at the MEFV locus. Hum. Mutat. 2004;23:e399. DOI: 10.1002/humu.9229

[35] Ece A, Çakmak E, Uluca Ü, Kelekçi S, Yolbaş İ, Güneş A, Yel S, Tan İ, Şen V. The MEFV mutations and their clinical correlations in children with familial Mediterranean fever in southeast Turkey. Rheumatol. Int. 2014;34(2):207-212. DOI: 10.1007/s00296-013-2858-1

[36] Yildirim ME, Kurtulgan HK, Ozdemir O, Kilicgun H, Aydemir DS, Baser B, Sezgin I. Prevalence of MEFV gene mutations in a large cohort of patients with suspected familial Mediterranean fever in Central Anatolia. Ann. Saudi. Med. 2019;39(6):382-387. DOI: 10.5144/0256-4947.2019.382

[37] Alzyoud R, Alsweiti M, Maittah H, Zreqat E, Alwahadneh A, Abu-Shukair M, Habahbeh L, Mutereen M. Familial Mediterranean fever in Jordanian Children: single centre experience. Mediterr. J. Rheumatol. 2018;29(4):211-216. DOI: 10.31138/mjr.29.4.211

[38] el-Garf A, Salah S, Iskander I, Salah H, Amin SN. MEFV mutations in Egyptian patients suffering from familial Mediterranean fever: analysis of 12 gene mutations. Rheumatol. Int. 2010;30(10):1293-1298. DOI: 10.1007/s00296-009-1140-z

[39] Mansour AR, El-Shayeb A, El Habachi N, Khodair MA, Elwazzan D, Abdeen N, Said M, Ebaid R, ElShahawy N, Seif A, Zaki N. Molecular Patterns of MEFV Gene Mutations in Egyptian Patients with Familial Mediterranean Fever: A Retrospective Cohort Study. Int. J. Inflam. 2019;2019:2578760. DOI: 10.1155/2019/2578760

[40] Al-Haggar MS, Yahia S, Abdel-Hady D, Al-Saied A, Al-Kenawy R, Abo-El-Kasem R. Phenotype-genotype updates from familial Mediterranean fever database registry of Mansoura University Children' Hospital, Mansoura, Egypt. Indian. J. Hum. Genet. 2014;20(1):43-50. DOI: 10.4103/0971-6866.132755

[41] Belmahi L, Cherkaoui IJ, Hama I, Sefiani A. MEFV mutations in Moroccan patients suffering from familial Mediterranean Fever. Rheumatol. Int. 2012;32(4):981-984. DOI: 10.1007/s00296-010-1732-7

[42] Mattit H, Joma M, Al-Cheikh S, El-Khateeb M, Medlej-Hashim M, Salem N, Delague V, Mégarbané A. Familial Mediterranean fever in the Syrian population: gene mutation frequencies, carrier rates and phenotype-genotype correlation. Eur. J. Med. Genet. 2006;49(6):481-486. DOI: 10.1016/j.ejmg.2006.03.002

[43] Sabbagh AS, Ghasham M, Abdel Khalek R, Greije L, Shammaa DM, Zaatari GS, Mahfouz RA. MEFV gene mutations spectrum among Lebanese patients referred for Familial Mediterranean Fever work-up: experience of a major tertiary care center. Mol. Biol. Rep. 2008;35(3):447-451. DOI: 10.1007/s11033-007-9105-3

[44] Gershoni-Baruch R, Shinawi M, Leah K, Badarnah K, Brik R. FMF: prevalence, penetrance and genetic drift. Eur. J. Hum. Genet. 2001;9(8):634-637. DOI: 10.1038/sj.ejhg.5200672

[45] Sharkia R, Mahajnah M, Zalan A, Athamna M, Azem A, Badarneh K, Faris F. Comparative screening of FMF mutations in various communities of the Israeli society. Eur. J. Med. Genet. 2013;56(7):351-355. DOI: 10.1016/j.ejmg.2013.04.002

[46] Touitou I. The spectrum of familial Mediterranean fever (FMF) mutations. Eur. J. Hum. Genet. 2001;9:47-483. DOI: 10.1038/sj.ejhg.5200658

[47] Celep G, Durmaz ZH, Erdogan Y, Akpinar S, Kaya SA, Guckan R. The Spectrum of MEFV Gene Mutations and Genotypes in the Middle Northern Region of Turkey. Eurasian. J. Med. 2019;51(3):252-256. DOI: 10.5152/eurasianjmed.2019.18396

[48] Debeljak M, Toplak N, Abazi N, Szabados B, Mulaosmanović V, Radović J, Perko D, Vojnović J, Constantin T, Kuzmanovska D, Avčin T. The carrier rate and spectrum of MEFV gene mutations in central and southeastern European populations. Clinical and Experimental Rheumatology. Clin. Exp. Rheumatol. 2015;33(6 Suppl 94):S19-S23.

[49] Giaglis S, Papadopoulos V, Kambas K, Doumas M, Tsironidou V, Rafail S, et al. MEFV alterations and population genetics analysis in a large cohort of Greek patients with familial Mediterranean fever. Clin. Genet. 2007;71(5):458-467. DOI: 10.1111/j.1399-0004.2007.00789.x

[50] Ritis K, Giaglis S, Spathari N, Micheli A, Zonios D, Tzoanopoulos D, Deltas CC, Rafail S, Mean R, Papadopoulos V, Tzioufas AG, Moutsopoulos HM, Kartalis G. Non-isotopic RNase cleavage assay for mutation detection in MEFV, the gene responsible for familial Mediterranean fever, in a cohort of Greek patients. Ann. Rheum. Dis. 2004;63(4):438-443. DOI: 10.1136/ard.2003.009258

[51] Constantinou. Deltas CC, Mean R, Rossou E, Costi C, Koupepidou P, Hadjiyanni I, Hadjiroussos V, Petrou P, et al. Familial Mediterranean fever (FMF) mutations occur frequently in the Greek-Cypriot population of Cyprus. Genet. Test. 2002;6(1):15-21. DOI: 10.1089/109065702760093861

[52] Güncan S, Bilge NŞ, Cansu DÜ, Kaşifoğlu T, Korkmaz C. The role of MEFV mutations in the concurrent disorders observed in patients with familial Mediterranean fever. Eur. J. Rheumatol. 2016;3(3):118-121. DOI: 10.5152/eurjrheum.2016.16012

[53] Azizlerli G, Köse AA, Sarica R, Gül A, Tutkun IT, Kulaç M, Tunç R, Urgancioğlu M, Dişçi R. Prevalence of Behçet's disease in Istanbul, Turkey. Int. J. Dermatol. 2003;42(10):803-806. DOI: 10.1046/j.1365-4362.2003.01893.x

[54] Gallizzi R, Pidone C, Cantarini L, Finetti M, Cattalini M, Filocamo G, Insalaco A, Rigante D, et al. A national cohort study on pediatric Behçet's disease: cross-sectional data from an Italian registry. Pediatr. Rheumatol. Online. J. 2017;15(1):84. DOI: 10.1186/s12969-017-0213-x

[55] Touitou I, Magne X, Molinari N, Navarro A, Quellec AL, Picco P, Seri M, Ozen S, Bakkaloglu A, Karaduman A, Garnier JM, Demaille J, Koné-Paut I. MEFV mutations in Behçet's disease. Hum. Mutat. 2000;16(3):271-272. DOI: 10.1002/1098-1004(200009)16:3<271::AID-HUMU16>3.0.CO;2-A

[56] Hamzaoui A, Houman MH, Massouadia M, Ben Salem T, Khanfir MS, Ben Ghorbel I, Miled M. Contribution of Hla-B51 in the susceptibility and specific clinical features of Behcet's disease in Tunisian patients. Eur. J. Intern. Med. 2012;23(4):347-349. DOI: 10.1016/j.ejim.2011.12.011

[57] Atagunduz P, Ergun T, Direskeneli H. MEFV mutations are increased in Behçet's disease (BD) and are associated with vascular involvement. Clin. Exp. Rheumatol. 2003;21(4 Suppl 30): S35-S37.

[58] Oztürk MA, Ertenli I, Kiraz S, C Haznedaroğlu I, Celik I, Kirazli S, Calgüneri M. Plasminogen activator inhibitor-1 as a link between pathological fibrinolysis and arthritis of Behçet's disease. Rheumatol. Int. 2004;24(2):98-102. DOI: 10.1007/s00296-003-0324-1

[59] Milenkovic J, Milojkovic M, Jevtovic Stoimenov T, Djindjic B, Miljkovic E. Mechanisms of plasminogen activator inhibitor 1 action in stromal remodeling and related diseases. Biomed. Pap. Med. Fac. Univ. Palacky. Olomouc. Czech. Repub. 2017;161(4):339-347. DOI: 10.5507/bp.2017.046

[60] Gurgey A, Balta G, Boyvat A. Factor V Leiden mutation and PAI-1 gene 4G/5G genotype in thrombotic patients with Behcet's disease. Blood. Coagul. Fibrinolysis. 2003;14(2):121-124. DOI: 10.1097/00001721-200302000-00001

[61] Zeidan MJ, Saadoun D, Garrido M, Klatzmann D, Six A, Cacoub P. Behçet's disease physiopathology: a contemporary review. Auto. Immun. Highlights. 2016;7(1):4. DOI: 10.1007/s13317-016-0074-1

[62] Nienhuis AW, Nathan DG. Pathophysiology and Clinical Manifestations of the β-thalassemias. Cold. Spring. Harb. Perspect. Med. 2012;2:a011726. DOI: 10.1101/cshperspect.a011726

[63] Origa R. β-Thalassemia. Gen. Med. 2017;19:609-619. DOI: 10.1038/gim.2016.173

[64] Galanello R, Origa R. Beta-thalassemia. Orphanet. J. Rare. Dis. 2010;5:2-15. DOI: 10.1186/1750-1172-5-11

[65] Modell B, Darlison M. Global epidemiology of haemoglobin disorders

and derived service indicators. Bull. World. Health. Organ. 2008;86:480-487. DOI: 10.2471/BLT.06.036673

[66] Weatherall DJ, Williams TN, Allen SJ, O'Donnell A. The population genetics and dynamics of the thalassemias. Hematol. Oncol. Clin. North. Am. 2010; 24:1021-1031. DOI: 10.1016/j.hoc.2010.08.010

[67] Weatherall DJ. The inherited diseases of hemoglobin are an emerging global health burden. Blood. 2010; 115: 4331-4336. DOI: 10.1182/blood-2010-01-251348

[68] Taher TA, Weatherall DJ, Maria Domenica Cappellini MD. Thalassemia. Lancet. 2018; 391:155-167. DOI: 10.1016/S0140-6736(17)31822-6

[69] Danjou F, Anni F, Galanello R. β-thalassemia: From genotype to phenotype. Haematologica. 2011;96:1573-1575. DOI: 10.3324/haematol.2011.055962

[70] Thein SL. 2013. Molecular basis of β-thalassemia. Cold. Spring. Harb. Perspect. Med. DOI: 10.1101/cshperspect.a011700.

[71] Weatherall DJ. 2001. Phenotype-genotype relationships in monogenic disease: Lessons from the thalassemias. Nat. Rev. Genet. 2001;2:245-255. DOI: 10.1038/35066048

[72] Musallam KM, Sankaran VG, Cappellini MD, Duca L,Nathan DG, Taher AT. Fetal hemoglobin levels and morbidity in untransfused patients with β-thalassemia intermedia. Blood. 2012;119:364-367. DOI: 10.1182/blood-2011-09-382408

[73] Karimi M, Haghpanah S, Farhadi A, Yavarian M. Genotype–phenotype relationship of patients with β-thalassemia taking hydroxyurea: A 13-year experience in Iran. Int. J. Hematol.2011;95:51-56. DOI: 10.1007/s12185-011-0985-6

[74] Musallam KM, Rivella S, Vichinsky E, Rachmilewitz EA. Non transfusion dependent thalassemias. Haematologica. 2013; 98: 833-844. DOI: 10.3324/haematol.2012.066845

[75] Musallam KM, Taher AT, Rachmilewitz EA. B-thalassemia intermedia: a clinical perspective. Cold. Spring. Harb. Perspect. Med. 2012; 2: a013182. DOI: 10.1101/cshperspect.a013482

[76] Weatherall D.William Allan award address. The thalassemias: the role of molecular genetics in an evolving global health problem. Am. J. Hum. Genet 2004;74:385-392. DOI: 10.1086/381402

[77] Rivella S. The role of ineffective erythropoiesis in non transfusion -dependent thalassemia. Blood. Rev. 2012;26:12-15. DOI: 10.1016/S0268-960X(12)70005-X

[78] Haidar R, Mhaidli H, Taher AT. Paraspinal extramedullary hematopoiesis in patients with thalassemia intermedia. Eur. Spine. J. 2010;19:871-878. DOI: 10.1007/s00586-010-1357-2

[79] Cappellini MD, Poggiali E, Taher AT, Musallam KM. Hypercoagulability in β-thalassemia: a status quo. Expert. Rev. Hematol. 2012;5:505-511. DOI: 10.1586/ehm.12.42

[80] Ginzburg Y, Rivella S. β-Thalassemia: A model for elucidating the dynamic regulation of ineffective erythropoiesis and iron metabolism. Blood. 2011;118:4321-4330. DOI: 10.1182/blood-2011-03-283614

[81] Tanno T, Bhanu NV, Oneal PA, Goh SH, Staker P, Lee YT,Moroney JW, Reed CH, Luban NL, Wang RH, et al. High levels of GDF15 in thalassemia suppress expression of the iron regulatory protein hepcidin. Nat. Med. 2007;13:1096-1101. DOI: 10.1038/nm1629

[82] Tanno T, Porayette P, Sripichai O, Noh SJ, Byrnes C,Bhupatiraju A, Lee YT,

Goodnough JB, Harandi O,Ganz T. Identification of TWSG1 as a second novel erythroid regulator of hepcidin expression in murine and human cells. Blood. 2009; 114:181-186. DOI: 10.1182/blood-2008-12-195503

[83] Stojanovic D, Mitic V, Stojanovic M, Petrovic D, Ignjatovic A, Stefanovic N, et al. The Partnership Between Renalase and Ejection Fraction as a Risk Factor for Increased Cardiac Remodeling Biomarkers in Chronic Heart Failure Patients. Curr. Med. Res. Opin. 2020;36(6):909-919. DOI: 10.1080/03007995.2020.1756233

[84] Camaschella C, Nai A. Ineffective erythropoiesis and regulation of iron status in iron loading anaemias. Br. J. Haematol. 2016; 172: 512-523. DOI: 10.1111/bjh.13820

[85] Preza GC, Ruchala P, Pinon R, Ramos E, Qiao B, Perallta MA, Sharma S, Waring G, Ganz T, Nameth E. Mini hepcidins are rationally designed small peptides that mimic hepcidin activity in mice and may be useful for treatment of iron overload. J. Clin. Invest. 2011;121:4880-4888. DOI: 10.1172/JCI57693

[86] Angastiniotis M, Lobitz S. Thalassemias: An Overview. Int. J. Neonatal. Screen. 2019;5:2-11. DOI: 10.3390/ijns5010016

[87] Najdanović JG, Cvetković VJ, Stojanović S, Vukelic-Nikolic M, Stanisavljevic M, et al. The Influence of Adipose-Derived Stem Cells Induced into Endothelial Cells on Ectopic Vasculogenesis and Osteogenesis. Cel. Mol. Bioeng. 2015;8:577-590. DOI: 10.1007/s12195-015-0403-x

[88] Morris CR, Vichinsky EP. 2010. Pulmonary hypertension in thalassemia. Ann. NY. Acad. Sci. 2010;1202:205-213. DOI: 10.1111/j.1749-6632.2010.05580.x

[89] Brancaleoni V, Pierro ED, Motta I, Cappellini MD. Laboratory diagnosis of thalassemia. Int. Jnl. Lab. Hem. 2016;38:32-34. DOI: 10.1111/ijlh.12527

[90] Kirk P, Roughton M, Porter JB, et al. Cardiac T2* magnetic resonance for prediction of cardiac complications in thalassemia major. Circulation. 2009;**120**:1961-1968. DOI: 10.1161/CIRCULATIONAHA.109.874487

[91] St Pierre TG, El Beshlawy A, Elalfy M, et al. Multicenter validation of spin density projection assisted R2MRI for the noninvasive measurement of liver iron concentration. Magn. Reson. Med. 2014;71:2215-2223. DOI: 10.1002/mrm.24854

[92] Carpenter JP, He T, Kirk P, et al. On T2* magnetic resonance and cardiac iron. Circulation. 2011;123:1519-1528. DOI: 10.1161/CIRCULATIONAHA.110.007641

[93] Baronciani D, Angelucci E, Potschger U, et al. Hemopoietic stem cell transplantation in thalassemia: a report from the European Society for Blood and Bone Marrow Transplantation Hemoglobinopathy Registry, 2000-2010. Bone. Marrow. Transplant. 2016;51:536-541. DOI:10.1038/bmt.2015.293

[94] Andersen O, Wetten OF, De Rosa MC, Andre C, Carelli Alinovi C, Colafranceschi M, Brix O, Colosimo A. Haemoglobin polymorphisms affect the oxygen-binding properties in Atlantic cod populations. Proc. Biol. Sci. 2009;276(1658):833-841. DOI: 10.1098/rspb.2008.1529

[95] Williams TN, Weatherall DJ. World distribution, population genetics, and health burden of the hemoglobinopathies. Cold. Spring. Harb. Perspect. Med. 2012;2(9):a011692. DOI: 10.1101/cshperspect.a011692

[96] Christianson A, Howson CP, Modell B. 2006. March of Dimes global report on birth defects. March of Dimes Birth Defects Foundation, New York.

[97] Taylor SM, Cerami C, Fairhurst RM. Hemoglobinopathies: slicing the Gordian knot of Plasmodium falciparum malaria pathogenesis. PLoS. Pathog. 2013;9(5):e1003327. DOI: 10.1371/journal.ppat.1003327

[98] Taylor SM, Fairhurst RM. Malaria parasites and red cell variants: when a house is not a home. Curr. Opin. Hematol. 2014;21(3):193-200. DOI: 10.1097/MOH.0000000000000039

[99] World Health Organization. World Malaria Report 2011 [Internet]. Geneva: 2012. Available from: https://www.who.int/malaria/world_malaria_report_2011/WMR2011_noprofiles_lowres.pdf?ua=1

[100] Williams TN. Human red blood cell polymorphisms and malaria. Curr. Opin. Microbiol. 2006;9(4):388-394. DOI: 10.1016/j.mib.2006.06.009

[101] Bunn HF. The triumph of good over evil: protection by the sickle gene against malaria. Blood. 2013;121(1):20-25. DOI: 10.1182/blood-2012-08-449397

[102] Haldane JBS. The rate of mutation of human genes. Hereditas. 1949;35:267-273. DOI: 10.1111/j.1601-5223.1949.tb03339.x

[103] Ingram VM. Abnormal human haemoglobins. III. The chemical difference between normal and sickle cell haemoglobins. Biochim. Biophys. Acta. 1959;36:402-411. DOI: 10.1016/0006-3002(59)90183-0

[104] Taylor SM, Parobek CM, Fairhurst RM. Haemoglobinopathies and the clinical epidemiology of malaria: a systematic review and meta-analysis. Lancet. Infect. Dis. 2012;12(6):457-468. DOI: 10.1016/S1473-3099(12)70055-5

[105] Williams TN, Mwangi TW, Wambua S, Alexander ND, Kortok M, Snow RW, Marsh K. Sickle cell trait and the risk of Plasmodium falciparum malaria and other childhood diseases. J. Infect. Dis. 2005a;192(1):178-186. DOI: 10.1086/430744

[106] Allison AC. Protection afforded by sickle-cell trait against subtertian malareal infection. Br. Med. J. 1954a;1(4857):290-294. DOI: 10.1136/bmj.1.4857.290

[107] Allison AC. Notes on sickle-cell polymorphism. Ann. Hum. Genet. 1954b;19(1):39-51. DOI: 10.1111/j.1469-1809.1954.tb01261.x

[108] Piel FB, Hay SI, Gupta S, Weatherall DJ, Williams TN. Global burden of sickle cell anaemia in children under five, 2010-2050: modelling based on demographics, excess mortality, and interventions. PLoS. Med. 2013;10(7):e1001484. DOI: 10.1371/journal.pmed.1001484

[109] Eridani S. Sickle cell protection from malaria. Hematol. Rep. 2011;3(3):e24. DOI: 10.4081/hr.2011.e24

[110] Valentine WN, Neil JV. Hematologic and genetic studies on the transmission of Thalassemia minor (Mediterranean anemia). Arch. Int. Med. 1944;74:185-196. DOI: 10.1001/archinte.1944.00210210032005

[111] Allison AC. Polymorphism and natural selection in human populations. Cold. Spring. Harb. Symp. Quant. Biol. 1964;29:137-149. DOI: 10.1101/sqb.1964.029.01.018

[112] Friedman MJ. Erythrocytic mechanism of sickle cell resistance to malaria. Proc. Natl. Acad. Sci. U S A. 1978;75(4):1994-1997. DOI: 10.1073/pnas.75.4.1994

[113] Luzzatto L, Pinching AJ. Commentary to R Nagel-Innate resistance to malaria: the intraerythrocytic cycle. Blood. Cells. 1990;16:340-347.

[114] Živković J, Najman S, Vukelić M, Stojanović S, Aleksić M,

Stanisavljević M, Najdanović J. Osteogenic effect of inflammatory macrophages loaded onto mineral bone substitute in subcutaneous implants. Arch. Biol. Sci. 2015;67(1):173-186. DOI:10.2298/ABS140915020Z

[115] Roth EF Jr, Friedman M, Ueda Y, Tellez I, Trager W, Nagel RL. Sickling rates of human AS red cells infected in vitro with Plasmodium falciparum malaria. Science. 1978;202(4368):650-652. DOI:10.1126/science.360396

[116] Baruch DI, Ma XC, Singh HB, Bi X, Pasloske BL, Howard RJ. Identification of a region of PfEMP1 that mediates adherence of Plasmodium falciparum infected erythrocytes to CD36: conserved function with variant sequence. Blood. 1997;90(9):3766-3775. DOI: 10.1182/blood.V90.9.3766

[117] Cholera R, Brittain NJ, Gillrie MR, Lopera-Mesa TM, Diakité SA, Arie T, Krause MA, Guindo A, Tubman A, Fujioka H, Diallo DA, Doumbo OK, Ho M, Wellems TE, Fairhurst RM. Impaired cytoadherence of Plasmodium falciparum-infected erythrocytes containing sickle hemoglobin. Proc. Natl. Acad. Sci. U S A. 2008;105(3):991-996. DOI: 10.1073/pnas.0711401105

[118] Archer NM, Petersen N, Clark MA, Buckee CO, Childs LM, Duraisingh MT. Resistance to Plasmodium falciparum in sickle cell trait erythrocytes is driven by oxygen-dependent growth inhibition. Proc. Natl. Acad. Sci. U S A. 2018;115(28):7350-7355. DOI: 10.1073/pnas.1804388115

[119] Baruch DI, Ma XC, Pasloske B, Howard RJ, Miller LH. CD36 peptides that block cytoadherence define the CD36 binding region for Plasmodium falciparum-infected erythrocytes. Blood. 1999;94(6):2121-2127. DOI: 10.1182/blood.V94.6.2121

[120] Urban BC, Ferguson DJ, Pain A, Willcox N, Plebanski M, Austyn JM, Roberts DJ. Plasmodium falciparum-infected erythrocytes modulate the maturation of dendritic cells. Nature. 1999;400(6739):73-77. DOI: 10.1038/21900

[121] Aitman TJ, Cooper LD, Norsworthy PJ, Wahid FN, Gray JK, Curtis BR, McKeigue PM, Kwiatkowski D, Greenwood BM, Snow RW, Hill AV, Scott J. Malaria susceptibility and CD36 mutation. Nature 2000;405(6790):1015-1016. DOI: 10.1038/35016636

[122] Duraisingh MT, Maier AG, Triglia T, Cowman AF. Erythrocyte-binding antigen 175 mediates invasion in Plasmodium falciparum utilizing sialic acid-dependent and -independent pathways. Proc. Natl. Acad. Sci. U S A. 2003;100(8):4796-4801. DOI: 10.1073/pnas.0730883100

[123] Milet J, Nuel G, Watier L, Courtin D, Slaoui Y, Senghor P, Migot-Nabias F, Gaye O, Garcia A. Genome wide linkage study, using a 250K SNP map, of Plasmodium falciparum infection and mild malaria attack in a Senegalese population. PLoS. One. 2010;5(7):e11616. DOI: 10.1371/journal.pone.0011616

[124] LaMonte G, Philip N, Reardon J, Lacsina JR, Majoros W, Chapman L, Thornburg CD, Telen MJ, Ohler U, Nicchitta CV, Haystead T, Chi JT. Translocation of sickle cell erythrocyte microRNAs into Plasmodium falciparum inhibits parasite translation and contributes to malaria resistance. Cell. Host. Microbe. 2012;12(2):187-199. DOI: 10.1016/j.chom.2012.06.007

[125] Piel FB, Patil AP, Howes RE, Nyangiri OA, Gething PW, Williams TN, Weatherall DJ, Hay SI. Global distribution of the sickle cell gene and geographical confirmation of the malaria hypothesis. Nat. Commun. 2010;1:104. DOI: 10.1038/ncomms1104

[126] Kilian N, Srismith S, Dittmer M, Ouermi D, Bisseye C, Simpore J,

Cyrklaff M, Sanchez CP, Lanzer M. Hemoglobin S and C affect protein export in Plasmodium falciparum-infected erythrocytes. Biol. Open. 2015;4(3):400-410. DOI: 10.1242/bio.201410942.

[127] Mackintosh CL, Beeson JG, Marsh K. Clinical features and pathogenesis of severe malaria. Trends. Parasitol. 2004;20(12):597-603. DOI: 10.1016/j.pt.2004.09.006

[128] Miller LH, Ackerman HC, Su XZ, Wellems TE. Malaria biology and disease pathogenesis: insights for new treatments. Nat. Med. 2013;19(2):156-167. DOI: 10.1038/nm.3073

[129] Roberts DJ, Williams TN. Haemoglobinopathies and resistance to malaria. Redox. Report. 2003; 8(5): 304-310. DOI: 10.1179/135100003225002998

[130] Baruch DI, Pasloske BL, Singh HB, Bi X, Ma XC, Feldman M, Taraschi TF, Howard RJ. Cloning the P. falciparum gene encoding PfEMP1, a malarial variant antigen and adherence receptor on the surface of parasitized human erythrocytes. Cell. 1995;82(1):77-87. DOI: 10.1016/0092-8674(95)90054-3

[131] Smith JD, Chitnis CE, Craig AG, Roberts DJ, Hudson-Taylor DE, Peterson DS, Pinches R, Newbold CI, Miller LH. Switches in expression of Plasmodium falciparum var genes correlate with changes in antigenic and cytoadherent phenotypes of infected erythrocytes. Cell. 1995;82(1):101-110. DOI: 10.1016/0092-8674(95)90056-x

[132] Baruch DI, Gormely JA, Ma C, Howard RJ, Pasloske BL. Plasmodium falciparum erythrocyte membrane protein 1 is a parasitized erythrocyte receptor for adherence to CD36, thrombospondin, and intercellular adhesion molecule 1. Proc. Natl. Acad. Sci. U S A. 1996;93(8):3497-3502. DOI: 10.1073/pnas.93.8.3497

[133] Carlson J, Helmby H, Hill AV, Brewster D, Greenwood BM, Wahlgren M. Human cerebral malaria: association with erythrocyte rosetting and lack of anti-rosetting antibodies. Lancet. 1990;336(8729):1457-1460. DOI: 10.1016/0140-6736(90)93174-n

[134] Kaul DK, Roth EF Jr, Nagel RL, Howard RJ, Handunnetti SM. Rosetting of Plasmodium falciparum-infected red blood cells with uninfected red blood cells enhances microvascular obstruction under flow conditions. Blood. 1991;78(3):812-819. DOI: 10.1182/blood.V78.3.812.812

[135] Raventos-Suarez C, Kaul DK, Macaluso F, Nagel RL. Membrane knobs are required for the microcirculatory obstruction induced by Plasmodium falciparum-infected erythrocytes. Proc. Natl. Acad. Sci. U S A. 1985;82(11):3829-3833. DOI:10.1073/pnas.82.11.382940

[136] Fried M, Duffy PE. Adherence of Plasmodium falciparum to chondroitin sulfate A in the human placenta. Science. 1996;272(5267):1502-1504. DOI: 10.1126/science.272.5267.1502

[137] Chen Q, Barragan A, Fernandez V, Sundström A, Schlichtherle M, Sahlén A, Carlson J, Datta S, Wahlgren M. Identification of Plasmodium falciparum erythrocyte membrane protein 1 (PfEMP1) as the rosetting ligand of the malaria parasite P. falciparum. J. Exp. Med. 1998;187(1):15-23. DOI: 10.1084/jem.187.1.15

[138] Tan X, Traore B, Kayentao K, Ongoiba A, Doumbo S, Waisberg M, Doumbo OK, Felgner PL, Fairhurst RM, Crompton PD. Hemoglobin S and C heterozygosity enhances neither the magnitude nor breadth of antibody responses to a diverse array of Plasmodium falciparum antigens. J. Infect. Dis. 2011;204(11):1750-1761. DOI: 10.1093/infdis/jir638

[139] Cabrera G, Cot M, Migot-Nabias F, Kremsner PG, Deloron P, Luty AJ. The sickle cell trait is associated with enhanced immunoglobulin G antibody responses to Plasmodium falciparum variant surface antigens. J. Infect. Dis. 2005;191(10):1631-1638. DOI: 10.1086/429832

[140] Cyrklaff M, Sanchez CP, Kilian N, Bisseye C, Simpore J, Frischknecht F, Lanzer M. Hemoglobins S and C interfere with actin remodeling in Plasmodium falciparum-infected erythrocytes. Science. 2011;334(6060):1283-1286. DOI: 10.1126/science.1213775

[141] Miller LH, Baruch DI, Marsh K, Doumbo OK. The pathogenic basis of malaria. Nature. 2002;415(6872):673-679. DOI: 10.1038/415673a

[142] O'Meara WP, Bejon P, Mwangi TW, Okiro EA, Peshu N, Snow RW, Newton CR, Marsh K. Effect of a fall in malaria transmission on morbidity and mortality in Kilifi, Kenya. Lancet. 2008;372(9649):1555-1562. DOI: 10.1016/S0140-6736(08)61655-4

[143] O'Shaughnessy DF, Hill AVS, Bowden DK, Weatherall DJ, Clegg JB. Globin genes in Micronesia: Origins and affinities of Pacific Island peoples. Am. J. Hum. Genet. 1990;46:144-155.

[144] Pasvol G, Weatherall DJ, Wilson RJ. Cellular mechanism for the protective effect of haemoglobin S against P. falciparum malaria. Nature. 1978;274(5672):701-703. DOI: 10.1038/274701a0

[145] Luzzatto L, Nwachuku-Jarrett ES, Reddy S. Increased sickling of parasitised erythrocytes as mechanism of resistance against malaria in the sickle-cell trait. Lancet. 1970;1(7642):319-321. DOI: 10.1016/s0140-6736(70)90700-2

[146] Shear HL, Roth EF Jr, Fabry ME, Costantini FD, Pachnis A, Hood A, Nagel RL. Transgenic mice expressing human sickle hemoglobin are partially resistant to rodent malaria. Blood. 1993;81(1):222-226. DOI: 10.1182/blood.V81.1.222.222

[147] Le Hesran JY, Personne I, Personne P, Fievet N, Dubois B, Beyemé M, Boudin C, Cot M, Deloron P. Longitudinal study of Plasmodium falciparum infection and immune responses in infants with or without the sickle cell trait. Int. J. Epidemiol. 1999;28(4):793-798. DOI: 10.1093/ije/28.4.793

[148] Williams TN, Mwangi TW, Roberts DJ, Alexander ND, Weatherall DJ, Wambua S, Kortok M, Snow RW, Marsh K. An immune basis for malaria protection by the sickle cell trait. PLoS. Med. 2005b;2(5):e128. DOI: 10.1371/journal.pmed.0020128

[149] Ferreira A, Marguti I, Bechmann I, Jeney V, Chora A, Palha NR, Rebelo S, Henri A, Beuzard Y, Soares MP. Sickle hemoglobin confers tolerance to Plasmodium infection. Cell. 2011;145: 398-409. DOI: 10.1016/j.cell.2011.03.049

[150] Pai GS, Sprenkle JA, Do TT, Mareni CE, Migeon BR. Localization of loci for hypoxanthine phosphoribosyltransferase and glucose-6-phosphate dehydrogenase and biochemical evidence of nonrandom X chromosome expression from studies of a human X-autosome translocation. Proc. Natl. Acad. Sci. U S A. 1980;77(5):2810-2813. DOI: 10.1073/pnas.77.5.2810

[151] Luzzatto L, Nannelli C, Notaro R. Glucose-6-Phosphate Dehydrogenase Deficiency. Hematol. Oncol. Clin. North. Am. 2016;30(2):373-393. DOI: 10.1016/j.hoc.2015.11.006.

[152] Verrelli BC, McDonald JH, Argyropoulos G, Destro-Bisol G, Froment A, Drousiotou A, Lefranc G, Helal AN, Loiselet J, Tishkoff SA. Evidence for balancing selection from nucleotide sequence analyses of human

G6PD. Am J Hum Genet. 2002;71(5):1112-1128. DOI: 10.1086/344345

[153] Frank JE. Diagnosis and management of G6PD deficiency. Am. Fam. Physician. 2005;72(7):1277-1282

[154] Ruwende C, Hill A. Glucose-6-phosphate dehydrogenase deficiency and malaria. J. Mol. Med. (Berl.). 1998;76(8):581-588. DOI: 10.1007/s001090050253

[155] Ruwende C, Khoo SC, Snow RW, Yates SN, Kwiatkowski D, Gupta S, Warn P, Allsopp CE, Gilbert SC, Peschu N, Newbold CI, Greenwood BM, Marsh K, Hill AVS. Natural selection of hemi- and heterozygotes for G6PD deficiency in Africa by resistance to severe malaria. Nature. 1995;376(6537):246-249. DOI: 10.1038/376246a0

[156] Ganczakowski M, Town M, Bowden DK, Vulliamy TJ, Kaneko A, Clegg JB, Weatherall DJ, Luzzatto L. Multiple glucose 6-phosphate dehydrogenase-deficient variants correlate with malaria endemicity in the Vanuatu archipelago (southwestern Pacific). Am. J. Hum. Genet. 1995;56(1): 294-301

[157] Siniscalco M, Bernini L, Latte B, Motulski AG. Favism and thalassaemia in Sardinia and their relationship to malaria. Nature. 1961;190:1179-1180. DOI: 10.1038/1901179a0

[158] Chen EY, Cheng A, Lee A, Kuang WJ, Hillier L, Green P, Schlessinger D, Ciccodicola A, D'Urso M. Sequence of human glucose-6-phosphate dehydrogenase cloned in plasmids and a yeast artificial chromosome. Genomics. 1991;10(3):792-800. DOI: 10.1016/0888-7543(91)90465-q

[159] Beutler E. Glucose-6-phosphate dehydrogenase deficiency. N. Engl. J. Med. 1991; 17;324(3):169-174. DOI: 10.1056/NEJM199101173240306

[160] Luzzatto L, Mehta A. Glucose-6-phosphate dehydrogenase deficiency," in The Metabolic of Inherited Disease, C. R. Scriver, A. L. Baudet, W. S. Sly, and D Valle, Eds., pp. 2237-2265, McGraw-Hill, New York, NY, USA, 6th edition, 1989.

[161] Vulliamy TJ, D'Urso M, Battistuzzi G, Estrada M, Foulkes NS, Martini G, Calabro V, Poggi V, Giordano R, Town M. Diverse point mutations in the human glucose-6-phosphate dehydrogenase gene cause enzyme deficiency and mild or severe hemolytic anemia. Proc. Natl. Acad. Sci. U S A. 1988;85(14):5171-5175. DOI: 10.1073/pnas.85.14.5171

[162] Beutler E. The genetics of glucose-6-phosphate dehydrogenase deficiency. Semin. Hematol. 1990;27(2):137-164.

[163] Barišić M, Korać J, Pavlinac I, Krželj V, Marušić E, Vulliamy T, Terzić J. Characterization of G6PD deficiency in southern Croatia: description of a new variant, G6PD Split. J. Hum. Genet. 2005;50(11):547-549. DOI: 10.1007/s10038-005-0292-2

[164] Calabrò V, Mason PJ, Filosa S, Civitelli D, Cittadella R, Tagarelli A, Martini G, Brancati C, Luzzatto L. Genetic heterogeneity of glucose-6-phosphate dehydrogenase deficiency revealed by single-strand conformation and sequence analysis. Am. J. Hum. Genet. 1993;52(3):527-536

[165] Nkhoma ET, Poole C, Vannappagari V, Hall SA, Beutler E. The global prevalence of glucose-6-phosphate dehydrogenase deficiency: a systematic review and meta-analysis. Blood. Cells. Mol. Dis. 2009;42(3):267-278. DOI: 10.1016/j.bcmd.2008.12.005.

[166] Beutler E. G6PD: population genetics and clinical manifestations. Blood. Rev. 1996;10(1):45-52. DOI: 10.1016/s0268-960x(96)90019-3

[167] World Health Organization. World Malaria Report 2017: 196. (2017).

[168] Dean L. Blood Groups and Red Cell Antigens [Internet]. Bethesda (MD): National Center for Biotechnology Information (US); 2005. Available from: https://www.ncbi.nlm.nih.gov/books/NBK2261/&usg=AL/

[169] Miller LH, Mason SJ, Clyde DF, McGinniss MH. The resistance factor to Plasmodium vivax in blacks. The Duffy-blood-group genotype, FyFy. N. Engl. J. Med. 1976;295(6):302-304. DOI: 10.1056/NEJM197608052950602

[170] Chaudhuri A, Polyakova J, Zbrzezna V, Williams K, Gulati S, Pogo AO. Cloning of glycoprotein D cDNA, which encodes the major subunit of the Duffy blood group system and the receptor for the Plasmodium vivax malaria parasite. Proc. Natl. Acad. Sci. U S A. 1993;90(22):10793-10797. DOI: 10.1073/pnas.90.22.10793

[171] VanBuskirk KM, Sevova E, Adams JH. Conserved residues in the Plasmodium vivax Duffy-binding protein ligand domain are critical for erythrocyte receptor recognition. Proc. Natl. Acad. Sci. U S A. 2004;101:15754-15759. DOI: 10.1073/pnas.0405421101

[172] Collins A, Keats BJ, Dracopoli N, Shields DC, Morton NE. Integration of gene maps: chromosome 1. Proc. Natl. Acad. Sci. U S A. 1992;89(10):4598-4602. DOI: 10.1073/pnas.89.10.4598

[173] Iwamoto S, Omi T, Kajii E, Ikemoto S. Genomic organization of the glycoprotein D gene: Duffy blood group Fya/Fyb alloantigen system is associated with a polymorphism at the 44-amino acid residue. Blood. 1995;85(3):622-626. DOI: 10.1182/blood.V85.3.622.bloodjournal853622

[174] Parasol N, Reid M, Rios M, Castilho L, Harari I, Kosower NS. A novel mutation in the coding sequence of the FY*B allele of the Duffy chemokine receptor gene is associated with an alterederythrocyte phenotype. Blood. 1998;92(7):2237-2243. DOI: 10.1182/blood.V92.7.2237

[175] Mallinson G, Soo KS, Schall TJ, Pisacka M, Anstee DJ. Mutations in the erythrocyte chemokine receptor (Duffy) gene: the molecular basis of the Fya/Fyb antigens and identification of a deletion in the Duffy gene of an apparently healthy individual with the Fy(a-b-) phenotype. Br. J. Haematol. 1995;90(4):823-829. DOI: 10.1111/j.1365-2141.1995.tb05202.x

[176] Tournamille C, Colin Y, Cartron JP, Le Van Kim C. Disruption of a GATA motif in the Duffy gene promoter abolishes erythroid gene expression in Duffy-negative individuals. Nat. Genet. 1995 Jun;10(2):224-228. doi: 10.1038/ng0695-224. PMID: 7663520

[177] Aldarweesh F. The Duffy Blood Group System. In: Erhabor O, Munshi A, editors. Human Blood Group Systems [Online First]. London: IntechOpen Limited; 2019. DOI: 10.5772/intechopen.89952

[178] Carter R. Speculations on the origins of Plasmodium vivax malaria. Trends. Parasitol. 2003;19:214-219. DOI: 10.1016/S1471-4922(03)00070-9

[179] Adams JH, Sim BK, Dolan SA, Fang X, Kaslow DC, Miller LH. A family of erythrocyte binding proteins of malaria parasites. Proc. Natl. Acad. Sci. U S A. 1992;89(15):7085-7089. DOI: 10.1073/pnas.89.15.7085

[180] Rayner JC, Vargas-Serrato E, Huber CS, Galinski MR, Barnwell JW. A Plasmodium falciparum homologue of Plasmodium vivax reticulocyte binding protein (PvRBP1) defines a trypsin-resistant erythrocyte invasion pathway.

J. Exp. Med. 2001;194(11):1571-1181. DOI: 10.1084/jem.194.11.1571

[181] Mendis K, Sina BJ, Marchesini P, Carter R. The neglected burden of Plasmodium vivax malaria. Am. J. Trop. Med. Hyg. 2001;64(1-2 Suppl):97-106. DOI: 10.4269/ajtmh.2001.64.97

[182] Kaschagen LJ, Mueller I, Kiniboro B, Bockarie MJ, Reeder JC, Kazura JW, Kastens W, McNamara DT, King CH, Whalen CC, Zimmerman PA. Reduced *Plasmodium vivax* erythrocyte infection in PNG Duffy-negative heterozygotes. PLoS. One. 2007; 28;2(3):e336. DOI: 10.1371/journal.pone.0000336

[183] Grimberg BT, Udomsangpetch R, Xainli J, McHenry A, Panichakul T, Sattabongkot J, Cui L, Bockarie M, Chitnis C, Adams J, Zimmerman PA, King CL. Plasmodium vivax invasion of human erythrocytes inhibited by antibodies directed against the Duffy binding protein. PLoS. Med. 2007;4(12):e337. DOI: 10.1371/journal.pmed.0040337

[184] Howes RE, Patil AP, Piel FB, Nyangiri OA, Kabaria CW, Gething PW, Zimmerman PA, Barnadas C, Beall CM, Gebremedhin A, Ménard D, Williams TN, Weatherall DJ, Hay SI. The global distribution of the Duffy blood group. Nat. Commun. 2011;2:266. DOI: 10.1038/ncomms1265.

[185] King CL, Adams JH, Xianli J, Grimberg BT, McHenry AM, Greenberg LJ, Siddiqui A, Howes RE, da Silva-Nunes M, Ferreira MU, Zimmerman PA. Fy(a)/Fy(b) antigen polymorphism in human erythrocyte Duffy antigen affects susceptibility to Plasmodium vivax malaria. Proc. Natl. Acad. Sci. U S A. 2011;108(50):20113-20118. DOI: 10.1073/pnas.1109621108

[186] Ryan JR, Stoute JA, Amon J, Dunton RF, Mtalib R, Koros J, Owour B, Luckhart S, Wirtz RA, Barnwell JW, Rosenberg R. Evidence for transmission of Plasmodium vivax among a duffy antigen negative population in Western Kenya. Am. J. Trop. Med. Hyg. 2006;75(4):575-581.

Chapter 10

Co-Evolution between New Coronavirus (SARS-CoV-2) and Genetic Diversity: Insights on Population Susceptibility and Potential Therapeutic Innovations

Mahmood A. Al-Azzawi and Moustafa A. Sakr

Abstract

The DNA sequences are different between the distinct individuals and these variations produce the species genetic diversity. SARS-CoV-2 virus is a zoonotic SARS-like coronavirus that spreads globally, causing the COVID-19 pandemic disease. The immune response genes are the most various and different in the human genome, correlating with infectious diseases. Genetic variants in the angiotensin-converting enzyme 2 (ACE2) receptor, TMPRSS2, HO-1, BCL11A, and CYP2D6 are predicted to either encourage or inhibit the interaction with the viral proteins and subsequently contribute to coronavirus genetic risk factors. The genetic susceptibility to SARS-CoV-2 was investigated by analyzing different genes' polymorphisms such as ACE2 and TMPRSS2, HO-1, and BCL11A. A specific genetic susceptibility to COVID-19 was found through different populations in TMPRSS2, ACE2, HO-1, and BCL11A genes. Particularly, ACE2 gene polymorphisms were shown to be correlated with pulmonary and cardiovascular conditions by modifying the angiotensinogen-ACE2 system, which recommends the possible explanations of COVID-19 susceptibility based on genetic diversity. Moreover, the COVID-19 treatment could be complicated by such genetic polymorphisms. In conclusion, a good characterization of functional polymorphisms and the host genetics can assist in identifying the pathophysiology of the disease pathway to stratify the risk evaluation and to personalize the treatment procedures.

Keywords: gene polymorphisms, infectious diseases, host genetics, SARS-CoV-2, TMPRSS2, HO-1, ACE-2, BCL11A, coronavirus, COVID-19

1. Introduction

Infectious diseases have been and continue to be a source of concern and intimidation for human and animal life, and due to the absence of effective strategies in disease control, epidemics appear and spread day after day and cause a significant increase in mortality. Over decades, genetic and genomic studies provided invulnerable evidence that the host showed a genetic variation in its response to infectious agents, that may otherwise affect epidemiological risks, morbidity, and survival [1–4]. Determining the host genetic implications in the risk of the epidemic and its severity remains the biggest obstacle to the infectious disease research progression [5, 6]. Because of the large size of the samples required by quantitative genetic studies, the definition of disease resistance based on individual mortality must be changed because it is easy in any case to know if the subjects' mortality was happening due to the exposition to infectious diseases or not. But, this is not true in the case of survival because it is multisided, and it may depend not only on an individual's resistance to infectious agents but also on his ability to survive after getting a disease or infection [7, 8].

Obviously, interest has increased in the infectivity genetic regulation, which can be described as the capability of a pathogen to infect an individual upon contact. Comprehension of the genetic regulation of infectivity is especially relevant if there are contrary genomic associations between these traits and elements of tolerance or resistance [9–11]. Such unfavorable genetic associations could be arising if subjects with much genetic survival not only come over with infection but also have a tendency to shed more pathogens [12]. Endurance and resistance infectivity may be controlled by several gene sets with variable contributions, both in degree and direction for survival [7, 13]. Despite this, no study has investigated these three traits at the same time. It is worth noting that plenty of quantitative genetic studies revealed variation in genetic resistance [2, 14–16], however, only a few studies showed a genetic difference in disease survival [7, 8]. In the context of infectious diseases, genomic selection may definitely restrict the spread of the disease by implementing a mechanism for determining high-risk people of infection [1].

Almost two decades after the outset of the Severe Acute Respiratory Syndrome (SARS), produced by a beta coronavirus, recently called SARS-CoV-1, the world was surprised by the emergence of a more virulent and infectious new virus in late 2019. This virus soon spread to almost all parts of the world and quickly reached the epidemic disease state [17]. The new coronavirus 2019 (COVID-19) outbreak originated from the SARS-CoV-2 virus suddenly became a major public health threat. COVID-19 is characterized by different types of clinical characterizations: affected patients can be asymptomatic, symptomatic with mild respiratory symptoms, or manifest severe pneumonia [18–21]. It is noted that these estimations are variable and began to approach accuracy as more cases are described, examined, and analyzed. Curiously enough, there is a clear difference in these estimations among different countries, worthy to mention that, the differences in the severity of the virus were recorded between the sexes and different age categories [18, 20, 22]. The infected cases have increased drastically [23]. Transmission from one person to another has been confirmed [24]. The virus was discovered in Bronchoalveolar lavage (BAL) [22], saliva and nasopharyngeal swabs [25], sputum [26], and throat [27, 28]. Even though the number of patients with COVID-19 was asymptomatic or mildly symptomatic still indecisive until now, but some studies have suggested that the percentage is between 40 and 80% [29, 30].

Among the most debatable characteristics in the clinical course and pathogenesis of COVID-19 is the heterogeneous hazard in the development to the acute form. Some significant clinical factors have been specified as severe disease predictors in different populations around the world, essentially include old age, male sex, obesity, and presence of multiple co-morbidities, such as diabetes mellitus, hypertension (HTN), cardiovascular disease, and impaired liver and renal function [20, 31–33]. In fact, some patients continue completely without symptoms until the final viral shedding, however, others experience a highly aggressive form of the disease [34–39]. These severe cases in the clinical picture of COVID-19 firmly propose that other co-factors may have a vital role in modifying disease development and progression. The suppressed immune response in the elders, co-morbidities, or smoking condition, may explain the variances in the COVID-19 disease severity between individuals and populations [40], but severe disease has also been detected in young persons, apparently free from these risk factors. This shows that most risk factors clarifying COVID-19 disease severity are yet mysterious. Therefore, to recognize the mechanisms beyond COVID-19 disease severity is critical to provide suitable protective measures and sufficient triage approaches, drug innovation processes, and eventually the pandemic control. The genetic diversity between hosts can be explained the big difference in the incidence of SARS CoV-2 rates and the severity of COVID 19.

In this chapter, we will focus on some genetic variants and their implications for the severity of COVID-19. From these genes, we will take the consideration of the ACE2, TPRSS2, HO-1, and BCL11A genes, and the association between the DNA polymorphisms of these genes with the genetic susceptibility of the COVID-19, Whereas, systematic investigation of the functional polymorphism in these genes among diverse populations could tile the way for reliable medicine and personalized treatment approaches for COVID-19, this will call genetics to take the initiative in combating the virus pandemic.

2. Pathways of cellular infection by SARS-CoV-2

SARS-CoV-1 and SARS-CoV-2 connect to a similar receptor on the surface of human cells, known as angiotensin-converting enzyme 2 (ACE2) [41]. This complex particularly includes the receptor-binding domain (RBD) positioned within the virus spike protein (S protein). However, recent laboratory studies have revealed that unlike SARS-CoV-1, the SARS-CoV-2 RBD favors creating a greater binding capacity (i.e. 1204 versus 998 Å) [41, 42]. The SARS-CoV-2 infects and enters the infected cell by binding the viral spike protein with ACE2 of the host cell through the RBD. Even so, the splitting of spike protein needs to be done by human protease, where S protein subunits (S1 and S2) are broken apart from each other, with the last domain undergoes considerable structural modifications necessary to bind with the cell membrane of the host cell [43]. The transmembrane serine

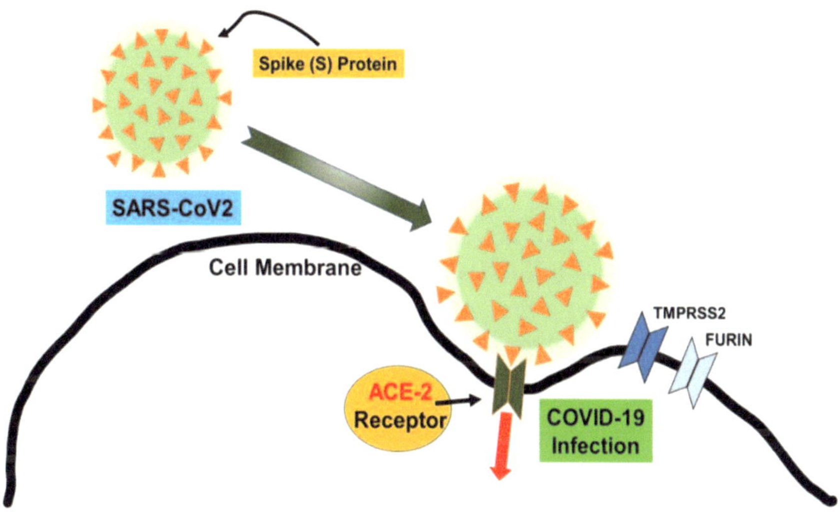

Figure 1.
Illustration of the COVID-19 virus spike protein. Across ACE-2 receptors, the spike invades the cell. Afterward, the spike is cleaved by the host cell proteases, membrane protease 2 (TMPRSS2), and furin, which results in COVID-19 infection activation [48].

protease 2 (TMPRSS2), together with lysosomal cathepsins, considers one of the most crucial proteases in this approach [44]. Moreover, a type 1 membrane-bound enzyme (furin), also splits the site between SARS-CoV-2 spike protein (both S1 and S2 subunits). Most significantly, furin can be expressed in numerous organs, involving the lungs. Furin stimulates the splitting of spike protein (S1/S2) after the binding of SARS-CoV-2 to ACE2 receptor, and this stimulation by itself is necessary to enter the virus into the cell [45]. This different pathway, which includes furin-mediate activation, would allow SARS-CoV-2 to be less dependent on co-expressions of TMPRSS2 on the cell surface of the infect cells. Hence, SARS-CoV-2 could be able to enter a wide range of low TMPRSS2 expressing cells. Lastly, disintegrin and metalloproteinase domain-containing protein 17 (ADAM17) stimulate the release of ectodomains for a number of transmembrane proteins, such as ACE2 [46]. Therefore, increased ADAM17 activity is thought to be correlated with increased shedding of ACE2 and eventually decreases the possibility of cellular entry by SARS-CoV-2 [47] (**Figure 1**).

3. ACE2 expression in human tissues

The expression of ACE2 in the different human tissues was controversial because ACE2 was newly identified as a major binding site across which SARS-CoV-2 enters human host cells. Recently, many studies were performed to detect the cell types where ACE2 receptor is mainly expressed, which could describe the possible SARS-CoV-2 targets. One study was conducted to address the expression of ACE2 in various natural human tissues, and the analysis of the results regarding age and sex. Highest ACE2 expression levels were detected in the tissues of the small intestine, testicle, thyroid heart, adipose tissues, and kidneys. Esophagus, pancreas, lungs, liver, adrenal gland bladder, and colon were found to express the intermediate level while the lowest expression was found in the stomach nerves, blood vessels, uterus, muscle, spleen, bone marrow, and brain. Regarding lungs, the levels of ACE2 expression were upregulated and downregulated in relation to the immune pattern of men and women respectively [49]. ACE2 also was expressed in certain types of epithelial cells in the airway, such as type II alveolar epithelial cells and ciliated nasal epithelium. Moreover, it was found to be highly co-expressed with the TMPRSS2 in the nasal epithelium, which explains their higher infectivity by COVID-19 [50]. ACE2 is localizing also on the oral cavity mucosa. For now, these results revealed the underlying mechanism that the oral cavity poses a significant potential risk for 2019-nCoV susceptibility, and ACE2 was also expressed in lymphocytes inside the oral mucosa [51]. These findings have reminded us that COVID-19 attacks the lymphocytes and causes lymphopenia, mostly in severe forms of the disease [52].

More importantly, ACE2 also are expressed in endothelial cells [53]. That explains why COVID-19 disease affects multiorgan in the patients [54]. these results indicate that SARS-CoV-2 virus promotes the initiation of endotheliitis in many organs as a direct result of the viral intervention and the inflammatory response of the host. Additionally, the triggering of pyroptosis and apoptosis may have an important role in endothelial cell injury in COVID-19 patients and can account for the weakened systemic microcirculatory performance in various blood vessels and their clinical consequences in COVID-19 patients [55]. This supposition affords justifications for treatments to stabilize the endothelium during viral reproduction, especially by anti-inflammatory cytokines drugs, cholesterol-lowering drugs, and ACE inhibitors [56–59]. This approach can be especially appropriate for weak patients with an earlier endothelial disorder, such as hypertension, diabetes mellitus, obesity, cardiovascular disease co-morbidities patients [55].

4. Implication of human polymorphism of ACE2 in disease susceptibility

A lot of ACE2 variants have been recognized in different databases [60, 61]. over the last decades, much focus has been assigned on some of ACE2 polymorphisms, due to their effects on the development of cardiovascular disease (CVD) and, more specifically, their association with hypertension (HT). ACE2 restricts the negative profibrotic and vasoconstrictor influences of AngII, as the breakdown of AngII to Ang (1-7) decreases the AngII oxidative stress of the cerebral arteries endothelium [62]. Ang (1-7) has been stated to have antifibrotic and vasodilation [63, 64]. Low cardiac expression of ACE2 levels has been notified in hypertension and diabetes heart failure [65, 66]. ACE2 gene polymorphisms were first detected in the Chinese people with different ACE2 variants (rs4830542, rs4240157, and rs4646155) linked to hypertension (HT) [67–70]. Also, ACE2 SNP rs21068809 (C > T) was found to be linked to the clinical features of HT [71]. In India, a study of 246 patients with HT and 274 normal subjects showed a connection of ACE2 rs21068809 SNP with HT [72]. in Brazilian cohorts, a study of genetic association of the combination of ACE2 G8790A and ACE I/D polymorphisms reveal susceptibility to HT [73]. ACE polymorphism has been described in African- Americans with HT [74].

5. Viral ACE2 receptor polymorphism and coronavirus infection

ACE2 gene variants are still possible to affect SARS-CoV-2 infectivity. In SARS-CoV, the function of the S1 domain of the S protein is to mediate the binding of ACE2 receptors while the S2 domain is potentially undergoing post binding trans-conformational modulations which activate the fusion to the cell membrane [75]. The viral (RBD) found in S1 has been adjusted to amino acid number 270 to 510 [76]. The Leu584Ala point mutation of ACE2 significantly weakened the shedding activity of the enzyme and promoted the entrance of SARS-CoV into the host cells [77]. An ACE2 soluble form lacks the transmembrane and cytoplasmic domain was stated able to prevent SARS-CoV S protein binding to ACE2 [46]. Recombinant SARS-CoV-2 spike proteins were observed to downregulated ACE2 expression by releasing sACE2 and thus enhancing injury of the lung [78]. SARS-CoV and SARSCoV-2 participate in the identity of 76% of the amino acid residues necessary for binding of ACE2 within the SARS-CoV-2 spike S1 domain. A lot of amino acid residues of the ten human ACE2 proteins were compared by multiple sequence alignment, a 100% identity among the ACE2 sequences was observed in four different ACE2 isoforms. The role of these ACE2 isoforms remains unpredictable in SARS-CoV-2 infection outcome. According to the work by Cao et al., [61] 32 polymorphisms of ACE2, including 7 hotspot variables (Ile486Val, Lys26Arg, Asn638Ser, Asn720Asp, Ser692Pro, Ala627Val, and Leu731Ile/Phe) were identified in different peoples, that make some individuals could be more or less susceptible to the virus than others.

In a preliminary study, the distribution of the allele frequency for 1700 polymorphisms in the ACE2 gene was conducted between various populations of the world. What is noteworthy is that 11 common and rare variants were detected linked to the high ACE2 expression. It was observed that their expression is irregularly distributed among different populations groups. This study found that the polymorphism of the ACE2 gene (variant 4,646,127) was closely related to the higher expression levels of the ACE2 gene in the East Asian population, and this paved the way to study this important issue more specifically [61]. These results were confirmed by a similar subsequent study by [79], which also evidenced that the allele frequency of these variants associated with overexpression of ACE2. Also, different ACE2 polymorphisms encoded a number of proteins for SARS-CoV-2 spike protein has been studied, and it was found that each variant differs in compatibility with RBD sequence. Specifically, although the majority of genetic variants exhibited high physical similarity. Specifically, the two ACE2 gene alleles (rs143936283 and rs73635825) showed a quite low binding strength for the SARS-CoV-2 spike protein, which could mean a lower possibility of viral binding and possible to infection resistance [80]. It has been observed that the probability of some natural genetic variants of ACE2, particularly those assigned to attach with the SARS-CoV-2 spike protein, may be linked with flexible virus-host interaction, thus likely modifying severity and pathogenicity. A large analysis of the genome data-set was performed and showed that no less than nine human ACE2 variants (E23K, S19P, I21V, N64K, K26R, H378R, T27A, T92I, and Q102P) are prospective to increase predisposition to viral binding, while 17 other variants of ACE2 (that is, E37K, K31R, H34R, N33I, E35K, Y50F, D38V, G326E, N51S, M62V, D355N, K68E, F72V, Y83H, D509Y, G352V, and Q388L) were thought to be protected from viral entry, where they demonstrated a lower binding tendency to SARS-CoV-2 spike protein [81].

In another study, from five separate Italian centers, the authors found that three variants of ACE2 can be specified (p. Gly211Arg, lys26Arg, and p. Asn720Asp). It was noted that these three polymorphisms were recurrently identified in the Italian population rather than the East Asian population. These variants are closely located in the SARS-CoV-2 essential sequence of spike protein binding sites and therefore viral entry and division expected to be modified (for example, Asn720Asp is located on only 4 amino acids of TMPRSS2 cleavage site) [82]. This may tell a partial explanation for the high case mortality rate registered in Italy by comparison to China. Despite ACE2 practically serve as a receptor for coronavirus SARS entry into human host cells, another does not support the correlation between its common gene polymorphisms and receptivity or consequence of SARS [83]. It has also been observed that some ACE2 variants show differential efficacy in stimulating neutrophils, monocytes, natural killer cells (NK), macrophages, and T helper cells, thus

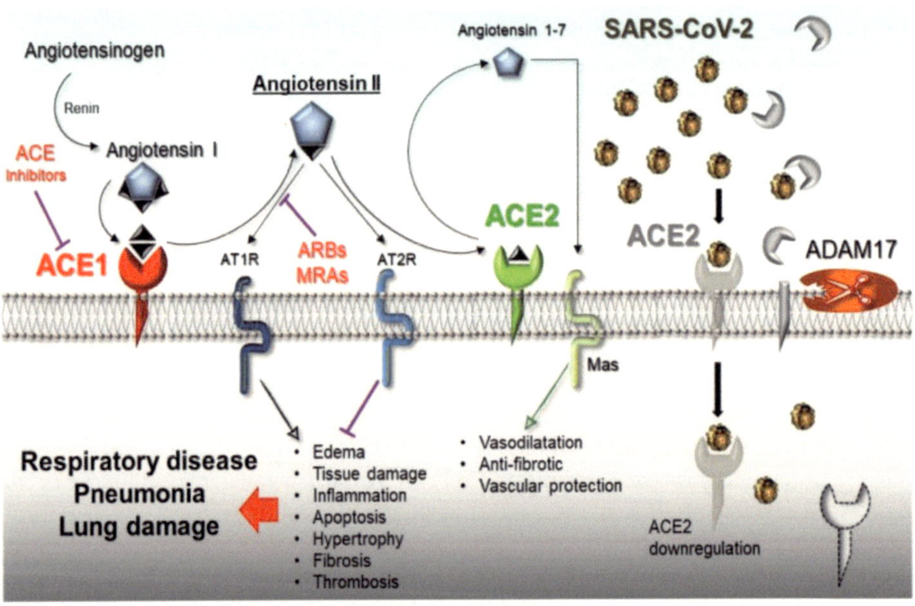

Figure 2.
Diagrammatic representation for the renin-angiotensin system (RAS) pathway. As ACE2/Ang 1-7/ Mas-axis and ACE1/Ang-II/ AT1R-axis occur, SARS-CoV-2 inhibition by cleavage of ACE2 by ADAM17 appears. ***ADAM17:*** *ADAM metallopeptidase domain 17;* ***ARBs:*** *angiotensin receptor blockers;* ***MRAs:*** *mineralocorticoid receptor antagonists [87].*

may probably either enhance or reduce the inflammatory or "cytokine storm" [84], in addition to stimulating the processing of Ang II, thereby improving or exacerbating vasoconstriction and participating to the improvement or exacerbation of topical or systemic tissue infection [85, 86] (**Figure 2**).

6. TMPRSS2 polymorphism analysis with COVID-19 disease

TMPRSS2 and ACE2 have been associated with SARS-corona (CoV) disease, influenza, and SARS-CoV-2 in facilitating viral entrance into the infected host cell TMPRSS2 considers as an androgen-reactive serine protease enzyme that cleaves SARS-CoV-2 Spike protein, mediating viral activation and entry [88]. Single-nucleotide polymorphisms of TMPRSS2 enzyme have been studied in several diseases such as in breast cancer, the rs2276205 (A > G) with low-frequency allele was correlated with increased patients' endurance [89]. In prostate cancer, the rs12329760 (C > T) of TMPRSS2 has a higher frequency in men with prostate cancer in his family, while ERG gene fusion [90, 91] Rs383510 (T > C) and rs2070788

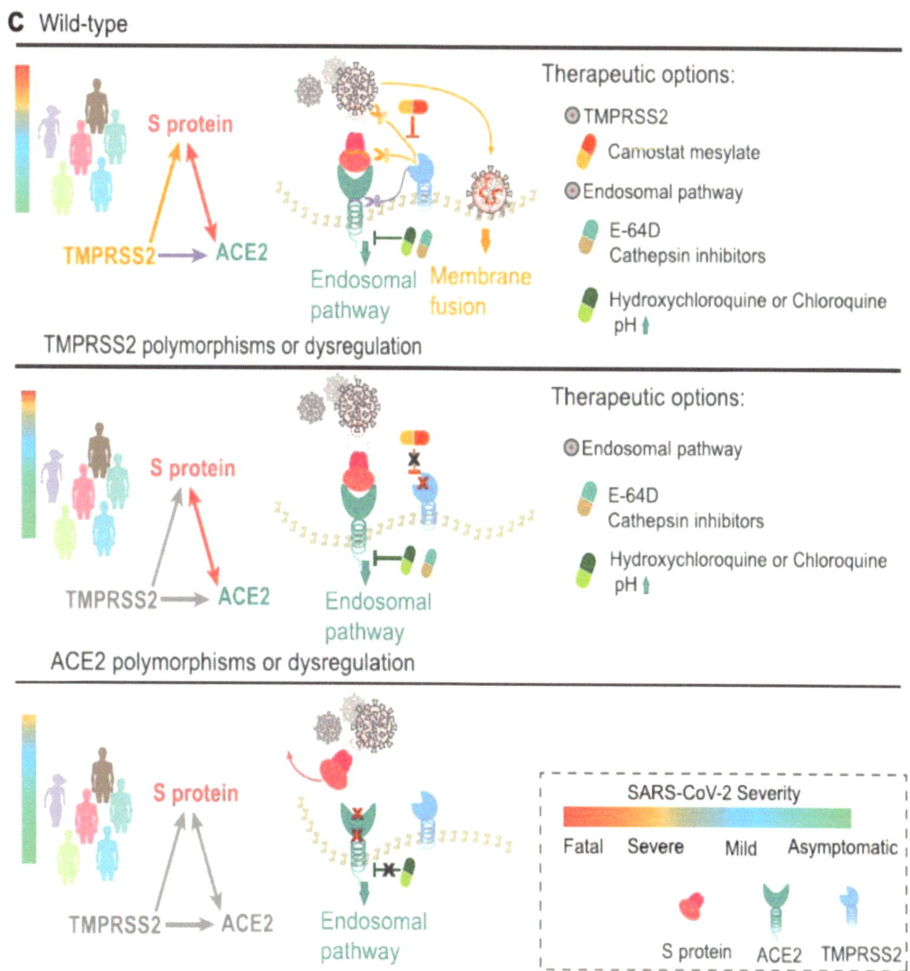

Figure 3.
A polymorphism and dysregulation of ACE2, and TMPRSS2 in COVID-19 and a suggested model for active compound medicines (e.g., hydroxychloroquine, Camostat mesylate, and E-64D [a protease inhibitor] for COVID-19) [93].

(G > A) were correlated with aggressive H7N9, H1N1, and increased lung expression of TMPRSS2 [92]. A study by Hou et al., indicated that 4% of nonidentical variants of TMPRSS2 are stop-codon mutations, Meanwhile, 59% are harmful mutations in TMPRSS2 coding regions [93]. The harmful variants (p.Arg240Cys, p.Val160Met, p.Gly181Arg, p.Pro335Leu, p.Gly432Ala, and p.Gly259Ser) in the coding region of TMPRSS2, are the same with somatic alterations arising in various types of cancer. In the same contest, Hou et al. found that, the p. Asp435Tyr which is a key site for catalytic residue binding of TMPRSS2 has unique low-frequency allele, but predominant SNPs in TMPRSS2 and offer possible descriptions for differential genetic infectivity to COVID-19 and for risk influences, such as those with tumor and male patients. By using the analysis of single-cell RNA-seq, Schuler et al. revealed that the expression of TMPRSS2 was upregulated in ciliated cells and alveolar epithelial type 1 cells and increased with humans aging [94]. This observation indicates that the developmental TMPRSS2 expression regulation may have a role in the relative protection of the children and infants from COVID-19 infection. Yet, it might be of great importance to investigate the link between TMPRSS2 polymorphisms and the age relationship with COVID-19 susceptibility (**Figure 3**).

7. Heme oxygenase-1 enzyme (HO-1) genetic polymorphisms and COVID-19 severity

Many studies demonstrated that the HO-1 gene polymorphisms, particularly the promoter region GT dinucleotide repeat mutation regulates the inducibility of HO-1 to ROS [95–101]. Subjects with more GT repeats have been believed to be more sensitive to cardiovascular endothelium diseases such as atherosclerosis coronary artery disease and aortic aneurysm s [95, 98, 99]. The lower Expression level of HO-1 in those with more GT repeats make the patients to be more affected to decrease endothelial hemostasis and inflammation [95–101]. While, GT sequences short alleles are correlated with increased HO-1 inducibility, which in turn reduced inflammation and enhanced cytoprotection [101]. Patients with COVID-19 complications perhaps have longer GT sequences and decreased vessel hemostasis.

COVID-19 disease has poor effects in diabetic and obese individuals, maybe because those people are already having high interleukin 6 levels of (IL-6) and they are in a proinflammatory state due to leptin and insulin resistance [102, 103]. As a result, the negative clinical outcomes of COVID-19 infection in obese patients was recorded [103]. Peterson et al. have revealed that obesity raises high-density lipoprotein (HDL) oxidation [104]. Oxidized HDL (Ox-HDL) is thought to produce pro-inflammatory cytokines by the direct action on adipocyte stem cells [105]. Ox-HDL initiates an inflammatory cascading with inflammatory cytokines, tumor necrosis factor (TNF), interleukins (IL-6, IL-1), and increasing of Angiotensin II (ANG II), a biomarker for early cardiovascular system disorders [104]. This made the obese individuals are more sensitive to heart failure due to infection of COVID-19 [106]. Up-regulation of HO-1-derived bilirubin may enhance the COVID-19 bad effect, this risk was reduced by an increased HO-1 level [107, 108]. Hence, up-regulation of the level of HO-1 with pharmacological treatment [109] may have valuable action in acute inflammation conditions.

8. BCL11A polymorphisms

BCL11A Genetic polymorphisms were correlating to produce fetal hemoglobin in overall population, and these genetic variants were later found to be able to modify the severity of β-thalassemia and sickle cell diseases. Although the elevation of fetal hemoglobin can ameliorate the severity of these disorders. In an attempt to best comprehend the genetic background of this heterogeneity, genome-wide surveys were performed with 362,129 joint SNPs on a large cohort population of β-thalassemia and sickle cell patients to explore the genetic linking and relationship with HbF levels, in addition to other traits related to red blood cells. Among the principal variants influencing HbF levels, BCL11A SNP rs11886868 in the was completely correlated with this trait. This BCL11A variant was correlated with raised fetal hemoglobin (HbF) production in beta-thalassemia patients. Also, the similar BCL11A variants were substantially correlated with sickle cell patients HbF levels. These findings show that modifying HbF levels by BCL11A variants, consider as an essential factor in improving the beta-thalassemia phenotype and may potentially help improve other hemoglobin disorders. These findings can help describe the molecular mechanisms for regulating fetal globin and may ultimately participate in the evolution of new therapeutic strategies for sickle cell anemia and beta-thalassemia [110–112]. Hence, these results can provide an explanation of why some individuals naturally exhibit diseases mild symptoms, while others have shown very acute clinical symptoms. Therefore, it is imperative to perceive the role of genetic polymorphisms of these genes in SARS-CoV-2 infection in human populations to interpret the observed heterogeneity in predisposition and COVID-19 infection severity [88, 113].

9. Genetic polymorphism and therapy effectiveness

COVID-19 may be inactivated or partially treated by the following approaches: ACE2 receptor attaching site blocking either by antibody or specific ligand or using ACE2 soluble form that can neutralize the virus by binding the virus spike protein, and, yet, cover ACE2 binding site on the host cell surface and reducing the tissue injury. The genetic polymorphisms of cytochrome (CYP) 2D6 can affect drug metabolism using this approach, which contains 50% currently using drugs [114]. The metabolism of these genes can be increased by these polymorphisms and in turn, reduce their efficiency or significantly decline their metabolism causing drug toxicity [115]. Slow drug metabolizers permit toxic effects of the medications as chloroquine to become accumulated and resulting in cardiac problems with an increased hazard of cardiac arrest, specifically in diabetes and obesity patients. CYP2D6 Polymorphism is much high in Asians and African Americans [116–118], which extremely influenced by this disorder. One Korea study studying Lupus disease demonstrated considerable variation in the level of hydroxychloroquine due to polymorphisms of CYP2D6 [119]. This may explain the clinical outcomes differences when using this drug. Because of the metabolism abnormalities due to these genetic polymorphisms, resistant malaria strains will be arising [120–122]. Heart failure patients can be affected by the same CYP 2D6 gene polymorphisms since it is accountable for metoprolol metabolism [123, 124]. These gene variants affect several other medications such as barbiturates, Isoniazid (INH), serotonin reuptake inhibitor (omeprazole hydralazine sulfasalazine, etc.) [125]. Individuals with CYP2D6 polymorphisms and the HO-1 GT allele make therapy and disease outcomes challenging. Some of the patients who carry these polymorphisms will respond perfectly to drugs and have a low risk of COVID-19 patients to develop complications such as multiorgan failure and ARDS, while other patients will express drug toxicity levels and multiorgan problems [115]. This can describe why clinicians are unable to predict the multiorgan failure with COVID -19 disease and different outcomes from using 4-aminoquinolones.

10. Personalized medicine guided by host genetic of COVID-19

SARS-CoV-2 inhibition can be done by spike protein and ACE2 differential glycosylation [126]. Several polymorphisms, such as p.Pro389His, p.Met383Thr and p.Asp427Tyr slightly inhibited by hydroxychloroquine. This can be clarifying why hydroxychloroquine treatment was not significantly in a different hospital than others [127]. However, more pharmacogenomics experiments between the genetic data and drug response from COVID-19 patients are extremely needed. The viral entry to the host cell by binding to the cell membrane through S protein can be blocked by TMPRSS2 [88]. The SARS-CoV-2 pathogenesis and infection depend on the TMPRSS2 presence, in a high pH environment [128, 129]. The inhibitor of endosomal acidification such as hydroxychloroquine and CatB/L inhibitors might work only in absence of TMPRSS2- in SARS-CoV-2 infected and may not work or has no or less effective in patients with TMPRSS2 wild-type [128]. So far, the populations with missense polymorphisms and stop-gained of TMPRSS2 polymorphisms may be good sensitive to treatment with hydroxychloroquine. Furthermore, the patients who carry TMPRSS2 and ACE2 wildtype, a mix of hydroxychloroquine or chloroquine with camostat may have the best clinical advantage. The ACE2 can be cleaved by TMPRSS2 at Arginine 697 to 716 [130], which improves viral entry. Thus, patients with, p.Arg710Cys p.Arg708Trp, p.Arg716Cys and p.Arg710His polymorphisms in ACE2 might have fewer symptoms of COVID-19 disease as the cleavage site of ACE2 gene loses by these polymorphisms (**Figure 3**) [113].

11. Conclusion

The pandemic COVID-19 by SARS-CoV-2 coronavirus is multifactorial in which human inheritances might play a pivotal role together with the co-morbidity diseases and other risk factors. The disease clinical course has been depending on the link between genetic variants, such as the CYP2D6 enzyme system, HO-1 (anti-inflammatory gene), and ACE-2 enzyme. Beside ACE2 polymorphisms, there is TMPRSS2 gene variance that possibly changes the pathogenicity of the virus by changing the interaction between ACE2 and SARS-CoV-2 virus. A good characterization of functional polymorphisms and the host genetics can assist in identifying the pathophysiology of the disease pathway to stratify the risk evaluation and to personalize the treatment procedures.

Conflict of interest

The authors declare no conflict of interest.

Author details

Mahmood A. Al-Azzawi[1*] and Moustafa A. Sakr[2]

1 College of Dentistry, Al-Ayen University, An-Nasiriyah, Iraq

2 Genetic Engineering Institute, University of Sadat City, Sadat City, Egypt

*Address all correspondence to: mmahmood41@yahoo.com

References

[1] Anacleto O, Cabaleiro S, Villanueva B, et al. Genetic differences in host infectivity affect disease spread and survival in epidemics. Scientific Reports. 2019;**9**(1):4924. DOI: 10.1038/s41598-019-40567-w

[2] Bishop SC, Woolliams JA. Genomics and disease resistance studies in livestock. Livestock Science. 2014;**166**:190-198. DOI: 10.1016/j.livsci.2014.04.034

[3] Yáñez JM, Houston RD, Newman S. Genetics and genomics of disease resistance in salmonid species. Frontiers in Genetics. 2014;**5**:415. DOI: 10.3389/fgene.2014.00415

[4] O'Brien SJ, Evermann JF. Interactive influence of infectious disease and genetic diversity in natural populations. Trends in Ecology & Evolution. 1988;**3**(10):254-259. DOI: 10.1016/0169-5347(88)90058-4

[5] Doeschl-Wilson AB, Davidson R, Conington J, et al. Implications of host genetic variation on the risk and prevalence of infectious diseases transmitted through the environment. Genetics. 2011;**188**(3):683-693. DOI: 10.1534/genetics.110.125625

[6] King KC, Lively CM. Does genetic diversity limit disease spread in natural host populations? Heredity (Edinb.). 2012;**109**(4):199-203. DOI: 10.1038/hdy.2012.33

[7] Saura M, Carabaño MJ, Fernández A, et al. Disentangling genetic variation for resistance and endurance to scuticociliatosis in turbot using pedigree and genomic information. Frontiers in Genetics. 2019;**10**:539. DOI: 10.3389/fgene.2019.00539

[8] Kause A, Odegård J. The genetic analysis of tolerance to infections: A review. Frontiers in Genetics. 2012;**3**:262. DOI: 10.3389/fgene.2012.00262

[9] Gopinath S, Lichtman JS, Bouley DM, et al. Role of disease-associated tolerance in infectious superspreaders. Proceedings of the National Academy of Sciences of the United States of America. 2014;**111**:15780-15785. DOI: 10.1073/pnas.1409968111

[10] Wong G, Liu W, Liu Y, et al. MERS, SARS, and Ebola: The role of super-spreaders in infectious disease. Cell Host & Microbe. 2015;**18**(4):398-401. DOI: 10.1016/j.chom.2015.09.013

[11] Leavy O. Infectious disease: The tolerance of superspreaders. Nature Reviews. Immunology. 2014;**14**:776-777. DOI: doi.org/10.1038/nri3776

[12] Rauw WM. Immune response from a resource allocation perspective. Frontiers in Genetics. 2012;**3**:267. DOI: 10.3389/fgene.2012.00267

[13] Nath M, Woolliams JA, Bishop SC. Assessment of the dynamics of microparasite infections in genetically homogeneous and heterogeneous populations using a stochastic epidemic model. Journal of Animal Science. 2008;**86**:1747-1757. DOI: doi.org/10.2527/jas.2007-0615

[14] Karlsson EK, Kwiatkowski DP, Sabeti PC. Natural selection and infectious disease in human populations. Nature Reviews. Genetics. 2014;**15**(6):379-393. DOI: 10.1038/nrg3734

[15] Barreiro LB, Quintana-Murci L. From evolutionary genetics to human immunology: How selection shapes host defence genes. Nature Reviews. Genetics. 2010;**11**(1):17-30. DOI: 10.1038/nrg2698

[16] Houston RD. Future directions in breeding for disease resistance in aquaculture species. Revista Brasileira de Zootecnia. 2017;**46**(6):545-551. DOI: doi.org/10.1590/s1806-92902017000600010

[17] Lippi G, Sanchis-Gomar F, Henry BM. Coronavirus disease 2019 (COVID-19): The portrait of a perfect storm. Annals of Translational Medicine. 2020;**8**(7):497. DOI: 10.21037/atm.2020.03.157

[18] Chen N, Zhou M, Dong X, et al. Epidemiological and clinical characteristics of 99 cases of 2019 novel coronavirus pneumonia in Wuhan, China: A descriptive study. Lancet. 2020;**395**(10223):507-513. DOI: 10.1016/S0140-6736(20)30211-7

[19] Huang C, Wang Y, Li X, et al. Clinical features of patients infected with 2019 novel coronavirus in Wuhan, China. Lancet. 2020;**395**(10223):497-506. DOI: 10.1016/S0140-6736(20)30183-5

[20] Wu Z, McGoogan JM. Characteristics of and important lessons from the coronavirus disease 2019 (COVID-19) outbreak in China: Summary of a report of 72 314 cases from the Chinese Center for Disease Control and Prevention. JAMA; 2020;**323**(13):1239-1242. DOI: 10.1001/jama.2020.2648

[21] Xu XW, Wu XX, Jiang XG, et al. Clinical findings in a group of patients infected with the 2019 novel coronavirus (SARS-Cov-2) outside of Wuhan, China: Retrospective case series. BMJ. 2020;**368**:m606. DOI: 10.1136/bmj.m606

[22] Zhou F, Yu T, Du R, et al. Clinical course and risk factors for mortality of adult inpatients with COVID-19 in Wuhan, China: A retrospective cohort study. Lancet. 2020;**395**(10229):1054-1062. DOI: 10.1016/S0140-6736(20)30566-3

[23] Velavan TP, Meyer CG. The Covid-19 epidemic. Tropical Medicine & International Health. 2020;**25**(3):278-280. DOI: 10.1111/tmi.13383

[24] Nishiura H, Linton NM, Akhmetzhanov AR. Initial cluster of novel coronavirus (2019-nCoV) infections in Wuhan, China is consistent with substantial human-to-human transmission. Journal of Clinical Medicine. 2020;**9**(2):488. DOI: 10.3390/jcm9020488

[25] To KK, Tsang OT, Yip CC, et al. Consistent detection of 2019 novel coronavirus in saliva. Clinical Infectious Diseases. 2020;**71**(15):841-843. DOI: 10.1093/cid/ciaa149

[26] Lin X, Gong Z, Xiao Z, et al. Novel coronavirus pneumonia outbreak in 2019: Computed tomographic findings in two cases. Korean Journal of Radiology. 2020;**21**(3):365-368. DOI: 10.3348/kjr.2020.0078

[27] Bastola A, Sah R, Rodriguez-Morales AJ, et al. The first 2019 novel coronavirus case in Nepal. The Lancet. Infectious Diseases. 2020;**20**(3):279-280. DOI: 10.1016/S1473-3099(20)30067-0

[28] Petrosillo N, Viceconte G, Ergonul O, et al. COVID-19, SARS and MERS: Are they closely related? Clinical Microbiology and Infection. 2020;**26**(6):729-734. DOI: 10.1016/j.cmi.2020.03.026

[29] Day M. Covid-19: Identifying and isolating asymptomatic people helped eliminate virus in Italian village. BMJ. 2020;**368**:m1165. DOI: 10.1136/bmj.m1165

[30] Lauretani F, Ravazzoni G, Roberti MF, et al. Assessment and treatment of older individuals with COVID 19 multi-system disease: Clinical and ethical implications. Acta Bio-Medica. 2020;**91**(2):150-168

[31] Onder G, Rezza G, Brusaferro S. Case-fatality rate and characteristics of patients dying in relation to COVID-19 in Italy. JAMA. 2020;**323**(18):1775-1776. DOI: 10.1001/jama.2020.4683

[32] CDC Covid-Response Team. Severe outcomes among patients with coronavirus disease 2019 (COVID-19)—United States. MMWR. Morbidity and Mortality Weekly Report. 2020;**69**(12):343-346. DOI: 10.15585/mmwr.mm6912e2

[33] Williamson EJ, Walker AJ, Bhaskaran K, et al. Factors associated with COVID-19-related death using OpenSAFELY. Nature. 2020;**584**(7821):430-436. DOI: 10.1038/s41586-020-2521-4

[34] Lippi G, Sanchis-Gomar F, Henry BM. Association between environmental pollution and prevalence of coronavirus disease 2019 (COVID-19) in Italy. medRxiv. 2020. DOI: 10.1101/2020.04.22.20075986

[35] Lippi G, Henry BM, Mattiuzzi C, et al. The death rate for COVID-19 is positively associated with gross domestic products. Acta Bio-Medica. 2020;**91**(2):224-225. DOI: 10.23750/abm.v91i2.9514

[36] Lippi G, Sanchis-Gomar F, Henry BM. Active smoking and COVID-19: A double-edged sword. European Journal of Internal Medicine. 2020;**77**:123-124. DOI: 10.1016/j.ejim.2020.04.060

[37] Yang J, Zheng Y, Gou X, et al. Prevalence of comorbidities and its effects in patients infected with SARS-CoV-2: A systematic review and meta-analysis. International Journal of Infectious Diseases. 2020;**94**:91-95. DOI: 10.1016/j.ijid.2020.03.017

[38] Rodriguez-Morales AJ, Cardona-Ospina JA, Gutiérrez-Ocampo E, et al. Clinical, laboratory and imaging features of COVID-19: A systematic review and meta-analysis. Travel Medicine and Infectious Disease. 2020;**34**:101623. DOI: 10.1016/j.tmaid.2020.101623

[39] Cascella M, Rajnik M, Cuomo A, et al. Features, evaluation and treatment coronavirus (COVID-19). In: StatPearls [Internet]. Treasure Island (FL): StatPearls Publishing; 2020. Available from: https://www.ncbi.nlm.nih.gov/books/NBK554776/ [Updated: 04 July 2020]

[40] Guan WJ, Liang WH, Zhao Y, et al. Comorbidity and its impact on 1590 patients with COVID-19 in China: A nationwide analysis. European Respiratory Journal. 2020;**55**(5):2000547. DOI: 10.1183/13993003.00547-2020

[41] Shang J, Ye G, Shi K, et al. Structural basis of receptor recognition by SARS-CoV-2. Nature. 2020;**581**(7807):221-224. DOI: 10.1038/s41586-020-2179-y

[42] Brielle ES, Schneidman-Duhovny D, Linial M. The SARS-CoV-2 exerts a distinctive strategy for interacting with the ACE2 human receptor. Viruses. 2020;**12**(5):497. DOI: 10.3390/v12050497

[43] Belouzard S, Millet JK, Licitra BN, Whittaker GR. Mechanisms of coronavirus cell entry mediated by the viral spike protein. Viruses. 2012;**4**(6):1011-1033. DOI: 10.3390/v4061011

[44] Shang J, Wan Y, Luo C, et al. Cell entry mechanisms of SARS-CoV-2. Proceedings of the National Academy of Sciences of the United States of America. 2020;**117**(21):11727-11734. DOI: 10.1073/pnas.2003138117

[45] Walls AC, Park YJ, Tortorici MA, et al. Structure, function, and antigenicity of the SARS-CoV-2 spike glycoprotein. Cell. 2020;**181**(2):281-292.e6. DOI: 10.1016/j.cell.2020.02.058

[46] Lambert DW, Yarski M, Warner FJ, et al. Tumor necrosis factor-alpha convertase (ADAM17) mediates regulated ectodomain shedding of the severe-acute respiratory syndrome-coronavirus (SARS-CoV) receptor, angiotensin-converting enzyme-2 (ACE2). Journal of Biological Chemistry. 2005;**280**(34):30113-30119. DOI: 10.1074/jbc.M505111200

[47] Rizzo P, Vieceli Dalla Sega F, Fortini F, et al. COVID-19 in the heart and the lungs: Could we "notch" the inflammatory storm? Basic Research in Cardiology. 2020;**115**(3):31. DOI: 10.1007/s00395-020-0791-5

[48] Fakhouri EW, Peterson SJ, Kothari J, et al. Genetic polymorphisms complicate COVID-19 therapy: Pivotal role of HO-1 in cytokine storm. Antioxidants. 2020;**9**(7):636. DOI: org/10.3390/antiox9070636

[49] Li MY, Li L, Zhang Y, Wang XS. Expression of the SARS-CoV-2 cell receptor gene ACE2 in a wide variety of human tissues. Infectious Diseases of Poverty. 2020;**9**(1):45. DOI: 10.1186/s40249-020-00662-x

[50] Sungnak W, Huang N, Bécavin C, et al. SARS-CoV-2 entry factors are highly expressed in nasal epithelial cells together with innate immune genes. Nature Medicine. 2020;**26**(5):681-687. DOI: 10.1038/s41591-020-0868-6

[51] Xu H, Zhong L, Deng J, et al. High expression of ACE2 receptor of 2019-nCoV on the epithelial cells of oral mucosa. International Journal of Oral Science. 2020;**12**:8. DOI: 10.1038/s41368-020-0074-x

[52] Henry BM, de Oliveira MHS, Benoit S, et al. Hematologic, biochemical and immune biomarker abnormalities associated with severe illness and mortality in coronavirus disease 2019 (COVID-19): A meta-analysis. Clinical Chemistry and Laboratory Medicine. 2020;**58**(7):1021-1028. DOI: 10.1515/cclm-2020-0369

[53] Ferrario CM, Jessup J, Chappell MC, et al. Effect of angiotensin-converting enzyme inhibition and angiotensin II receptor blockers on cardiac angiotensin-converting enzyme 2. Circulation. 2005;**111**(20):2605-2610. DOI: 10.1161/CIRCULATIONAHA.104.510461

[54] Monteil V, Kwon H, Prado P, et al. Inhibition of SARS-CoV-2 infections in engineered human tissues using clinical-grade soluble human ACE2. Cell. 2020;**181**(4):905-913.e7. DOI: 10.1016/j.cell.2020.04.004

[55] Varga Z, Flammer AJ, Steiger P, et al. Endothelial cell infection and endotheliitis in COVID-19. Lancet. 2020;**395**(10234):1417-1418. DOI: 10.1016/S0140-6736(20)30937-5

[56] Taddei S, Virdis A, Ghiadoni L, et al. Effects of angiotensin converting enzyme inhibition on endothelium-dependent vasodilatation in essential hypertensive patients. Journal of Hypertension. 1998;**16**(4):447-456. DOI: 10.1097/00004872-199816040-00006

[57] Flammer AJ, Sudano I, Hermann F, et al. Angiotensin-converting enzyme inhibition improves vascular function in rheumatoid arthritis. Circulation. 2008;**117**(17):2262-2269. DOI: 10.1161/CIRCULATIONAHA.107.734384

[58] Hürlimann D, Forster A, Noll G, et al. Anti-tumor necrosis factor-alpha treatment improves endothelial function in patients with rheumatoid arthritis. Circulation. 2002;**106**(17):2184-2187. DOI: 10.1161/01.cir.0000037521.71373.44

[59] Feldmann M, Maini RN, Woody JN, et al. Trials of anti-tumour necrosis factor therapy for COVID-19 are urgently needed. Lancet.

2020;**395**(10234):1407-1409. DOI: 10.1016/S0140-6736(20)30858-8

[60] Leiden Open Variation Database. ACE2 Gene Homepage. Available from: https://databases.lovd.nl/shared/genes/ACE2 [Last accessed: 12 May 2020]

[61] Cao Y, Li L, Feng Z, et al. Comparative genetic analysis of the novel coronavirus (2019-nCoV/SARS-CoV-2) receptor ACE2 in different populations. Cell Discovery. 2020;**6**:11. DOI: 10.1038/s41421-020-0147-1

[62] Peña Silva RA, Chu Y, Miller JD, et al. Impact of ACE2 deficiency and oxidative stress on cerebrovascular function with aging. Stroke. 2012;**43**(12):3358-3363. DOI: 10.1161/STROKEAHA.112.667063

[63] Tallant EA, Clark MA. Molecular mechanisms of inhibition of vascular growth by angiotensin-(1-7). Hypertension. 2003;**42**(4):574-579. DOI: 10.1161/01.HYP.0000090322.55782.30

[64] Crackower MA, Sarao R, Oudit GY, et al. Angiotensin-converting enzyme 2 is an essential regulator of heart function. Nature. 2002;**417**(6891):822-828. DOI: 10.1038/nature00786

[65] Díez-Freire C, Vázquez J, Correa de Adjounian MF, et al. ACE2 gene transfer attenuates hypertension-linked pathophysiological changes in the SHR. Physiological Genomics. 2006;**27**(1):12-19. DOI: 10.1152/physiolgenomics.00312.2005

[66] Tikellis C, Pickering R, Tsorotes D, et al. Interaction of diabetes and ACE2 in the pathogenesis of cardiovascular disease in experimental diabetes. Clinical Science (London). 2012;**123**(8):519-529. DOI: 10.1042/CS20110668

[67] Niu W, Qi Y, Hou S, et al. Correlation of angiotensin-converting enzyme 2 gene polymorphisms with stage 2 hypertension in Han Chinese. Translational Research. 2007;**150**(6):374-380. DOI: 10.1016/j.trsl.2007.06.002

[68] Fan XH, Wang YB, Wang H, et al. Polymorphisms of angiotensin-converting enzyme (ACE) and ACE2 are not associated with orthostatic blood pressure dysregulation in hypertensive patients. Acta Pharmacologica Sinica. 2009;**30**(9):1237-1244. DOI: 10.1038/aps.2009.110

[69] Chen YY, Liu D, Zhang P, et al. Impact of ACE2 gene polymorphism on antihypertensive efficacy of ACE inhibitors. Journal of Human Hypertension. 2016;**30**(12):766-771. DOI: 10.1038/jhh.2016.24

[70] Luo Y, Liu C, Guan T, et al. Association of ACE2 genetic polymorphisms with hypertension-related target organ damages in south. Hypertension Research. 2019;**42**(5):681-689. DOI: 10.1038/s41440-018-0166-6

[71] Chen Q, Tang X, Yu CQ, et al. Correlation of angiotensin-converting enzyme 2 gene polymorphism with antihypertensive effects of benazepril. Beijing Da Xue Xue Bao. Yi Xue Ban. 2010;**42**(3):293-298

[72] Patnaik M, Pati P, Swain SN, et al. Association of angiotensin-converting enzyme and angiotensin-converting enzyme-2 gene polymorphisms with essential hypertension in the population of Odisha, India. Annals of Human Biology. 2014;**41**(2):145-152. DOI: 10.3109/03014460.2013.837195

[73] Pinheiro DS, Santos RS, Jardim PCBV, et al. The combination of ACE I/D and ACE2 G8790A polymorphisms revels susceptibility to hypertension: A genetic association study in Brazilian patients. PLOS One. 2019;**14**(8):e0221248. DOI: 10.1371/journal.pone.0221248

[74] Duru K, Farrow S, Wang JM, et al. Frequency of a deletion polymorphism in the gene for angiotensin converting enzyme is increased in African-Americans with hypertension. American Journal of Hypertension. 1994;7(8):759-762. DOI: 10.1093/ajh/7.8.759

[75] Devaux CA, Rolain JM, Raoult D. ACE2 receptor polymorphism: Susceptibility to SARS-CoV-2, hypertension, multi-organ failure, and COVID-19 disease outcome. Journal of Microbiology, Immunology, and Infection. 2020;**53**(3):425-435. DOI: 10.1016/j.jmii.2020.04.015

[76] Babcock GJ, Esshaki DJ, Thomas WD Jr, et al. Amino acids 270 to 510 of the severe acute respiratory syndrome coronavirus spike protein are required for interaction with receptor. Journal of Virology. 2004;**78**(9):4552-4560. DOI: 10.1128/jvi.78.9.4552-4560.2004

[77] Xiao F, Zimpelmann J, Agaybi S, et al. Characterization of angiotensin-converting enzyme 2 ectodomain shedding from mouse proximal tubular cells. PLOS One. 2014;**9**(1):e85958. DOI: 10.1371/journal.pone.0085958

[78] Glowacka I, Bertram S, Herzog P, et al. Differential downregulation of ACE2 by the spike proteins of severe acute respiratory syndrome coronavirus and human coronavirus NL63. Journal of Virology. 2010;**84**(2):1198-1205. DOI: 10.1128/JVI.01248-09

[79] Chen J, Jiang Q, Xia X, et al. Individual variation of the SARS-CoV-2 receptor ACE2 gene expression and regulation. Aging Cell. 2020;**19**(7);e13168. DOI: 10.1111/acel.13168

[80] Hussain M, Jabeen N, Raza F, et al. Structural variations in human ACE2 may influence its binding with SARS-CoV-2 spike. Journal of Medical Virology. 2020. DOI: 10.1002/jmv.25832

[81] Stawiski EW, Diwanji D, Suryamohan K, et al. Human ACE2 receptor polymorphisms predict SARS-CoV-2 susceptibility. bioRxiv. 2020; 2020.04.07.024752. DOI:10.1101/2020.04.07.024752

[82] Benetti E, Tita R, Spiga O, et al. ACE2 gene variants may underlie interindividual variability and susceptibility to COVID-19 in the Italian population. European Journal of Human Genetics. 2020:1-13. DOI: 10.1038/s41431-020-0691-z

[83] Chiu RW, Tang NL, Hui DS, et al. ACE2 gene polymorphisms do not affect outcome of severe acute respiratory syndrome. Clinical Chemistry. 2004;**50**(9):1683-1686. DOI: 10.1373/clinchem.2004.035436

[84] Li G, He X, Zhang L, et al. Assessing ACE2 expression patterns in lung tissues in the pathogenesis of COVID-19. Journal of Autoimmunity. 2020;**112**:102463. DOI: 10.1016/j.jaut.2020.102463

[85] Yang M, Zhao J, Xing L, et al. The association between angiotensin-converting enzyme 2 polymorphisms and essential hypertension risk: A meta-analysis involving 14,122 patients. Journal of the Renin-Angiotensin-Aldosterone System. 2015;**16**(4):1240-1244. DOI: 10.1177/1470320314549221

[86] Liu D, Chen Y, Zhang P, et al. Association between circulating levels of ACE2-Ang-(1-7)-MAS axis and ACE2 gene polymorphisms in hypertensive patients. Medicine (Baltimore). 2016;**95**(24):e3876. DOI: 10.1097/MD.0000000000003876

[87] Gemmati D, Bramanti B, Serino ML, et al. COVID-19 and individual genetic susceptibility/receptivity: Role of ACE1/ACE2 genes, immunity, inflammation and coagulation. Might the double X-chromosome in females be protective against SARS-CoV-2 compared to

the single X-chromosome in males. International Journal of Molecular Sciences. 2020;**21**(10):3474. DOI: 10.3390/ijms21103474

[88] Hoffmann M, Kleine-Weber H, Schroeder S, et al. SARS-CoV-2 cell entry depends on ACE2 and TMPRSS2 and is blocked by a clinically proven protease inhibitor. Cell. 2020;**181**(2):271-280.e8. DOI: 10.1016/j.cell.2020.02.052

[89] Luostari K, Hartikainen JM, Tengstrom M, et al. Type II transmembrane serine protease gene variants associate with breast cancer. PLOS One. 2014;**9**(7):e102519. DOI: 10.1371/journal.pone.0102519

[90] FitzGerald LM, Agalliu I, Johnson K, et al. Association of TMPRSS2-ERG gene fusion with clinical characteristics and outcomes: Results from a population-based study of prostate cancer. BMC Cancer. 2008;**8**:230. DOI: 10.1186/1471-2407-8-230

[91] Giri VN, Ruth K, Hughes L, et al. Racial differences in prediction of time to prostate cancer diagnosis in a prospective screening cohort of high-risk men: Effect of TMPRSS2 Met160Val. BJU International. 2011;**107**(3):466-470. DOI: 10.1111/j.1464-410X.2010.09522.x

[92] Cheng Z, Zhou J, To KK, et al. Identification of TMPRSS2 as a susceptibility gene for severe 2009 pandemic A(H1N1) influenza and A(H7N9) influenza. The Journal of Infectious Diseases. 2015;**212**(8): 1214-1221. DOI: 10.1093/infdis/jiv246

[93] Hou Y, Zhao J, Martin W, et al. New insights into genetic susceptibility of COVID-19: An ACE2 and TMPRSS2 polymorphism analysis. BMC Medicine. 2020;**18**(1):216. DOI: 10.1186/s12916-020-01673-z

[94] Schuler BA, Habermann AC, Plosa EJ, et al. Age-related expression of SARS-CoV-2 priming protease TMPRSS2 in the developing lung. Biorxiv: The Preprint Server for Biology. 2020. DOI: 10.1101/2020.05.22.111187

[95] Pechlaner R, Willeit P, Summerer M, et al. Heme oxygenase-1 gene promoter microsatellite polymorphism is associated with progressive atherosclerosis and incident cardiovascular disease. Arteriosclerosis, Thrombosis, and Vascular Biology. 2015;**35**(1):229-236. DOI: 10.1161/ATVBAHA.114.304729

[96] Yamada N, Yamaya M, Okinaga S, et al. Microsatellite polymorphism in the heme oxygenase-1 gene promoter is associated with susceptibility to emphysema. American Journal of Human Genetics. 2000;**66**(1):187-195. DOI: 10.1086/302729

[97] Okamoto I, Krögler J, Endler G, et al. A microsatellite polymorphism in the heme oxygenase-1 gene promoter is associated with risk for melanoma. International Journal of Cancer. 2006;**119**(6):1312-1315. DOI: 10.1002/ijc.21937

[98] Hirai H, Kubo H, Yamaya M, et al. Microsatellite polymorphism in heme oxygenase-1 gene promoter is associated with susceptibility to oxidant-induced apoptosis in lymphoblastoid cell lines. Blood. 2003;**102**(5):1619-1621. DOI: 10.1182/blood-2002-12-3733

[99] Guénégou A, Leynaert B, Bénessiano J, et al. Association of lung function decline with the heme oxygenase-1 gene promoter microsatellite polymorphism in a general population sample. Results from the European Community Respiratory Health Survey (ECRHS), France. Journal of Medical Genetics. 2006;**43**(8):e43. DOI: 10.1136/jmg.2005.039743

[100] Exner M, Schillinger M, Minar E, et al. Heme oxygenase-1 gene

promoter microsatellite polymorphism is associated with restenosis after percutaneous transluminal angioplasty. Journal of Endovascular Therapy. 2001;**8**(5):433-440. DOI: 10.1177/152660280100800501

[101] Bao W, Song F, Li X, et al. Association between heme oxygenase-1 gene promoter polymorphisms and type 2 diabetes mellitus: A HuGE review and meta-analysis. American Journal of Epidemiology. 2010;**172**(6):631-636. DOI: 10.1093/aje/kwq162

[102] Zhou Y, Rui L. Leptin signaling and leptin resistance. Frontiers of Medicine. 2013;**7**(2):207-222. DOI: 10.1007/s11684-013-0263-5

[103] Peterson SJ, Dave N, Kothari J. The effects of heme oxygenase upregulation on obesity and the metabolic syndrome. Antioxidants & Redox Signaling. 2020;**32**(14):1061-1070. DOI: 10.1089/ars.2019.7954

[104] Peterson SJ, Shapiro JI, Thompson E, et al. Oxidized HDL, adipokines, and endothelial dysfunction: A potential biomarker profile for cardiovascular risk in women with obesity. Obesity (Silver Spring). 2019;**27**(1):87-93. DOI: 10.1002/oby.22354

[105] Peterson SJ, Vanella L, Bialczak A, et al. Oxidized HDL and isoprostane exert a potent adipogenic effect on stem cells: Where in the lineage? Journal of Cell, Stem cells and Regenerative Medicine. 2016;**2**(1). DOI: 10.16966/2472-6990.109

[106] Aghagoli G, Gallo Marin B, Soliman LB, et al. Cardiac involvement in COVID-19 patients: Risk factors, predictors, and complications: A review. Journal of Cardiac Surgery. 2020;**35**(6):1302-1305. DOI: 10.1111/jocs.14538

[107] Singh SP, McClung JA, Thompson E, et al. Cardioprotective heme oxygenase-1-PGC1α signaling in epicardial fat attenuates cardiovascular risk in humans as in obese mice. Obesity (Silver Spring). 2019;**27**(10):1634-1643. DOI: 10.1002/oby.22608

[108] Peterson SJ, Yadav R, Iacobellis G. Cardioprotective heme oxygenase 1-PGC1α signaling in epicardial fat attenuates cardiovascular risk in humans as in obese mice. Obesity (Silver Spring). 2019;**27**(10):1560-1561. DOI: 10.1002/oby.22629

[109] Peterson SJ, Rubinstein R, Faroqui M, et al. Positive effects of heme oxygenase upregulation on adiposity and vascular dysfunction: Gene targeting vs. pharmacologic therapy. International Journal of Molecular Sciences. 2019;**20**(10):2514. DOI: 10.3390/ijms20102514

[110] Menzel S, Garner C, Gut I, et al. A QTL influencing F cell production maps to a gene encoding a zinc-finger protein on chromosome 2p15. Nature Genetics. 2007;**39**(10):1197-1199. DOI: 10.1038/ng2108

[111] Lettre G, Sankaran VG, Bezerra MA, et al. DNA polymorphisms at the BCL11A, HBS1L-MYB, and beta-globin loci associate with fetal hemoglobin levels and pain crises in sickle cell disease. Proceedings of the National Academy of Sciences of the United States of America. 2008;**105**(33):11869-11874. DOI: 10.1073/pnas.0804799105

[112] Uda M, Galanello R, Sanna S, et al. Genome-wide association study shows BCL11A associated with persistent fetal hemoglobin and amelioration of the phenotype of beta-thalassemia. Proceedings of the National Academy of Sciences of the United States of America. 2008;**105**(5):1620-1625. DOI: 10.1073/pnas.0711566105

[113] Yan R, Zhang Y, Li Y, et al. Structural basis for the recognition

of SARS-CoV-2 by full-length human ACE2. Science. 2020;**367**(6485):1444-1448. DOI: 10.1126/science.abb2762

[114] Ingelman-Sundberg M. Genetic polymorphisms of cytochrome P450 2D6 (CYP2D6): Clinical consequences, evolutionary aspects and functional diversity. The Pharmacogenomics Journal. 2005;5(1):6-13. DOI: 10.1038/sj.tpj.6500285

[115] Haertter S. Recent examples on the clinical relevance of the CYP2D6 polymorphism and endogenous functionality of CYP2D6. Drug Metabolism and Drug Interactions. 2013;**28**(4):209-216. DOI: 10.1515/dmdi-2013-0032

[116] Wan YJ, Poland RE, Han G, et al. Analysis of the CYP2D6 gene polymorphism and enzyme activity in African-Americans in southern California. Pharmacogenetics. 2001;**11**(6):489-499. DOI: 10.1097/00008571-200108000-00004

[117] Yee MM, Josephson C, Hill CE, et al. Cytochrome P450 2D6 polymorphisms and predicted opioid metabolism in African American children with sickle cell disease. Journal of Pediatric Hematology/Oncology. 2013;**35**(7):e301-e305. DOI: 10.1097/MPH.0b013e31828e52d2

[118] Gaedigk A, Bhathena A, Ndjountché L, et al. Identification and characterization of novel sequence variations in the cytochrome P4502D6 (CYP2D6) gene in African Americans. The Pharmacogenomics Journal. 2005;**5**(3):173-182. DOI: 10.1038/sj.tpj.6500305 [published correction appears in Pharmacogenomics J. 2005;5(4):276. Rogan, PK (added)]

[119] Lee JY, Vinayagamoorthy N, Han K, et al. Association of polymorphisms of cytochrome P450 2D6 with blood hydroxychloroquine levels in patients with systemic lupus erythematosus. Arthritis & Rheumatology. 2016;**68**(1):184-190. DOI: 10.1002/art.39402

[120] He X, Pan M, Zeng W, et al. Multiple relapses of plasmodium vivax malaria acquired from West Africa and association with poor metabolizer CYP2D6 variant: A case report. BMC Infectious Diseases. 2019;**19**(1):704. DOI: 10.1186/s12879-019-4357-9

[121] Haraya K, Kato M, Chiba K, et al. Prediction of inter-individual variability on the pharmacokinetics of CYP2C8 substrates in human. Drug Metabolism and Pharmacokinetics. 2017;**32**(6):277-285. DOI: 10.1016/j.dmpk.2017.09.001

[122] Silvino AC, Costa GL, Araújo FC, et al. Variation in human cytochrome P-450 drug-metabolism genes: A gateway to the understanding of plasmodium vivax relapses. PLOS One. 2016;**11**(7):e0160172. DOI: 10.1371/journal.pone.0160172

[123] Wang B, Yang LP, Zhang XZ, et al. New insights into the structural characteristics and functional relevance of the human cytochrome P450 2D6 enzyme. Drug Metabolism Reviews. 2009;**41**(4):573-643. DOI: 10.1080/03602530903118729

[124] Mottet F, Vardeny O, de Denus S. Pharmacogenomics of heart failure: A systematic review. Pharmacogenomics. 2016;**17**(16):1817-1858. DOI: 10.2217/pgs-2016-0118

[125] Zhao RZ, Jiang S, Zhang L, et al. Mitochondrial electron transport chain, ROS generation and uncoupling (review). International Journal of Molecular Medicine. 2019;**44**(1):3-15. DOI: 10.3892/ijmm.2019.4188

[126] Savarino A, Di Trani L, Donatelli I, et al. New insights into the antiviral effects of chloroquine. The Lancet. Infectious Diseases. 2006;**6**(2):67-69. DOI: 10.1016/S1473-3099(06)70361-9

[127] Rosenberg ES, Dufort EM, Udo T, et al. Association of treatment with hydroxychloroquine or azithromycin with in-hospital mortality in patients with COVID-19 in New York state. Journal of the American Medical Association. 2020;**323**(24):2493-2502. DOI: 10.1001/jama.2020.8630

[128] Shulla A, Heald-Sargent T, Subramanya G, et al. A transmembrane serine protease is linked to the severe acute respiratory syndrome coronavirus receptor and activates virus entry. Journal of Virology. 2011;**85**(2):873-882

[129] Simmons G, Gosalia DN, Rennekamp AJ, et al. Inhibitors of cathepsin L prevent severe acute respiratory syndrome coronavirus entry. Proceedings of the National Academy of Sciences of the United States of America. 2005;**102**(33):11876-11881. DOI: 10.1073/pnas.0505577102

[130] Heurich A, Hofmann-Winkler H, Gierer S, et al. TMPRSS2 and ADAM17 cleave ACE2 differentially and only proteolysis by TMPRSS2 augments entry driven by the severe acute respiratory syndrome coronavirus spike protein. Journal of Virology. 2014;**88**(2):1293-1307. DOI: 10.1128/JVI.02202-13

Chapter 11

Sequencing of Norovirus in Southern, Nigeria: Prevalent Genotypes and Putative GII.4 Novel Recombinants among Children

Favour Osazuwa

Abstract

Norovirus is now known to be the leading cause of gastroenteritis among children worldwide. This present report highlights the genetic diversity of norovirus among children less than 5 years in Southern, Nigeria. Stool specimens were collected from 300 children with diarrhea and analyzed for norovirus using conventional reverse transcriptase-Polymerase Chain Reaction. Sequencing of the capsid region was performed to genotype the strains. Norovirus was detected in 45 (11.1%) of children with diarrhea. Genogroup II norovirus was detected in 38/45 (84.4%) patients, while genogroup I (GI) noroviruses were identified in 7/38 (15.6%) patients. Genotype diversity was large, as demonstrated by the nine identified genotypes (2 GI and 7 GII). GII.4 was the most predominant genotype. Two norovirus GII.4 variants, New Orleans_2009 and Sydney_2012 were identified in this study. A putative novel GII.4 recombinant was also detected. This study report for the first time the detection of norovirus GII.17 Kawasaki strain in South–South, region of Nigeria.

Keywords: norovirus, genetic diversity, children, RT-PCR

1. Introduction (Virology of Norovirus)

Norovirus has been identified to constitute a key biological cause of gastroenteritis worldwide [1]. It is also the most common cause of gastroenteritis world over [1]. Norovirus causes an estimated 1.1 million hospitalizations and up to 218,000 deaths among children less than 5 years annually [1]. Clinical manifestations of norovirus infection are characterized by non-bloody diarrhea, vomiting and stomach pain [2]. Incubation period is usually within 12 to 48 hrs after infection with the virus [3]. Complications are uncommon, but may include dehydration, especially in the young, the old, and those with other health problems [3]. Norovirus is typically spread by faeco-oral route, through contaminated food or water or from person to person [4]. Potential risk factors are poor environmental hygiene and overcrowded living quarters [5]. Vomiting, in particular, transmits infection effectively and appears to allow airborne transmission [5].

2. Norovirus genome and genetic diversity

Taxonomically, norovirus is a positive sense RNA, non-enveloped virus in the family caliciviridae [6]. The genus has one species, known as Norwalk virus [6]. Norovirus are highly genetically diverse, Phylogenetically, they can be classified into 10 different genogroups (GI-GX) and several genotypes in this order; 60 P-types (14 GI, 37 GII, 2 GIII, 1 GIV, 2 GV, 2 GVI, 1 GVII and 1 GX), with each of these genotypes having several genetic clusters and sub groups [7].

The genome is organized into three open reading frames (ORF) [8]. ORF 1 encoded six non-structural proteins (NS1-NS6) and the RNA dependent RNA Polymerase RdRP, while ORF 2 encodes the capsid proteins-the major structural proteins VP1 containing the shell (S) and protruding (P) domains. The S domain surrounds the viral RNA and the P domain, which consists of the P2 domain, is linked to the S domain through a flexible hinge [9]. ORF 3 encodes a minor structural protein [9]. In spite of the large genetic diversity of norovirus, it is noteworthy that norovirus of the genotype GII.4 are responsible for a majority of infections [10, 11].

3. Norovirus recombinants

Recombination are mechanism in the evolution of RNA viruses, this creates changes in virus genomes by exchanging sequences, thereby generating genetic variation and producing new viruses [12]. RNA recombination is very common among RNA viruses belonging to the family *picornaviridae, coronaviridae, retroviridae and caliciviridae* [13]. Recombination commonly occurs at the ORF1-ORF2 junction [14] although other recombination sites have been reported. Recombination event is high among the GII.4 noroviruses [14]. Norovirus recombination has been recognized to be a major tool it uses to evade host immune recognition [15]. Monitoring the incidence of rate of generation of new norovirus recombinants is a vital tool in the understanding of norovirus evolution and continuous global spread. Norovirus recombination has been linked to increased rate of generation of new norovirus genotypes and subtypes [16], this has also hampered the possibility of a possible vaccine.

4. Justification for this study

The introduction of rotavirus vaccines throughout the world has made norovirus the most common aetiologic agent of gastroenteritis world over [17]. An update on the predominant norovirus genotype in a given population is needed for the development of effective vaccine. There are no data on the genetic diversity of norovirus among children in South–South, Nigeria. Also, there are no data on the prevalence of norovirus in Nigeria. Against this background, this study was conducted to determine the prevalent norovirus genotypes and existence of possible GII.4 recombinants among children under 5 years with diarrhea in South–South, Nigeria.

5. Method

5.1 Study area and study population

This cross-sectional study was conducted in the period, March, 2018 to February, 2019. 405 children with clinical symptoms of diarrhea/gastroenteritis from a pool of 2813, attending outpatient clinics of four secondary health facilities (Central Hospital, Warri, Central Hospital, Benin, Primary Health Centre Pessu and Federal Medical Centre, Yenagoa) in Delta, Edo and Bayelsa States, Niger-Delta region, Nigeria were randomly included. Inclusion criteria where at least 3 clinical episodes of diarrhea- with an onset of 1 to 7 days whose parents or guardians consented for their ward/children to participate were included in this study. One hundred (100) asymptomatic apparently healthy age and sex matched children who served as controls (**Figure 1**).

5.2 Sample collection and processing

Stool specimens were collected into clean universal containers. Supernatant obtained from stool suspension of 50% in 1 ml sterile phosphate buffered saline were stored at -20°C for RT-PCR analysis of norovirus.

5.2.1 RNA extraction

RNA extraction was from thawed frozen samples were performed using AccuPrep® Viral RNA Extraction Kit (Bioneer, Daejon South Korea), following manufacturers instruction.

5.2.2 cDNA synthesis

cDNA synthesis was carried out on a 20 μl reverse transcription reaction of 1.0ug of extracted RNA on 0.2 ml tubes of Accupower Cycle script RT Premix (Bioneer Corporation, South Korea). Standard protocols as recommended by manufactures were followed for cDNA synthesis.

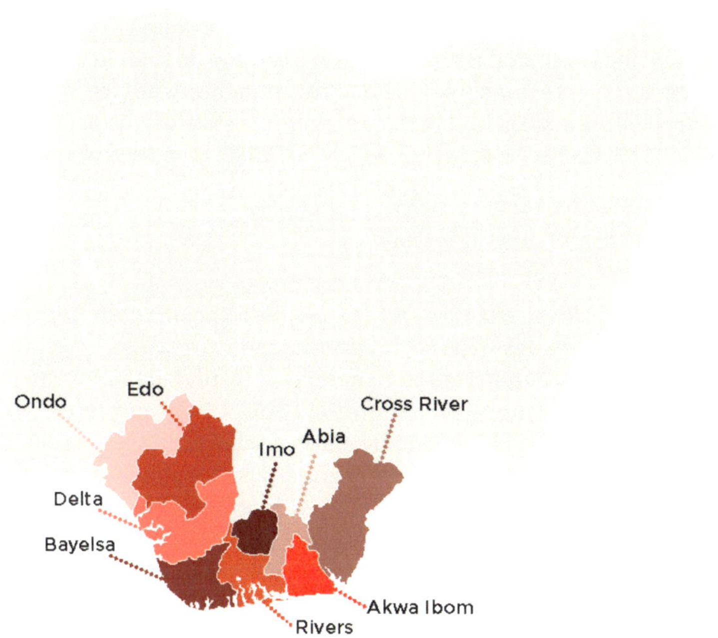

Figure 1.
Map of Southern Nigeria. States included in this study were Bayelsa, Delta and Edo.

5.2.3 Polymerase chain reaction

The cDNA generated was then amplified by PCR in a 45 μL reaction mixture as described in a previous study [18]. Specific primers (G1SKRCAACCCARCCA TTRTACA) and G1FFN (GGAGATCGCAATCTCCTGCCC) were used for GI genotyping, while for genotyping GII noroviruses, primers GIIFBN (TGGGAGGGCGATCGCAATCT) and GIISKR (CCRCCNGCATRHCCRTTRTACAT), respectively, were used in an RT-PCR analysis. The products were visualized on UV illuminator and photographed using Polaroid camera [19]. The RT-PCR used is a very sensitive method, it can detect as few as 5 x 10^6 copies per gram of stool sample. U-TaQ DNA polymerase (SBS genetech, Beijing, China), a high-fidelity thermostable enzyme that can withstand prolonged incubation at high temperature up to 95°C without significant loss of activity was used for this RT-PCR protocol.

5.2.4 Norovirus sequencing

The amplicons from the partial gene regions of the viral capsid genes were purified using QIAquick PCR purification kit (Qiagen Inc., Valencia, CA). Nucleotide sequencing was done using Big Dye ® Terminator v 3.1 Cycle sequencing kit (Applied Biosystems, Carlsbad, CA) on 3130 DNA genetic analyzer (Applied Biosystems, Carlsbad, CA), Sequences were edited using sequencher® Version 5.4.6 DNA sequence analysis software (Gene codes Corporation, Ann Arbor, MI, USA). Norovirus genotypes were determined by comparison of corresponding sequences of norovirus strains using the online norovirus genotyping tool version 1.0. available at (www.rivm.nl/mpf/norovirus/typing tool).

5.2.5 Phylogenetic analysis

For confirmation of genotyping, nucleotide sequences obtained were aligned with reference sequences using MUSCLE [20]. The evolutionary history was inferred by using Maximum Likelihood method and Kimura 2-Parameter model. The tree with the highest log likelihood is shown. The percentage of trees in which the associated taxa clustered together is shown next to the branches. The tree is drawn to scale with the branch lengths measured in the number of substitutions per site. Accession Number KF307755, KF3077756, MN271357, MN271380 were used as outgroup.

6. Results

Out of the 405 children enrolled, 45 (11.1%) were positive for norovirus using RT-PCR (**Table 1**). Only norovirus genogroup I and II were recovered in this study, norovirus genogroup IV were not detected. Norovirus genogroup II was the most prevalent (84.4%) among the children (**Table 2**). GII noroviruses where also most commonly encountered among all centers included in this study (**Table 3**). Based on capsid gene sequences recovered from the 45 norovirus positive samples, two GI noroviruses genotype was detected in this study GI.3, 71.4% and genotype G1.5, 28.6% respectively. Seven genotypes of GII noroviruses were detected, genotypes GII.4 (63%), GII.12 (7.9%), GII.17 (7.9%), GII.6 (5.3%), GII.7 (5.3%), GII.14 (5.3%) and GII.2 (5.3%) (**Table 4**). GII.4 was found to be the most prevalent genotypes in all States studied, Delta (53.4%), Bayelsa (75.0%) and Edo State (46.2%) (**Table 5**).

Subtyping analysis with the norovirus genotyping tool demonstrated that the majority of the norovirus GII.4 recovered in this study where homologous to the

DOI: http://dx.doi.org/10.5772/intechopen.94389

Location	No. Tested	No. Infected (%)	P value
Central Hospital, Warri	83	13 (15.7)	0.0032
PHC, Pessu/ Ugbuwangwe	68	15 (22.1)	
FMC Yenagoa	80	4 (5.0)	
Central Hospital, Benin	99	3 (3.0)	
Stella Obasanjo Hospital	75	10 (13.3)	

$P < 0.05$.

Table 1.
Norovirus infection detected by RT-PCR among children under 5 years from different health facilities in south–south region of Nigeria.

Norovirus genogroup	Frequency (%)
GI	7 (15.6)
GII	38 (84.4)
Total	**45**

Table 2.
Frequency of occurrence of genogroup of norovirus in study population.

Study Centre	No. Tested	Nov GI	Nov GII
Central Hospital, Warri	13	2 (15.4)	11 (84.6)
PHC, Pessu/Uguwangwe	15	2 (13.3)	13 (86.7)
FMC, Yenagoa	4	1 (25.0)	3 (75.0)
Central Hospital, Benin	3	1 (33.3)	2 (66.7)
Stella Obasanjo CWH, Benin	10	1 (10.0)	9 (90.0)

Table 3.
Distribution of norovirus genogroups among the health facilities in the study.

Genotypes	Frequency
GI	
GI.3	5 (71.4%)
GI. 5	2 (28.6%)
GII	
GII.2	2 (5.3%)
GII.4	24 (63.0%)
GII.6	2 (5.3%)
GII.7	2 (5.3%)
GII. 12	3 (7.9%)
GII. 14	2 (5.3%)
GII.17	3 (7.9%)

Table 4.
Distribution of genotypes of norovirus in study population.

Genotypes	Delta	Bayelsa	Edo
GI.3	3 (10.7)	0	2 (15.4)
GI.5	1 (3.6)	1 (25.0)	0
GII.2	1 (3.6)	0	1 (7.7)
GII.4	15 (53.4)	3 (75.0)	6 (46.2)
GII.6	1 (3.6)	0	1 (7.7)
GII.7	2 (7.1)	0	0
GII.12	2 (7.1)	0	0
GII.14	1 (3.6)	0	1 (7.7)
GII.17	1 (3.6)0	1 (7.7)	
Total	**28**	**4**	**13**

Table 5.
Distribution of norovirus genotypes among study centers.

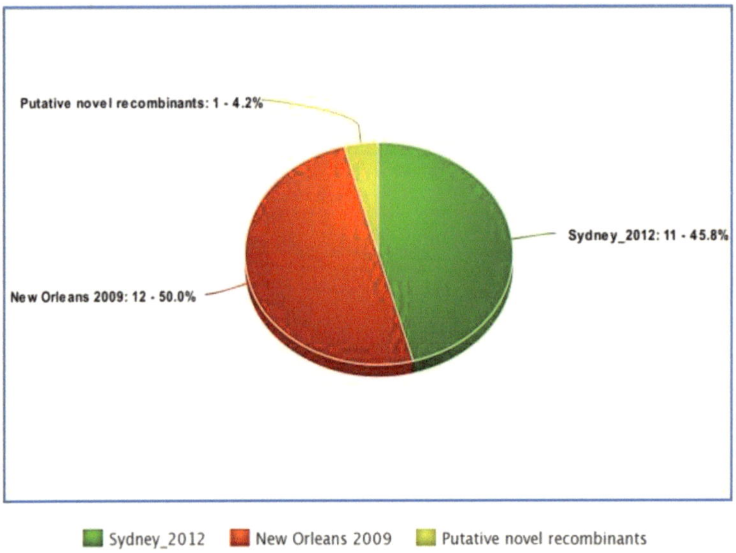

Figure 2.
Norovirus GII.4 variants among patients in this study.

New Orleans 2009 variant 12 (50.0%), while 10 (41.7%) isolates showed homology to the Sydney 2012 strains. Two GII.4 sequences were unassigned. All sequences where further subjected to phylogenetic analysis, genotypic assignment of all sequences was based on bootstrap cut off values >70%. The result showed that norovirus GII.4 were more genetically diverse (**Figure 2**).

Three genetic clusters of Sydney 2009 variants where found to circulate among study participants, with varying sequence identity (range 74–100%). The two sequences that were unassigned using the genotyping tool were assigned following phylogenetic analysis. One GII.4 isolate (MN271364) did not cluster with either New Orleans or Sydney 2009 reference strains despite been assigned to be GII.4 Sydney strain by the genotyping tool, but had genetic sequence similarity to a norovirus strain isolated in 1993 in Bristol (X76716), which itself is a recombinant strain which was neither New Orleans nor Sydney 2009 strain (**Figure 3**). This strain is however deemed a putative novel recombinant.

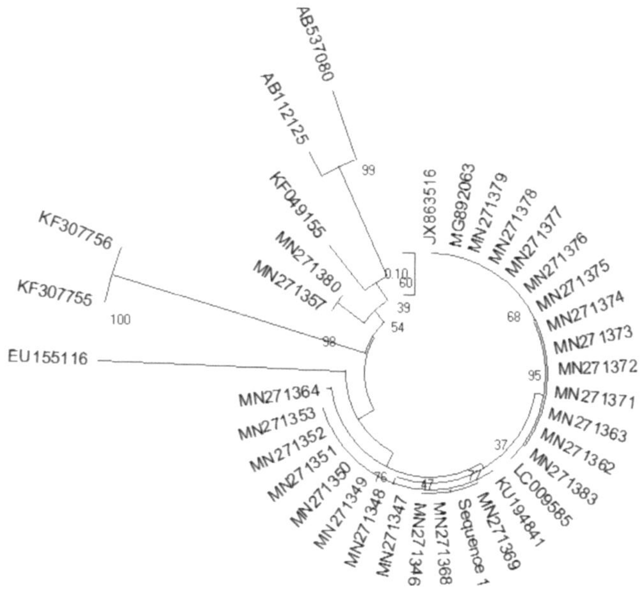

Figure 3.
Phylogenetic tree of norovirus GII.4 identified in this study: There were a total of 1487 positions in the final dataset. Sequences in this study are represented with filled triangles, reference sequences are not marked. MN271364 had low bootstrap values, as such classified to be a putative novel recombinant.

7. Discussion

A prevalence of 11.1% RT-PCR confirmed norovirus infection among children under 5 years of age was observed. This prevalence report is higher than the 6.7% in Ile –Ife, Osun State [21], but less than the 37.3% among children in Lagos, Nigeria [22]. This report supports the hypothesis that the prevalence of norovirus within Nigeria is not homogenous across communities and health centers. Hence, community and hospital-based surveillance are needed to provide an estimate of norovirus burden in Nigeria. The finding of this study might imply that norovirus may be one of the major etiologic agents of diarrhea among children less than 5 years in South–South region of Nigeria.

Only norovirus belonging to GI and GII genogroups were identified in this study, this concurs with findings from previous norovirus molecular epidemiologic studies [23, 24]. Norovirus genogroup II was the most commonly recovered norovirus in this study. This report concurs with findings from other parts of the World [25, 26]. This study further highlights the superior role of norovirus genogroup II in cases of norovirus induced gastroenteritis among children. In this study, GII.4 norovirus strains were most commonly detected. It has been well established that GII. 4 noroviruses are responsible for the majority of outbreaks worldwide [27].

Phylogenetic analysis revealed two GII.4 sequence variants, two Sydney 2012 clusters and one cluster for the New Orleans 2009 strain and a putative recombinant GII.4 virus. It is established that new variants of GII.4 norovirus emerge every 1–2 years, due to genetic recombination and point mutations resulting in the generation of new genetic clusters, recombinants/genotypes allowing increasing genetic fitness and continuous spread in populations by evading host immune responses [28]. GII.4 Possess the largest number of intra-genotypic variants and recombinants [29]. It is also possible for two or more variants to co-circulate at the same time in a geographical location [29]. Notable GII.4 variants causing majority of pandemic diseases are the Sydney 2012, New Orleans 2009, Farmington Hills 2002, US95/961995, Hunter 2004 and Den Haag 2009 [30]. Certain variants cause localized epidemics, Cairo 2007, Japan 2008 and Asia 2003 [30].

This study reports on the finding of norovirus GII.17 among children in Southern, Nigeria for the first time, this genotype is known to be very virulent [17]. First documented evidence of emergence of GII.17 occurred in the winter of 2014–2015 in Asia [31].

The finding of this study provides evidence of the existence of diverse genetic subtypes of norovirus in our locality. This data has illuminated the epidemiological profile of norovirus induced diarrhea/gastroenteritis in our locality.

7.1 Study limitations

Whole genome sequencing of both the major capsid protein and polymerase protein (RDrP) of the putative novel recombinant identified could not be performed because of lack of funds. This would have help to inform on the proper assignment of the putative novel recombinants as a novel genotype. Analysis of recombination breakpoints sites, homology model, evolutionary and phylogeographic relationships with reference will help provide greater information on the novel recombinants identified in this study.

7.2 In conclusion

Prevalence of RT-PCR confirmed norovirus infection among children with gastroenteritis in our locality was 11.1%. Norovirus genogroups I and II were the norovirus recovered in this study. G11.4 noroviruses were more prevalent, rare norovirus genotypes GII.2 and GII.17 were also encountered. The prevailing GII.4 variants in our study area belong to the New Orleans 2009 and Sydney 2012 strain, a putative novel GII.4 recombinant was encountered among study participants.

8. Published articles related to this research

Get more details on my research from these publications

1. Osazuwa, F., Grobler, H.S. & Johnson, W. Phylogenetic lineage of GII.17 norovirus identified among children in South–South, Nigeria. *BMC Res Notes* **13,** 347 (2020). https://doi.org/10.1186/s13104-020-05185-0 (SNIP: 0.864)

2. Osazuwa F, Johnson, W, GroblerHS. Genetic lineage of genogroup I norovirus identified among children with diarrhea in Niger-Delta region, Nigeria. Infectious Dis 2019, 3: 213–215. (Impact Factor: 2.494)

3. Osazuwa F, Okojie R, Akinbo FO, Johnson W, Grobler HS. Genetic diversity of norovirus among children under 5 years in the South–South region of Nigeria. New Zealand J Medical Laboratory Sciences 2020, 74:(1) 39–43.

Conflict of interest

The authors declare that they have no competing interests.

Ethics approval and consent to participate

This study was approved by the Ethics Committees of the Ministry of Health of Delta, Bayelsa and Edo States with reference numbers CHW/VOL14/130, FMCY/REC/ECC/2017/OC/046 and 732/T/89 respectively.

Footnote

The author is a doctoral student in Medical Molecular Biology at University of Benin.

Abbreviations

MUSCLE	**Mu**ltiple **S**equence **C**omparison by **L**og- **E**xpectation
dNTPS	Dideoxynucleotide triphosphates
RNA	Ribonucleic Acid

Author details

Favour Osazuwa
MDS Molecular Services, Sub-Saharan African Circuit Office, 36 Mike Akhigbe Way, Jabi, Abuja, Nigeria

*Address all correspondence to: favourdesires@gmail.com

References

[1] Ahmed SM, Hall AJ, Robinson AE, Verhoef L, Premkumar P, Parashar UM, Koopmans M Lopman, BL. Global prevalence of Norovirus in cases of gastroenteritis: a systematic review and meta-analysis. *Lancet Infect Dis* 2014; 14: 725-730.

[2] Brunette GW (2017). *CDC Yellow Book 2018: Health Information for International Travel*. Oxford University Press. p. 269. ISBN 9780190628611.

[3] "Norovirus Symptoms". *CDC*. 24 June 2016. Archived from the originalon 6 December 2018.

[4] Bok, K. and Green, K. Y (2013-03-16). "Norovirus Gastroenteritis in Immunocompromised Patients". *New England Journal of Medicine*. 368 (10): 971.

[5] Heijne JC, Teunis P, Morroy G, Wijkmans C, Oostveen S, Duizer E, Kretzschmar M, Wallinga J (2009). "Enhanced Hygiene Measures and Norovirus Transmission during an Outbreak". *Emerg. Infect. Dis*. 15 (1): 24-30.

[6] International committee on taxonomy of viruses *"ICTV Report Caliciviridae"*.

[7] Chhabra P, de Graaf M, Parra GI, Chan MC, Green K, Martella V *et al*. Updated classification of norovirus genogroups and genotypes. *J Gen Virol* 2019; 100 (10): 1393-1406.

[8] Green K. Caliciviridae: The noroviruses. In Knipe DM, Howley PM. (editors) *Fields virology* Philadelphia, PA: Lippincott Williams & Wilkins; 2013 pp 586-608.

[9] Lee S, Liu H, Wilen CB, Sychev ZE, Desai C *et al.* A secreted viral nonstructural protein determines intestinal norovirus pathogenesis. *Cell Host Microbe* 2019; 25:e845845-857.

[10] Noel JS, Fankhauser RL, Ando T, Monroe SS, Glass RI (2000). "Identification of a distinct common strain of "Norwalk-like viruses" having a global distribution". *J. Infect. Dis*. 179(6): 1334-44.

[11] Lu, L., Zhong, H., Xu, M. *et al.* Genetic diversity and epidemiology of Genogroup II noroviruses in children with acute sporadic gastroenteritis in Shanghai, China, 2012-2017. *BMC Infect Dis* 19, 736 (2019). https://doi.org/10.1186/s12879-019-4360-1.

[12] Worobey, M., and E. C. Holmes. 1999. Evolutionary aspects of recombination in RNA viruses. *J. Gen. Virol.* 80:2535-2543

[13] Lole, K. S., R. C. Bollinger, R. S. Paranjape, D. Gadkari, S. S. Kulkarni, N. G. Novak, R. Ingersoll, H. W. Sheppard, and S. C. Ray. 1999. Full-length human immunodeficiency virus type 1 genomes from subtype C-infected seroconverters in India, with evidence of intersubtype recombination. *J. Virol.*73:152-160.

[14] Bull RA, Tanaka MM, White PA: Norovirus recombination. *J Gen Virol* 2007, 88: 3347-3359. 10.1099/vir.0.83321-0

[15] Kaewkanya Nakjarung, Ladaporn Bodhidatta, Pimmnapar Neesanant, Paphavee Lertsethtakarn, Orntipa Sethabutr, Ket Vansith, Chhour Y. Meng, Brett E. Swierczewski, Carl J. Mason, "Molecular Epidemiology and Genetic Diversity of Norovirus in Young Children in Phnom Penh, Cambodia", *Journal of Tropical Medicine*, vol. 2016, Article ID 2707121,

[16] Fumian, Tulio Machado et al. "Norovirus Recombinant Strains Isolated from Gastroenteritis Outbreaks in Southern Brazil, 2004-2011." *PloS one* vol. 11,4 e0145391. 26 Apr. 2016, doi:10.1371/journal.pone.0145391

[17] Payne DC, Vinje J, Szilagyi PG, Edwards MD, Staat MA, Weinberg GA et al., Norovirus and medically attended gastroenteritis in US children. *The N Engl J Med*. 2013; 368:1121-1130.

[18] Kroneman A, Vennema H, Deforche K, v d Avoort H, Peñaranda S, Oberste MS, Vinjé J KM. An automated genotyping tool for enteroviruses and noroviruses.*JClin Virol* 2011; 51:121-125.

[19] Lee PY, Costumbrado J, Hsu CY, Jim YH. Agarose gel electrophoresis for the separation of DNA fragments. J visualized experiments. 2012: 62 doi 3791/39.

[20] Edgar RC. MUSCLE: Multiple sequence alignment with high accuracy and high throughput. *Nucleic Acids Res* 2004; 32 :1792-1797.

[21] Japhet, O.M, Adeyemi A. O., Famurewa, O., Svensson, L.and Nordgren, J. Molecular epidemiology of rotavirus and norovirus in Ile-Ife, Nigeria: high prevalence of G12P [8] rotavirus strains and detection of a rare norovirus genotype. *J Med Virol*; 2012; 84:1489-1496.

[22] Ayolabi CI, Ojo DA, Armah GE, Akpan I, Mafiana CF. Detection and partial characterization of norovirus among children with acute gastroenteritis in Lagos, Nigeria. *Int J Med. Med. Sci*.; 2010; 2(7):216-221.

[23] Maslin J, Nicand E, Ambert-Balay K, Fouet C, Kaplon J, Haus R, Pothier P, KohliE.Detection and characterization of Human caliciviruses associated with sporadic acute diarrhea in adults in Djibouti (horn of Africa). *Am J Trop Med Hyg*;2008;78:522-526.

[24] Armah GE, Gallimore CI, BinkaFN, Asmah RH, Green J, Ugoji U, Anto F, Brown DWG, Gray JJ. Characterization of Norovirus Strains in Rural Ghanaian Children with Acute Diarrhoea. *J. Med. Virol* 2006; 78: 1480-1485.

[25] Mattison K, Sebunya TK, Shukla A, Noliwe LN, Bidawid S. Molecular detection and characterization of noroviruses from children in Botswana. *J Med Virol*; 2010; 82(2):321-324.

[26] Huynen P, Mauroy A, Martin C, Savadogo LG, Boreux R, Thiry E, Melin P, De Mol P. Molecular epidemiology of norovirus infections in symptomatic and asymptomatic children from BoboDioulasso, Burkina Faso. *J ClinVirol*. 2013;58(3):515-521.

[27] Kabue JP, MeaderE, Hunter PR, Potgieter N. Genetic characterization of Norovirus strain in outpatient children of rural community of Vhembe district South Africa 2014-2016. 2017; *J Clin Virology* 94: 100-106.

[28] Xu J, Liu T, Wang H, Shen H: Intergenotype Recombination among New Norovirus GII.4 Variants in the Complete Genome. Intervirology 2017;60:138-143. doi: 10.1159/000484048.

[29] Siebenga JJ, Vennema H, Zheng DP, Vinje J, Lee BE, Pang XL *et al*. norovirus is a global problem:emergence and spread of norovirus GII.4 variants, 2001-2007. *J infect Dis* 2009; 200 (5), 800-8122.

[30] Hasing ME, Hazes B, Lee BE, Preiksaitis JK, Pang XL. Detection and analysis of recombination in GII.4 norovirus strains causing gastroenteritis causing gastroenteritis outbreaks in Alberta.*Infect Genet Evol* 2014; 27: 181-192.

[31] Lu JH, Sun L, Fang L, yang F, Mo Y, Lao J, Zhng H *et al*., Gastroenteritis outbreaks caused by norovirus GII.17, Guangdong. Province, China 2014-2015. Emerging Infect Dis 2015: 21 7: 1240-1242

Chapter 12

Genetic Diversity of Insulin Resistance and Metabolic Syndrome

Sanghoo Lee, Jinwoo Ahn, Jimyeong Park, Hyeonkyun Na, Youngkee Lee, Yejin Kim, Gayeon Hong and Kyoung-Ryul Lee

Abstract

A key in the etiology of a cluster of metabolic syndrome such as hyperglycemia, dyslipidemia, and obesity is known for insulin resistance, which is becoming a major global public health problem. Extensive studies have revealed many genetic factors for both insulin resistance and the components of metabolic syndrome. Advanced modern genotyping methods including genome-wide association studies and next-generation sequencing have allowed for the identification of both common and rare genetic variants related to these chronic disease-associated traits. Multiple genotype–phenotype studies are also needed to identify new and accurate genetic biomarkers in these conditions. The purpose of this chapter is to present genetic variants related to the pathogenesis of metabolic syndrome and insulin resistance and is to review the relevance between insulin resistance and metabolic syndrome clusters in terms of genetic diversity.

Keywords: metabolic disorders, genetic variation, genetic biomarker, genetic analysis

1. Introduction

Metabolic syndrome (MetS), known as syndrome X, Deadly Quartet, or insulin resistance syndrome is characterized by a cluster of metabolic risk factors such as obesity, hypertension, dyslipidemia, and elevated fasting plasma glucose [1]. The metabolic risk factors can result in type 2 diabetes (T2D) and cardiovascular disease (CVD) that are due to both genetic and environmental factors [2, 3]. For these reasons, MetS is becoming a global epidemic. The prevalence of MetS is estimated at 11.9–37.1% in Asia-Pacific region [4], 11.6–26.3% in Europe [5], and 22–24% in North America [6].

One of the primary mediators of MetS is known for insulin resistance (IR), which is a pathological state of improper cellular response to the hormone insulin in insulin-dependent cells such as skeletal muscle and adipose tissue [7]. IR is present in the majority of people with many metabolic disorders such as MetS and T2D. IR plays a crucial role in the pathophysiology of both T2D and CVD [7] but inversely related to insulin sensitivity in insulin-dependent tissues [8]. Clinical risk factors such as obesity, dyslipidemia, inflammation, hyperinsulinemia, and dysglycemia are also known to affect IR.

 IntechOpen

Although environmental factors such as lifestyle and socioeconomic status contribute to the development of IR and MetS, both IR and MetS are also being determined by genetic factors, as strongly evidenced by early familial genetic studies [9–11]. Based on these studies, advanced genetic analysis technologies such as genome-wide association studies (GWAS) and next-generation sequencing (NGS) are extensively being used to identify both common and rare genetic variants related to these metabolic disorder-associated traits.

This chapter is to present an overview of genetic variants involved in the pathogeneses of MetS and IR and to review the relevance between IR and MetS clusters in terms of genetic diversity.

2. Heritability of MetS and IR

The pieces of evidence for the heritability and co-occurrence of the metabolic traits have been revealed through early familial and twin genetic studies. The heritability of MetS, as defined by NCEP:ATPIII (National Cholesterol Education Program Adult Treatment Panel III) criteria, was estimated to be 24% (p = 0.006) in the Northern Manhattan Family Study, which was conducted in 803 subjects from 89 Caribbean-Hispanic families [9]. Each component of MetS has also an important genetic basis. The heritability was estimated at 46% for waist circumference (WC), 24% for fasting glucose, 60% for HDL-cholesterol, 47% for triacylglycerol (TAG), and 16% for systolic and 21% for diastolic blood pressure (BP). In the Linosa Study including 293 Caucasian native subjects from 51 families (123 parents and 170 offsprings), the heritability of MetS, as defined by NCEP:ATPIII, was estimated to be 27% (p = 0.0012) [12]. Among its components, the heritability for blood glucose and high-density lipoprotein (HDL)-cholesterol was 10% and 54%, respectively. The highest heritability was observed in the clustering of central obesity, hypertriglycemia, and low HDL-cholesterolemia (31%, $p < 0.001$). In an early study including 2508 adult male twin pairs, the accordance for the clustering of hypertension, diabetes, and obesity in the same individuals was 31.6% in monozygotic pairs and 6.3% in dizygotic pairs [13]. These early pieces of evidence have spurred many studies to find genetic determinants of MetS.

Although common genetic variants related to IR have been identified, these variants are known to make up only 25–44% of the heritability of IR [14–16]. For this reason, it is necessary to find low-frequency and rare genetic variations that affect the heritability of MetS and IR.

3. Genetic variants of MetS and IR

Significant progress has been made over the past decade to identify the genetic risk factors associated with the various traits of MetS. Although the complexity of MetS makes the identification of a genetic component of the disorder difficult, pieces of evidence for genetic determinants of MetS have been revealed through the linkage analysis approach, candidate gene association studies, GWAS, epigenetic studies, microRNAs, long-non-coding RNAs, system biological studies, and more recently NGS and whole-exome sequencing.

3.1 Linkage analysis approach

Many chromosomes and locus associated with MetS or its components or a combination of some of its components have been identified through linkage analysis. This approach has identified candidate quantitative trait loci (QTL). In 2209 subjects from 507 Caucasian families, a QTL associated with body mass index (BMI), WC, and fasting plasma insulin on chromosome 3q27 was identified, which includes genes such as the solute carrier family 2 of the facilitated glucose transporter (*GLUT2*) [17].

In a study including 456 Caucasian (white) and 217 African-American (black) subjects from 204 families, evidence of linkage for increased body fat, abdominal visceral fat, TAG, fasting glucose, fasting plasma insulin, blood pressure, and decreased HDL-cholesterol was identified on chromosome 10p11.2 and 19q13.4 and 10q13.4 in white [18]. In black subjects, the linkage was identified on chromosome 1p34.1 [18].

In a study including four ethnic groups (Caucasian, Mexican-American, African-American, and Japanese-American), evidence of linkage of MetS traits (weight/waist, lipid factor, and BP) was identified, where there is a strong linkage on chromosome 2q12.1-2q12 for Caucasian subjects and 3q26.1-3q29 for Mexican-American subjects [19].

Genetic data were obtained for 2467 subjects from 387 three-generation families and 1082 subjects from 256 sibships, where a genomic region on chromosome 2 included a pleiotropic locus contributing to the clustering of multiple metabolic syndrome (MMS)-related phenotypes (BMI, waist-to-hip ratio (WHR), subscapular skinfold, TAG, HDL-cholesterol, homeostasis model assessment (HOMA) index, plasminogen activator inhibitor-1-antigen, and serum uric acid) [20].

In a study including 250 German families, a genome-wide linkage scan for T2D supports the existence of MetS locus on chromosome 1p36.13 and T2D locus on chromosome 16p12.2 [21].

In a study with 715 individuals in 39 low-income Mexican American families, strong evidence of a major locus near markers *D1S1597* and *D1S407* on chromosome 1p36.21 that influences variation in symptomatic or clinical gallbladder disease through a genome-scan and linkage approach was revealed [22].

3.2 Candidate gene association studies

Candidate gene association studies identify and investigate many candidate genes that regulate biological processes related to MetS. Analysis of the mutation burden of candidate genes is among the first methods used to uncover MetS genes. Especially, the association of MetS and single nucleotide polymorphisms (SNPs) in related genes has been examined in many studies.

An association with MetS for 8 SNPs that are mostly in 25 genes involved in lipid metabolism was revealed in 88 studies with 4000 subjects. In these studies, the minor allele of C56G (*APOA5*), T1131C (*APOA5*), rs9939609 (*FTO*), C455T (*APOC3*), rs7903146 (*TCF7L2*), C482T (*APOC3*), and 174G > C (*IL6*) were more prevalent in subjects with MetS but the minor allele of Taq-1B (*CETP*) was less prevalent in those [23].

The association of *HSD11B1* variants and *HSD11B1* expression in abdominal adipose tissue with T2D, MetS, and obesity was identified in 802 studies. Especially, a polymorphic variant was identified to be related to T2D in a study including Pima Indians, and an association between MetS with another polymorphic variant at the *HSD11B1* gene was identified in an Indian study. However, most studies did not find an association between *HSD11B1* polymorphic variants and T2D, MetS, and obesity, suggesting that the variants may play a minor role to develop obesity, T2D, and MetS [24].

A meta-analysis study including 25 reports revealed an association of *ADIPOQ* rs2241766 and rs266729 polymorphisms with MetS in the Chinese population, where the G allele of rs2241766 increased the risk of MetS but no relevance to rs266729 was found [25].

In a study including 442 adults with MetS, it was revealed that *APOE* genotype affected IR, apolipoprotein (apo) CII, and CIII depending on plasma fatty acid (FA) levels in MetS. Elevated n-3 polyunsaturated FA (PUFA) was related to lower concentration of apo CIII in *E2* carriers and elevated C16:0 was related to IR in *E4* carriers. Decreased long-chain n-3 PUFA was associated with reduced apo CII level in *E2* carriers, after FA intervention. These results suggest that subjects with MetS may benefit from personalized dietary interventions based on *APOE* genotype [26].

A meta- and gene-based analysis including 18 studies was carried out to investigate the association of fat mass and obesity-related *FTO* gene polymorphisms with MetS, suggesting that FTO is strongly related to MetS ($p < 10^{-5}$) [27].

BALB/c mice are known to be resistant to a high-fat diet (HFD)-induced obesity. A recent study demonstrated that *Nod2−/−*BALB/c mice developed HFD-dependent obesity and risk factors of MetS such as hyperglycemia and hyperlipidemia. Interestingly, *Nod2−/−* HFD mice showed changes in the composition of gut flora and also delivered sensitivity to hyperglycemia, steatosis, and weight gain to wild type germ-free mice. Therefore, these results suggest that not only *Nod2* plays a novel role in obesity but also that *Nod2* and *Nod2*-regulated gut flora protect BALB/c mice from diet-induced obesity and metabolic disorders [28].

More recently, a multiple-genotype and multiple-phenotype analysis of a gene-based SNP set has been performed to identify new susceptible variants associated with MetS in 10,049 Korean individuals [29]. In this study, 27 SNP pairs were associated with MetS in the discovery stage and also replicated. Of these SNPs, 3 SNP pairs in each SIDT2, UBASH3B, and CUX2 gene were significant in the multiple-SNP and multiple-phenotype analysis rather than in the single-SNP and multiple-phenotype analysis. Especially, an association of MetS with an intronic SNP pair, rs7107152 ($p = 3.89 \times 10^{-14}$) and rs1242229 ($p = 3.64 \times 10^{-13}$), in *SIDT2* gene at 11q23.3 was found. These 2 SNPs are also associated with the expression of *SIDT2* and *TAGLN* that promote insulin secretion and lipid metabolism, respectively. These results suggest the usefulness of the multiple-genotype and multiple-phenotype analysis platform to identify new genetic loci in complex metabolic disorders such as MetS.

Although candidate genetic association studies have reported many genetic variations associated with MetS, often these results have not been replicated in other populations and been identified through GWAS. These examples include polymorphisms in or near genes encoding GAD2, ENPP1, and SCL6A14. Moreover, most of the identified genes underlie only one MetS trait. Few exceptions contain mutations in *ADIPOQ* related to hypertension, T2D, and dyslipidemia. Other examples contain mutations in *FOXC2*, *SREBP1*, *NR3C1*, and *GNB3* genes.

3.3 GWAS

GWA studies are an approach used to analyze an association of SNPs in subjects with MetS or IR and to date, being carried out by many researchers.

3.3.1 Genetic diversity of MetS

Over the past 10 years, GWAS have identified many genetic variants associated with each trait of MetS. Many genetic loci associated with lipid levels were discovered and refined by GWAS which identified 157 loci related to lipid levels at $p < 5 \times 10^{-8}$, including 62 loci not previously related to lipid levels [30]. Among the loci, 39 loci were associated with TAG levels and 71 with HDL-cholesterol.

Several loci associated with each component of MetS have pleiotropic effects on two or more traits related to MetS.

A GWA meta-analysis including 76,150 subjects showed that the rs2943634 variant near *IRS1* was associated with an elevated visceral to subcutaneous fat ratio, IR, dyslipidemia (higher TAG and lower HDL-cholesterol), risk of T2D, and reduced adiponectin levels [31]. Genetic variants in the *GCKR* gene were linked to fasting glucose levels [32], TAG [33], and non-alcoholic fatty liver disease [34]. Variants for obesity in/near *FTO* and *MC4R* genes were associated with specific measures of adiposity such as WC [35], HDL-cholesterol levels [30], IR [36, 37], and risk of T2D [35]. Variants in the *GRB14* gene were also linked to BMI-adjusted WHR [38], T2D [39], and fasting insulin levels.

In a GWAS comparing T2D subjects (n = 1924) and control (n = 2938) for autosomal SNPs (n = 490,032), SNPs in *FTO* gene region on chromosome 16 were identified to be strongly associated with T2D (e.g., rs9939609, OR = 1.27, $p = 5 \times 10^{-8}$). This strong association was furthermore reproduced by analyzing SNP rs9939609 in T2D subjects (n = 3757) and controls (n = 5346) (OR = 1.15, $p = 9 \times 10^{-6}$) [35]. However, some of these variants were also associated with MetS, suggesting that genes such as *FTO*, *MC4R*, and *IRS1* play important roles in the progression of MetS [40]. Especially, among several obesity-related loci found to be related to MetS-related traits in the GWAS studies, *FTO* and *MC4R* genes are considered to be the strongest candidates for body weight control, and *IRS1* is known to have an important effect on IR. These results may provide valuable information to understand the role of genetic control of adiposity and IR in the development of MetS.

GWA studies of MetS as a whole or a combination of its traits have also identified a number of both common and rare genetic variants. A GWA study was conducted to identify common genetic variants of MetS and its related components in 4560 Indian Asian men with a high prevalence of these conditions. In this study, no genetic variation showed an association with MetS as a whole. However, several variations were related to single components. Especially, 2 SNPs near *CETP*, 2 at 8p21.3 near *LPL*, 2 at 11q12.2 near *FADS1* and *FADS2*, and 1 at 21q22.3 near *FLJ41733* associated with HDL-cholesterol ($p < 10^{-6}$), and 1 SNP near *TCF7L2* associated with T2D ($p < 10^{-6}$) were identified [41].

A study by the STAMPEDE Consortium included 13 independent studies, comprising a total of 22,161 subjects of European ancestry, was conducted to find genetic determinants contributing to the correlated architecture of MetS traits, using MetS as a whole or pairs of its components as phenotypes [42]. In this study, the 5 SNPs in *LPL*, *APOA5* cluster (*ZNF259*, *BUD13*, and *APOA5*), and *CETP* genes were found to be associated with MetS. Especially, a total of 27 genetic variants in or near 16 genes were associated with bivariate combinations of 5 MetS traits, including variants in *LIPC* (chromosome 15q21-q23) associated with HDL-cholesterol-fasting glucose (rs2043085, $p = 1.3 \times 10^{-8}$) and with WC-HDL-cholesterol (rs10468017, $p = 5.5 \times 10^{-8}$), *ABCB11* (chromosome 2q24) associated with HDL-cholesterol-fasting glucose (rs569805, p = 8.5×10^{-8}) and with HDL-cholesterol-TAG (rs2954026, $p = 7.9 \times 10^{-9}$), *TRIB1* (chromosome 8q24.13) associated with TAG-BP (rs2954033, $p = 8.5 \times 10^{-9}$), *TFAP2B* (chromosome 6p12) associated with WC-fasting glucose (rs2206277, $p = 1.3 \times 10^{-7}$), *LOC100128354* (chromosome 11q21) and *MTNR1B* associated with BP-fasting glucose (rs1387153, $p = 8.1 \times 10^{-9}$), HDL-cholesterol-fasting glucose (rs1387153, p = 2.4×10^{-9}), and TAG-fasting glucose (rs10830956, $p = 4.8 \times 10^{-11}$), *LOC100129500* (chromosome 19q13.2) associated with HDL-cholesterol-TAG (rs439401, $p = 1.0 \times 10^{-8}$), and LOC100129150 variants with HDL-cholesterol-TAG (rs9987289, $p = 1.1 \times 10^{-8}$) and HDL-cholesterol-WC (rs9987289, $p = 3.7 \times 10^{-8}$) [42]. These common genetic variations can partly explain the covariation in the MetS traits.

In a study for susceptibility loci associated with MetS and its traits was conducted in four Finnish cohorts consisting of 2637 MetS cases and 7927 controls. One genetic variant (rs964184) in A *APOA1/C3/A4/A5* gene cluster region on chromosome 11, known as lipid locus was found to be associated with MetS in all 4 study samples ($p = 7.23 \times 10^{-9}$ in meta-analysis) and significantly associated with several very low-density lipoprotein (VLDL), TAG, and HDL metabolites ($p = 0.024–1.88 \times 10^{-5}$). Several genetic variants in or near 4 known loci related to lipids (LPL, CEPT, APOA1/C3/A4/A5, and APOB) were strongly associated with TAG/HDL/WC factors [43], but none was associated with 2 or more uncorrelated MetS traits. A polygenetic risk score (PRS), which was calculated as the number of alleles in loci associated with individual MetS traits, was significantly associated with MetS traits. These results suggest that genes associated with lipid metabolism pathways have crucial roles in the development of MetS. However, in this study, little evidence for pleiotropy associating obesity and dyslipidemia with the other MetS traits (hyperglycemia and hypertension) was found.

Genetic loci associated with the clustering of 6 MetS-related phenotypes (atherogenic dyslipidemia, vascular dysfunction, vascular inflammation, prothrombotic state, central obesity, and elevated plasma glucose) including 19 quantitative traits were identified by GWAS in 19,486 European American and 6287 African American Candidate Gene Association Resource Consortium participants [44]. In this study, 606 significant SNPs in and near 19 loci ($p = 2.13 \times 10^{-7}$) were identified in European Americans. Many of these loci were associated with at least one MetS-related trait domain and consistent with results in African Americans. Three new pleiotropic loci in or near *APOC1*, *BRAP*, and *PLCG1*, which were associated with multiple phenotype domains were identified. Several loci previously identified by GWAS for each trait of MetS, including *LPL*, *ABCA1*, and *GCKR*, were also associated with at least 2 trait domains. These results support the presence of genetic variants with pleiotropic effects on adiposity, inflammation, glucose regulation, dyslipidemia, vascular dysfunction, and thrombosis. Such loci could apply to uncover metabolic dysregulation and identify targets for early intervention.

3.3.2 Genetic diversity of IR

To date, many of the loci related to risks of developing IR have been identified and found to be associated with measures such as insulin sensitivity and secretion.

In an early meta-analysis, genetic variants related to IR were identified in 21 cohorts consisting of a non-diabetic group, which includes 46,186 subjects with measures of fasting glucose and 38,238 subjects with measures of fasting glucose and HOMA-IR. In additional 76,558 subjects, 25 SNPs were followed up with this approach, identifying 16 loci related to fasting glucose and 2 loci related to fasting insulin. In this study, several loci near *GCKR* including a new locus near *IGF1* were found to be associated with IR [32]. These results were replicated in a further 14 cohorts, which included 29,084 non-diabetic subjects with measures of fasting proinsulin, insulin secretion, and sensitivity [45]. Association of 37 risk loci for T2D with measures of insulin secretion, sensitivity, and processing and clearance was examined in 58,614 non-diabetic subjects and 17,327 subjects with measures of glycemic traits, revealing that the risk loci were grouped into 5 major categories including one cluster with 4 loci (*PPARG*, *KLF14*, *IRS1*, and *GCKR*) associated with IR [46].

A joint meta-analysis (JMA) approach has been developed to identify genetic variants associated with either fasting glucose and/or fasting insulin. This approach identified 6 loci that include 5 new variants associated with levels of fasting insulin (*IRS1*, *COBLL1-GRB14*, *PPP1R3B*, *PDGFC*, *UHRF1BP1*, and *LYPLAL1*) [47].

A large-scale meta-analysis including 133,010 subjects identified 17 loci significantly associated with fasting insulin. These loci included genes associated with other metabolic traits (*FTO*, *TCF7L2*, *PPARG*, ARL15, *RSPO3*, and *ANKRD55-MAP3K1*) and newly identified loci (*YSK4*, *FAM13A*, *TET2*, *PEPD*, and *HIP1*) [48]. In 2 further studies, these loci were used to make an IR PRS identify the relationship between variants associated with fasting insulin and the risk of each individual developing IR and T2D [49, 50]. The 2 studies identified that the IR GRSs were associated with decreased insulin sensitivity and lower BMI. In one of these 2 studies, a PRS was generated from 10 genetic loci that were related to lower HDL-cholesterol and higher TAG (*PPARG*, *IRS1*, *GRB14*, *PEPD*, *FAM13A1*, *PDGFC*, *LYPLAL1*, *RSPO3*, *ARL15*, and *ANKRD55-MAP3K1*) [49]. In the other study, 19 loci were used to generate their IR PGS and 11 risk variants were identified to be related to increased TAG and lower HDL-cholesterol along with a lower BMI [50]. In these studies, IR PRSs were used to highlight that IR can develop without obesity and high BMI.

IRS1 is a signaling adapter protein that is encoded by the *IRS1* gene in humans and a key factor of the insulin signaling pathway initiating the activation of phosphoinositide 3-kinase (PI3K) in response to insulin. The C allele at the SNP (rs2943641) near the *IRS1* gene was found to be associated with IR and hyperinsulinemia in a European population. Through functional studies, the risk allele was found to be associated with lower levels of basal IRS1 protein and decreased PI3K activity during insulin infusion, indicating a causative role for the genetic variant on risk of IR [51]. The SNP (rs2943650) near *IRS1* was also associated with lower HDL-cholesterol, elevated TAG, IR, and lower body fat percentage [31]. Significant associations of the variants in *FTO* with fasting insulin and insulin sensitivity were identified [37]. The risk variant in or near *TCF7L2* was found to be associated with both impaired β-cell function and IR [52]. A variant in *NAT2* was also found to be associated with a measure of insulin sensitivity in four European cohorts of 2764 non-diabetic individuals [53], supporting a role for *NAT2* in insulin sensitivity. In this study, a variant of *NAT2* was found to be strongly associated with reduced insulin sensitivity that was independent of BMI. The A allele at the SNP (rs1208) was significantly associated with IR-related traits, including increased fasting glucose, total cholesterol and LDL-cholesterol, hemoglobin A1C (HbA1c), TAG, and coronary artery disease (CHD). IGF1 is functionally similar to insulin and controls growth and development. Lower levels of IGF1 were found to be associated with decreased insulin sensitivity [54], and the SNP (rs35767) in the *IGF1* gene suggested that the G allele has lower levels of IGF-1 compared to the A allele [55].

In a GWA study of a UK cohort of Indian-Asian and European ancestry, *MC4R* was found to be associated with both IR with measures of HOMA-IR and WC, and with higher frequencies of risk alleles found in the Indian-Asian cohort [36].

In a GWA study of a cohort with Indian ancestry, 2 loci near *TMEM163* were found to be associated with both reduced plasma insulin and HOMA-IR [56].

In a GWA study of an African-American cohort, the SNP (rs7077836) near *TCERG1L* and the SNP (rs17046216) in *SC4MOL* were found to be associated with both fasting insulin and HOMA-IR [57]. *ARL15* belongs to a family of intracellular vesicle trafficking, and its exact function remains unknown. However, variants in *ARL15* were found to be associated with decreased levels of adiponectin and risk of T2D, CHD, and IR as measured by HOMA-IR [58].

To date, approximately 60 loci related to the risk of IR have been identified through GWAS, and among them, the top 10 IR-related loci have been replicated in 2 GWA studies [48, 59]. They are in and near the noncoding regions of *IRS1* (rs2943645), *PPARG* (rs17036328), *GRB14* (rs10195252), *PEPD* (rs731839), *PDGFC* (rs6822892), *MAP3K1* (rs459193), *ARL15* (rs4865796), *FAM13A* (rs3822072),

RSPO3 (rs2745353) and *LYPLAL1* (rs4846565). The PRS including the risk alleles of the 10 loci was associated with the cardiometabolic phenotypes such as lower BMI, lower body fat percentage, smaller hip circumference, and decreased leg fat mass as well as the risk phenotypes such as higher fasting insulin and higher TAG levels. These results suggest that limited storage capacity of subcutaneous adipose tissue (SAT) and consequently the elevation of ectopic fat deposition may be associated with the genetic link with IR [48, 49].

3.4 Low-frequency and rare variants

Whole-genome and exome sequencing approaches as relatively new genetic analysis technologies are being used to pinpoint the effects of minor allele frequencies (MAF $\leq$ 5%) and rare variants (MAF $\leq$ 0.5%) on the heritability of metabolic disorders such as MetS and IR.

The genomes of 1092 individuals from 14 populations were analyzed by using both the whole-genome and exome sequencing methods to identify low-frequency and rare genetic variants across 14 populations in the 1000 Genome Project [60]. The reference panels gained from this project can capture up to 98% accessible SNPs at a frequency of 1% in related populations and also enable researchers to analyze common and low-frequency variants in each individual from various populations. The 38 million SNP panels from the 1000 Genomes Project gave near complete coverage of common and low-frequency genetic variation with MAF $\geq$0.5% across European ancestry populations.

The European Network for Genetic and Genomic Epidemiology (ENGAGE) consortium carried out 22 GWAS to examine associations of genetic variants with WHR, fasting glucose, BMI, and fasting insulin in 87,048 individuals of European ancestry. This study identified two new loci for BMI, and fasting glucose and new lead SNPs at 29 loci including the SNP (rs1260326) near *GCKR* for fasting insulin [61].

Whole exome sequencing in a Danish cohort of 1000 individuals with T2D, BMI >27.5 kg/m^2, and hypertension and of 1000 controls identified 70,182 SNPs with MAF > 1%. Subsequent exome sequencing was performed in a two-stage follow-up in 15,989 Danes and a further 63,896 Europeans. This study showed associations of two common SNPs in *COBLL1* (MAF = 12.5%, OR = 0.88, p = 1.2 × 10^{-11}) and *MACF1* (MAF = 23.4%, OR = 1.10, p = 8.2 × 10^{-10}) with T2D and a low frequency variant in *CD300LG* (MAF = 3.5%, p = 8.5 × 10^{-14}) with fasting HDL-cholesterol [62].

Although physiological functions of risk variants in *COBLL1* and *MACF1* remain still unclear, a risk variant rs72836561 at *CD300LG* was found to be associated with the decreased mRNA expression level of *CD300LG* in both skeletal muscle and adipose tissue, elevated intramyocellular lipid, and decreased insulin sensitivity, through a functional study. These results suggest an association between this variant and MetS traits [63].

Exome sequencing in an Icelandic population revealed that a low-frequency (1.47%) variant (rs76895963) in *CCND2* decreased the risk of T2D (OR = 0.53, p = 5.0 × 10^{-21}) and was associated with elevated *CCND2* expression [64]. However, this variant was also associated with both greater height and higher BMI (1.17 cm per allele, p = 5.5 × 10^{-12} and 0.56 kg/m^2 per allele, p = 6.5 × 10^{-7}, respectively).

In 2733 individuals from the Greenlandic population that were historically isolated, combination analyses of Cardio-Metabochip based genotyping and exome sequencing revealed that a common variant in *TBC1D4* was associated with higher fasting glucose and decreased insulin sensitivity, resulting in decreased insulin-stimulated glucose uptake due to the variant [65].

Recently, whole-genome sequencing in a UK10K-cohort project consisting of 3781 healthy individuals with exome sequencing of 6000 individuals with either rare disease, severe obesity, or neurodevelopmental disorders has been performed to identify low-frequency and rare variants [66]. This project identified 24 million novel genetic variants including novel alleles associated with levels of TAG (*APOB*), adiponectin (*ADIPOQ*), and LDL-cholesterol (*LDLR* and *RGAG1*) from single-marker and rare variant aggregation tests and provided reference panels with increased coverage of low-frequency and rare variants. These panels are now being used to identify associations of low-frequency and rare variants with various traits related to health and disease.

3.5 Epigenetic determinants

Fatty acid-binding proteins (FABPs) play important roles in lipid metabolism and signaling. Dyslipidemia often occurs along with IR, obesity, and hypertension in individuals with MetS. The methylation status of CpG islands of a key regulator of lipid homeostasis, *FABP3*, is known as a quantitative trait associated with MetS phenotypes in humans. To identify if CpG methylation of *FABP3* affects MetS traits in 517 Northern European family populations, the CpG islands in the *FABP3* gene were profiled by a quantitative methylation analysis method. In this study, regional methylation was found to be strongly associated with plasma total cholesterol ($p = 0.00028$) and associated with LDL-cholesterol ($p = 0.00495$) [67]. Methylation at individual units was significantly associated with MetS traits such as insulin sensitivity and diastolic BP ($p < 0.0028$). These results suggest that DNA methylation of *FABP3* strongly affects MetS and might have important implications for insulin, lipids, and cardiovascular phenotypes of MetS.

Meanwhile, malnutrition in childhood, infancy, or fetus affects the prevalence of MetS in adults and their offspring [68], suggesting that maternal malnutrition affects gene expression in offspring through epigenetic mechanisms.

To date, most studies examining epigenetic changes related to MetS or IR have been conducted in animals and few studies have been conducted in humans. Therefore, further studies in humans are needed in the future.

4. CRISPR screen for genes affecting MetS or IR

Although many GWA studies are widely used to identify genetic loci associated with IR, it remains challenging to identify the causal gene in each locus [69]. Recently, structural and functional connections between GWAS loci and vicinal or distal genes were identified by chromosome conformation capture (3C) technology and expression quantitative trait loci (eQTL) studies [70, 71]. However, the 3C experiments are expensive and the eQTL studies cannot identify all the causal genes for a locus. Moreover, the 2 methods cannot pinpoint the causal genes and mechanisms related to the risk loci of IR. More recently, clustered regularly interspaced short palindromic repeats (CRISPR) knockout screening platform as an alternative method has been applied to pinpoint functions of new candidate causal genes at IR-associated loci in human preadipocytes and adipocytes [72]. This screening platform successfully characterized the functions of 10 new candidate causal genes at IR-associated loci. The 10 candidate genes (*PPARG*, *IRS-1*, *FST*, *PEPD*, *PDGFC*, *MAP3K1*, *GRB14*, *ARL15*, *ANKRD55*, and *RSPO3*) showed diverse phenotypes in the 3 insulin-sensitizing mechanisms, including lipid metabolism, adipogenesis, and insulin signaling, and the first 7 of these genes could affect all the 3 mechanisms. Additionally, 5 of 6 eQTL genes were identified as the top candidate causal genes (*IRS-1*, *GRB14*, *FST*, *PEPD*, and *PDGFC*), and expression levels of these 5 genes in human subcutaneous adipose tissue were found to be associated with increased risk of IR. Interestingly, it was first revealed in this study that the *FST, PEPD*, and *PDGFC* are involved in the functions of adipose in IR. Despite these findings, little is known about other functions of these 3 genes in adipose tissue, which may include novel molecular mechanisms for cardiometabolic disease. In this regard, studies will be needed to uncover new functions of these 3 genes in adipose tissue.

5. Conclusions

MetS and IR are central risk factors for the development of dyslipidemia, T2D, and CVD as well as complex metabolic traits. Many of the genetic variations implicated in the development of the MetS and IR are associated with glucose and lipid metabolism, respectively. Significant progress has been made in the identification of common and rare genetic variations associated with the MetS and IR in different populations, driven by the advent of GWAS and more recently, genome and exome sequencing approaches.

Despite many scientific efforts in identifying many genetic loci associated with the MetS and IR, their exact molecular pathogenesis remains unclear. Further studies are needed to identify functional links between the genetic variants and the phenotypes and subsequently to uncover the underlying molecular mechanisms of both metabolic disorders.

Clinical validation of the variants identified by several genetic analysis approaches is challenging for reasons resulting from implications by an individual's lifestyle and environmental factors as well as by genetic factors. In this aspect, studies including larger and more homogeneous populations are needed to identify genetic variants that underlie the association of the various traits of MetS and/or IR. However, results obtained from these studies should be replicated in different populations with a sufficient sample size to avoid false-positive associations and to reduce systematic biases and technical errors.

Approaches such as CRISPR, 3C, and eQTL are being used to identify structural and functional associations between genetic loci discovered by GWAS or exome sequencing and regional or distal genes. Among them, CRISPR as an *in vitro* screening platform may be used effectively to pinpoint causal genes at loci associated with MetS and IR in the near future. Currently, MetS and IR have been becoming a health and financial burden worldwide. The exact identification of validated variants that affect the MetS and IR might provide new preventive and treating strategies for the 2 metabolic disorders and related diseases.

Author details

Sanghoo Lee*, Jinwoo Ahn, Jimyeong Park, Hyeonkyun Na, Youngkee Lee, Yejin Kim, Gayeon Hong and Kyoung-Ryul Lee
Companion Biomarker Center, SCL Healthcare Co., Ltd., Gyeonggi-do, 16954, Republic of Korea

*Address all correspondence to: sprout30@scllab.co.kr

References

[1] Third report of the National Cholesterol Education Program (NCEP). Expert panel on detection, evaluation, ad treatment of high blood cholesterol in adults (adult treatment panel III) final report. Circulation. 2002;106:3143-21.

[2] Brown AE, Walker M, Genetics of insulin resistance and the metabolic syndrome. Curr. Cardiol. Rep. 2016;18:75.

[3] Dragsbaek K, Neergaard JS, Laursen JM, et al. Metabolic syndrome and subsequent risk of type 2 diabetes and cardiovascular disease in elderly women: challenging the current definition. Medicine(Baltimore). 2016;95:e4806.

[4] Ranasinghe P, Mathangasinghe Y, Hills AP, Misra A. Prevalence and trends of metabolic syndrome among adults in the Asia-Pacific region: a systematic review. BMC Public Health. 2017;17:101.

[5] van Vliet-Ostaptchouk JV, Nuotio ML, Slagter SN, et al. The prevalence of metabolic syndrome and metabolically healthy obesity in Europe: a collaborative analysis of ten large cohort studies. BMC Endocrinol Disorder. 2014;14:9.

[6] Lovre D, Mauvais-Jarvis F. Trends in prevalence of the metabolic syndrome. JAMA. 2015;314(9):950.

[7] Samuel VT, Shulman GI. The pathogenesis of insulin resistance: Integrating signaling pathways and substrate flux. J. Clin. Invest. 2016;126(1):12-22.

[8] Perry RJ, Samuel VT, Petersen KF, Shulman GI. The role of hepatic lipids in hepatic insulin resistance and type 2 diabetes. Nature. 2014;510(7503):84-91.

[9] Lin HF, Boden-Albala B, Juo SH, et al. Heritabilities of the metabolic syndrome and its components in the northern Manhattan family study. Diabetologia. 2005;48(10):2006-12.

[10] Lehtovirta M, Kaprio J, Forsblom C, et al. Insulin sensitivity and insulin secretion in monozygotic and dizygotic twins. Diabetologia. 2000;43(3):285-93.

[11] Elbein SC, Hasstedt SJ, Wegner K, Kahn SE. Heritability of pancreatic beta-cell function among nondiabetic members of Caucasian familial type 2 diabetic kindreds. J Clin Endocrinol Metab. 1999;84(4):1398-403.

[12] Bellia A, Giardina E, Lauro D, et al. "The Linosa Study": epidemiological and heritability data of the metabolic syndrome in a Caucasian genetic isolate. Nutr Metab Cardiovasc Dis. 2009;19:455-61.

[13] Carmelli D, Cardon LR, Fabsitz R. Clustering of hypertention, diabetes, and obesity in adult mail twins: same genes or same environments? Am J Hum Genet. 1994;55:566-73.

[14] Maher B. Personal genomes: the case of the missing heritability. Nature. 2008;456(7218):18-21.

[15] Eichler EE, Flint J, Gibson G, et al. Missing heritability and strategies for finding the underlying causes of complex disease. Nat Rev Genet. 2010;11(6):446-50.

[16] Manolio TA, Collins FS, Cox NJ, et al. Finding the missing heritability of complex diseases. Nature. 2009;461(7265):747-53.

[17] Kissebah AH, Sonnenberg GE, Myklebust J, et al. Quantitative trait loci on chromosomes 3 and 17 influence phenotypes of the metabolic

syndrome. Proc Natl Acad Sci. 2000;97(26):14478-83.

[18] Loos RJF, Katzmarzyk PT, Rao DC, et al. HERITAGE Family Study. Genome-wide linkage scan for the metabolic syndrome in the HERITAGE Family Study. J Clin Endocrinol Metab. 2003;88:5935-43.

[19] Edwards KL, Hutter CM, Wan JY, et al. Genome-wide linkage scan for the metabolic syndrome: the GENNID study. Obesity (Silver Spring). 2008;16:1596-1601.

[20] Tang W, Miller MB, Rich SS, et al. National Heart, Lung, and Blood Institute Family Heart Study. Diabetes. 2003;52(11):2840-47.

[21] Puppala S, Dodd GD, Fowler S, et al. A genomewide search finds major susceptibility loci for gallbladder disease on chromosome 1 in Mexican Americans. Am J Hum Genet. 2006;78(3):377-92.

[22] Cai G, Cole SA, Freeland-Graves JH, et al. Principal component for metabolic syndrome risk maps to chromosome 4p in Mexican Americans: the San Antonio Family Heart study. Hum Biol. 2004;76 (5):651-65.

[23] Povel CM, Boer JM, Reiling E, Feskens EJ. Genetic variants and the metabolic syndrome: a systematic review. Obesity Rev. 2011;12(11):952-67.

[24] do Nascimento FV, Piccoli V, Beer MA, et al. Association of *HSD11B1* polymorphic variants and adipose tissue gene expression with metabolic syndrome, obesity and type 2 diabetes mellitus: a systematic review. Diabetol Metab Syndr. 2015;7:38.

[25] Zhou JM, Zhang M, Wang S, et al. Association of the *ADIPOQ* rs2241766 and rs266729 polymorphisms with metabolic syndrome in the Chinese population: a meta-analysis. Biomed Environ Sci. 2016;29(7):505-15.

[26] Fallaize R, Carvalho-Wells AL, Tierney AC, et al. *APOE* genotype influences insulin resistance, apolipoprotein CII and CIII according to plasma fatty acid profile in the metabolic syndrome. Sci Rep. 2017;7:6274.

[27] Wang H, Dong S, Xu H, Qian J, Yang J. Genetic variants in *FTO* associated with metabolic syndrome: a meta- and gene-based analysis. Mol Biol Rep. 2012;39 (5):5691-98.

[28] Rodriguez-Nunez I, Caluag T, Kirby K, et al. *Nod2* and *Nod2*-regulated microbiota protect BALB/c mice from diet-induced obesity and metabolic dysfunction. Sci Rep. 2017;7(1):548.

[29] Moon S, Lee Y, Won S, Lee J. Multiple genotype–phenotype association study reveals intronic variant pair on *SIDT2* associated with metabolic syndrome in a Korean population. Hum Genomics. 2018;12:48.

[30] Willer CJ, MSchmidt E, Sengupta S, et al. Discovery and refinement of loci associated with lipid levels. Nat Genet. 2013;45(11):1274-83.

[31] Kilpeläinen TO, Zillikens MC, Stančáková A, et al. Genetic variation near *IRS1* associates with reduced adiposity and an impaired metabolic profile. Nat Genet. 2011;43(8):753-60.

[32] Dupuis J, Langenberg C, Prokopenko I, et al. New genetic loci implicated in fasting glucose homeostasis and their impact on type 2 diabetes risk. Nat Genet. 2010;42(2):105-16.

[33] Willer CJ, Sanna S, Jackson AU, et al., Newly identified loci that influence lipid concentrations and risk of coronary artery disease. Nat Genet. 2008;40:16199.

[34] Speliotes EK, Yerges-Armstrong LM , Wu J, et al. GOLD Consortium. Genome-wide association analysis identifies variants associated with nonalcoholic fatty liver disease that have distinct effects on metabolic traits. PLoS Genet. 2011;7:e1001324.

[35] Frayling TM, Timpson NJ, Weedon MN, et al. A common variant in the *FTO* gene is associated with body mass index and predisposes to childhood and adult obesity. Science. 2007; 316:889-94.

[36] Chambers JC, Elliott P, Zabaneh D, et al. Common genetic variation near *MC4R* is associated with waist circumference and insulin resistance. Nat Genet. 2008;40:716-8.

[37] Do R, Bailey SD, Desbiens K, et al. Genetic variants of *FTO* influence adiposity, insulin sensitivity, leptin levels, and resting metabolic rate in the Quebec Family Study. Diabetes. 2008;57:1147-50.

[38] Heid IM, Jackson AU, Randall JC, et al., Meta-analysis identifies 13 new loci associated with waist-hip ratio and reveals sexual dimorphism in the genetic basis of fat distribution. Nat Genet. 2010;42:949-60.

[39] Kooner JS, Saleheen D, Sim X, et al., Genome-wide association study in individuals of South Asian ancestry identifies six new type 2 diabetes susceptibility loci. Nat Genet. 2011;43:984-9.

[40] Stancáková A, Laakso M. Genetics of metabolic syndrome. Rev Endocr Metab Disord. 2014;15(4):243-52.

[41] Zabaneh D, Balding DJ. A genome-wide association study of the metabolic syndrome in Indian Asian nen. PLoS One. 2010;5:e11961.

[42] Kraja AT, Vaidya D, Pankow JS, et al. A bivariate genome-wide approach to metabolic syndrome. Diabetes. 2011;60:1329-39.

[43] Kristiansson K, Perola M, Tikkanen E, et al. Genome-wide screen for metabolic syndrome susceptibility loci reveals strong lipid gene contribution but no evidence for common genetic basis for clustering of metabolic syndrome traits. Circ Cardiovasc Genet. 2012; 5(2): 242-249.

[44] Avery CL, He Q, North KE, et al. A phenomics-based strategy identifies loci on *APOC1*, *BRAP*, and *PLCG1* associated with metabolic syndrome phenotype domains. PLoS Genet. 2011;7:e1002322.

[45] Ingelsson E, Langenberg C, Hivert MF, et al. Detailed physiologic characterization reveals diverse mechanisms for novel genetic loci regulating glucose and insulin metabolism in humans. Diabetes. 2010;59(5):1266-75.

[46] Dimas AS, Lagou V, Barker A, et al. Impact of type 2 diabetes susceptibility variants on quantitative glycemic traits reveals mechanistic heterogeneity. Diabetes. 2014;63:2158-71.

[47] Manning AK, HivertMF, Scott RA, et al. A genome-wide approach accounting for body mass index identifies genetic variants influencing fasting glycemic traits and insulin resistance. Nat Genet. 2012;44(6):659-69.

[48] Scott RA, Lagou V, Welch RP, et al. DIAbetes Genetics Replication and Meta-analysis (DIAGRAM) Consortium. Large-scale association analyses identify new loci influencing glycemic traits and provide insight into the underlying biological pathways. Nat Genet. 2012;44:991-1005.

[49] Scott RA, Fall T, Pasko D, et al. Common genetic variants highlight the role of insulin resistance and body

fat distribution in type 2 diabetes, independent of obesity. Diabetes. 2014;63(12):4378-87.

[50] Yaghootkar H, Scott RA, White CC, et al. Genetic evidence for a normal-weight "metabolically obese" phenotype linking insulin resistance, hypertension, coronary artery disease, and type 2 diabetes. Diabetes. 2014;63(12):4369-77.

[51] Rung J, Cauchi S, Albrechtsen A, et al. Genetic variant near *IRS1* is associated with type 2 diabetes, insulin resistance and hyperinsulinemia. Nat Genet. 2009;41(10):1110-15.

[52] Liu PH, Chang YC, Jiang YD, et al. Genetic variants of *TCF7L2* are associated with insulin resistance and related metabolic phenotypes in Taiwanese adolescents and Caucasian young adults. J Clin Endocrinol Metab. 2009;94(9):3575-82.

[53] Knowles JW, Xie E, Zhang Z, et al. Identification and validation of N-acetyltransferase 2 as an insulin sensitivity gene. J Clin Invest. 2015;125(4):1739-51.

[54] Succurro E, Andreozzi F, Marini MA, et al. Low plasma insulin like growth factor-1 levels are associated with reduced insulin sensitivity and increased insulin secretion in nondiabetic subjects. Nutr Metab Cardiovasc Dis. 2009;19(10):713-9.

[55] Mannino GC, Greco A, De Lorenzo C, et al. A fasting insulin raising allele at *IGF1* locus is associated with circulating levels of IGF-1 and insulin sensitivity. PLoS One. 2013;8(12):e85483.

[56] Tabassum R, Chauhan G, Dwivedi OP, et al. Genome-wide association study for type 2 diabetes in Indians identifies a new susceptibility locus at 2q21. Diabetes. 2013;62(3):977-86.

[57] Chen G, Bentley A, Adeyemo A, et al. A genome-wide association study reveals variants in *ARL15* that influence adiponectin levels. PLoS Genet. 2009;5(12):e1000768.

[58] Abecasis GR, Auton A, Brooks LD, et al. An integrated map of genetic variation from 1,092 human genomes. Nature. 1000 Genomes Project Consortium; 2012;**491**(7422):56-65

[59] Lotta LA, Gulati P, Day FR, Payne F, Ongen H, van de Bunt M, Gaulton KJ, Eicher JD, Sharp SJ, Luan J, et al; EPIC-InterAct Consortium; Cambridge *FPLD1* Consortium. Integrative genomic analysis implicates limited peripheral adipose storage capacity in the pathogenesis of human insulin resistance. *Nat Genet.* 2017;49:17-26.

[60] Abecasis GR, Auton A, Brooks LD, et al. An integrated map of genetic variation from 1,092 human genomes. Nature. 2012;491(7422):56-65.

[61] Horikoshi M, Mägi R, van de Bunt M, et al. Discovery and finemapping of glycaemic and obesity-related trait loci using high density imputation. PLoS Genet. 2015;11(7):e1005230.

[62] Albrechtsen A, Grarup N, Li Y, et al. Exome sequencing-driven discovery of coding polymorphisms associated with common metabolic phenotypes. Diabetologia. 2013;56(2):298-310.

[63] Støy J, Kampmann U, Mengel A, et al. Reduced *CD300LG* mRNA tissue expression, increased intramyocellular lipid content and impaired glucose metabolism in healthy male carriers of Arg82Cys in *CD300LG*: a novel genometabolic cross-link between *CD300LG* and common metabolic phenotypes. BMJ Open Diab Res Care. 2015;3(1):e000095.

[64] Steinthorsdottir V, Thorleifsson G, Sulem P, et al.

Identification of low-frequency and rare sequence variants associated with elevated or reduced risk of type 2 diabetes. Nat Genet. 2014;46(3):294-8.

[65] Moltke I, Grarup N, Jørgensen ME, et al. A common Greenlandic *TBC1D4* variant confers muscle insulin resistance and type 2 diabetes. Nature. 2014;512(7513):190-3.

[66] UK10K Consortium; Walter K, Min JL, Huang J, et al. The UK10K project identifies rare variants in health and disease. Nature. 2015;526(7571): 82-90.

[67] Zhang Y, Kent JW, Lee A, Cerjak D, et al. Fatty acid binding protein 3 (*fabp3*) is associated with insulin, lipids and cardiovascular phenotypes of the metabolic syndrome through epigenetic modifications in a Northern European family population.

[68] Kaati G, Bygren LO, Pembrey M, Sjostrom M. Transgenerational response to nutrition, early life circumstances and longevity. Eur J Hum Genet. 2007;15(7):784-90.

[69] Flannick J, Florez JC. Type 2 diabetes: genetic data sharing to advance complex disease research. Nat Rev Genet. 2016;17:535-549.

[70] Hakim O, Misteli T. SnapShot: chromosome confirmation capture. *Cell.* 2012;148:1068.e1-1068.e2.

[71] Nica AC, Dermitzakis ET. Expression quantitative trait loci: present and future. *Philos Trans R Soc Lond B Biol Sci.* 2013;368:20120362.

[72] Chen Z, Yu H, Shi X, Warren CR, Lotta LA, Friesen M, Meissner TB, Langenberg C, Wabitsch M, Wareham N, Benson MD, Gerszten RE, Cowan CA. Functional screening of candidate causal genes for insulin resistance in human preadipocytes and adipocytes. Circ Res. 2020;126:330-6.

Chapter 13

Genetics in Osteoarthritis Knee

Rajeshwar Nath Srivastava, Amar Chandra Sharma, Sudeepti Ratan Srivastava, Saloni Raj and Lavini Raj

Abstract

Osteoarthritis (OA) is a debilitating joint disorder with a complex pathogeny wherein diverse factors interact, causing a process of deterioration of the articular cartilage and the subchondral bone. It can be primary or secondary but has common clinical, radiological, and pathological manifestations. Unfortunately, there are no curative or preventive options available for this disease. The knee is the most common site to develop OA among all synovial joints. Both environmental and genetic factors play an essential role in the initiation of the disease. Identifying the genes underlying the genetic background could give new insights into the pathophysiology of knee osteoarthritis (KOA) and could potentially lead to new drug targets. Several genes involving developmental processes or maintenance of cartilage and bone are found to be associated with KOA susceptibility and progression. Understanding the gene functions has improved the knowledge towards the disease pathogenesis. So, it will be of interest to investigate the role of gene-gene interaction in the disease.

Keywords: KL grade, knee osteoarthritis, single nucleotide polymorphism, VAS, WOMAC

1. Introduction

Osteoarthritis (OA) is the most common degenerative arthritis caused by the breakdown of articular cartilage [1]. The prevalence of OA is high and expected to increase in the coming years [2]. Results of some epidemiologic studies indicate that the incidence of symptomatic OA is about 8–9% in China [3]. OA is a multi-factorial joint disorder in which growing age, genetic factors; hormonal as well as mechanical factors are significant contributors to its onset and progression. The molecular mechanism underlying the cartilage degeneration is poorly understood [4]. American College of Rheumatology defines Osteoarthritis as a heterogonous group of condition that leads to joint symptoms and signs which are associated with defective integrity of articular cartilage, in addition to related changes in the underlying bone at joint margin [5]. OA is primarily a non-inflammatory disorder of movable joint characterized by an imbalance between the synthesis and degradation of articular cartilage, leading to the classic pathological change of wearing away and destruction of cartilage [6]. OA affects nearly 21 million people in the United States, accounting for 25% of the visits to primary care physicians and half of all having NSAID (Non- Steroid Anti Inflammatory drugs) prescriptions. It is estimated that about 80% of the population was having radiographic evidence of Osteoarthritis by the age of 65 years, although only about 60% of these were symptomatic [7]. Epidemiological profile of Osteoarthritis in India is not clear, but it is estimated that more than 30-40% of the Indian population suffers from Osteoarthritis over the age of 50 years (www.wrongdiagnosis.com).

Osteoarthritis is considered to be of two types:

1. Primary

2. Secondary

2. Primary osteoarthritis

Primary OA is a chronic degenerative condition of mobile joints due to an unknown cause. This may result due to aging because few people do not show any clinical or functional signs of the diseases in the late 90s. The proteoglycan and water content of the cartilage reduce with the advancement of age, hence the toughness of cartilage and increasing the susceptibility of collagen fibers to degenerate [8]. Mild inflammation around the joint capsule may occur in OA as compared to rheumatoid arthritis. This inflammation is in response to the small particles of the debris produced by this cartilage breakage and then attempted clearance by the scavengers cells located in joint lining [8]. New bone outgrowths called 'spurs' or osteophytes may form on the margins of the joint, possibly in an attempt to improve the congruence of the articular cartilage surfaces. Some of these bone changes, along with low-grade inflammation, may cause pain and mobility.

3. Secondary osteoarthritis

This type of OA is caused by other factors or disease, but the resulting pathological changes are the same as for primary OA.

Leading causes of secondary Osteoarthritis:

- Accidental injury to joints [9].

- Inflammatory diseases [7] (such as Perthes disease, Lyme disease) and all chronic form of arthritis (e.g. gout, pseudogout and rheumatoid arthritis). In gout, uric acid crystals cause cartilage degradation at a faster pace [10].

- Healed infection of the joints.

- Sports injuries [11]

According to Creamer *et al.* [12], Knee Osteoarthritis (OA) is a significant cause of disability, particularly in older people. The factors determining disability remain unclear. A study was conducted to assess the impact of clinical and psychosocial variables on function in knee Osteoarthritis and to develop models to account for observed variance in self-reported disability. It was conducted that function in symptomatic knee Osteoarthritis is determined more by pain and obesity than by structural changes as seen on plain X-ray. Hunter and Felson [13] said that Osteoarthritis had been known believed to be a disease of articular cartilage since ages, but the current concept is entirely different. Hunter and Felson explained that OA is a structural and functional impairment of synovial joints resulting in a range of disease. OA entails the whole joint including the subchondral bone, menisci, ligaments, periarticular muscle, capsule and synovium.

Osteoarthritis (OA) is a chronic degenerative disorder of multifactorial etiology characterized by loss of articular cartilage, hypertrophy of bone at the margins, subchondral sclerosis and range of biochemical and morphological alterations of the synovial membrane and joint capsule. It may be either Primary Osteoarthritis or Secondary Osteoarthritis. Primary OA is mostly related to aging. It can present as localized, generalized or as erosive Osteoarthritis. Another disease or condition causes secondary OA [14].

Osteoarthritis (OA) is a disease of the musculoskeletal system that primarily involves the joints of the knee, hip, spine, hand and foot. OA is estimated to affect 40% of people >70 years of age [15], making it more prevalent than any other form of arthritis [16]. The Framingham Knee Osteoarthritis study suggests that knee osteoarthritis increases in prevalence throughout the elderly, more so in women than in men [17]. Females are found to have more severe OA, more number of joints are involved, and have more symptoms and increased hand and knee OA [18]. Osteoarthritis (OA) is the most common cause of musculoskeletal disability related to aging and is characterized by late-onset degeneration of articular cartilage [19]. OA is one of the leading causes of disability and dysfunction in the elderly population [20]; it has been estimated that the total cost for arthritis, including OA, is over 2% of the United States gross domestic product [21].

Prevalence of OA: Osteoarthritis (OA) is the second most common rheumatological problem and is the most frequent joint disease with a prevalence of 80% in the population having radiographic evidence. About 60% of radiographically evidenced subjects are symptomatic [22]. In India, more than 41.1% of the population suffers from Osteoarthritis beyond the age of 50 years [23]. This is the most common cause of locomotor disability in the elderly [24]. More than 20 million people are affected with OA in the United States, including 10% of adults of age 50 years. It has been estimated that 2% of women and 1.4% of men develop radiographic OA per year, but approx half of these individual show symptoms [19].

Etiology of OA: OA is a multifactorial disease with both genetic and environmental determinants, and all cases are probably affected by both, with a continuous distribution between the extremes of genetic or predominantly environmental causes [15]. The pathophysiology of OA is complex and do not comprehend with its clinical feature. The disease is rare before the age of 40 years but frequency increases with age as large no. of individual with ≥70 years demonstrate radiographic evidence of OA in some joints. All cases are probably affected by both genetics and environment, with a continuous distribution between the extremes of predominantly genetic or predominantly environmental causes [15].

Role of genes in OA: Role of genetics is emerging as an important etiological factor in recent times. More than 65 genes are associated with knee osteoarthritis (KOA) in different populations, and Indian population candidate genes like CALM-1 [25], VDR gene polymorphism [26], GDF-5 [22], SMAD-3, BMP-5, CCL2, COL2A1 [27] and COL2A1, CRTL1 [28] are associated with KOA. In recent years many studies have been conducted on KOA to investigate its association with SNPs. Genome-Wide Association Stu is performed on a large scale to identify the role of different loci in the development of the disease depending on the number of variants used. To date, more than ten loci (LSP1P3, GDF5, CHST11, FTO, GNL3, ASTN2, SENP6, PTHLH, TP63, CDC5L and CHST11) have been found associated with KOA through GWAS in European, Asian and Caucasians populations [29, 30]. Candidate gene studies have been responsible for identifying several susceptible loci for OA. GDF5, ASPN, FRZB and PTGS2 are few other genes which have been identified this way. These genes continue to be the subject of functional studies and further genetic replication in independent populations [31–36].

Valdev et al. [37] reported an association between an amino acid variant in the TRPV1 gene and risk of symptomatic KOA for the first time. This amino acid has been implicated in pain sensitivity previously. The observation that the genotype implicated in lower pain sensitivity is significantly associated with a lower risk of painful OA. After adjustment for confounding variables (age, sex, BMI and radiographic severity) the difference between symptomatic and asymptomatic OA also achieves statistical significance.

Knee OA subjects showed individual characteristics in their expression of PBMC gene. A set of 173 genes was identified to diagnose Knee OA cases. The sensitivity and specificity were 89% and 76% respectively [38]. Besides, they observed that patients with symptomatic KOA could be categorized into two distinct groups based on the level of inflammatory gene expression (e.g., IL-1β, IL-8, COX-2). The differential overexpression of these inflammatory genes in KOA subclasses was validated using qPCR ($P < 0.0001$) in 2 cohorts from NYUHJD and one cohort from Duke University.

Yerges Armstrong et al. [39] identified four SNPs significantly associated with radiographic KOA. The strongest signal ($p = 0.0009$) maps to 12q3, which contains a gene coding for SP7. Additional loci map to 7p14.1 (TXNDC3), 11q13.2 (LRP5) and 11p14.1 (LIN7C). The allele associated with higher BMD was also associated with higher odds of KOA in all four loci. This meta-analysis demonstrated that several GWAS identified BMD SNPs are nominally associated with prevalent radiographic KOA and further supports the hypothesis that BMD or its determinants may be a risk factor contributing to OA development.

A study by Shi-Xing Luo et al. [40] on IL-16 showed a significant association of SNP rs4778889 with altered gene expression levels as well as two other SNPs (rs11556218 and rs4072111). The latter two SNPs are located in an exon region, and their single nucleotide change results in an amino acid substitution. This was the first study to investigate the association of IL16 polymorphism with KOA risk, and a significant effect was observed IL16 rs1155218. Polymorphism represented an Asn to Lys substitution in exon 6 of the gene. It was mentioned that individual with rs11549465 C allele was at lower risk to develop the disease than those with T allele. Javier Fernández-Torres et al. [41] found that the SNP rs11549465 located in the exon 12 within the HIF1A gene was associated with KOA in Mexican patients. Their results showed that the presence of the *CC* homozygous variant or *C* allele represents potential risk factors for development of KOA. On the contrary to this, they detected that the heterozygous variant of CT or *T* allele of the rs11549465 polymorphism of the HIF1A gene (in comparison with the homozygous carriers) play a protective role against the disease.

Rui Zhang et al. [32] demonstrated that SNP rs143383 of GDF5 is a compelling risk factor for both knees and hand OA and provide further support for GDF5 in etiology of OA. A recent study by Kwo Wei Ho et al. [42] suggested that the COL11A2, a collagen-encoding gene, may play a role in pain sensitization after the development of OA. In a case-control study, Haohuan Li et al. [43] investigated the association between EN1 rs4144782 and susceptibility to KOA in a Chinese population. A significant association of SNP was observed with increased risk of KOA. The results reconfirmed the close connection between BMD and OA.

Indian scenario: Besides our studies, not much work has been done on the association of SNPs in the Indian population. Studies published from our laboratory on genetic polymorphism in KOA demonstrated an association of BMP5, CCL2, COL2A1, IL1B, SMAD3, GDF5, ESR-α, CALM1 and COMP genes with the development of KOA. Background of these genes is given below.

4. Bone morphogenic protein (BMP5) gene

BMP5 is a member of the TGF-β superfamily of secreted proteins whose family members are involved in synovial joint development and joint tissue homeostasis [44]. Hahn *et al.* [45] reported that BMP5 was found on human chromosome 6p12.1.

BMPs were originated as protagonists of bone formation and growth. They have a major role in morphogenesis of a variety of vertebrate tissues and organs. They stimulate all proliferation and matrix synthesis for differentiation of chondrocytes. BMP5, in particulate, are known to regulate ovarian development [8, 46], cardiac development [47], and limb bud development [48] with a well-defined role in the differentiation of chondrocytes through the promotion of cell proliferation and matrix synthesis [49, 50]. Southam *et al.* [51] found that Polymorphisms located within the transcribed region of *BMP5* and its proximal promoter had previously been excluded for association with OA. Nevertheless, there is increasing evidence; however, that polymorphisms in regulatory elements involved in gene transcription play an important role in conferring susceptibility to complex disease traits [52].

The studies carried out in multiple organisms demonstrated its role in the regulation of various episode of embryonic development which includes dorsal-ventral and left-right axis formation, mesenchymal-epithelial interactions, and differentiation of many specific tissues including lung, gut, kidney, hair, teeth, cartilage, and bone [7]. BMPs can trigger the entire process of cartilage and bone formation when implanted at ectopic sites in adult animals [53] and are usually expressed in and around early cartilage and bone precursors during embryonic development [54–57]. Moreover, mutations in different BMP genes block the formation of particular skeletal features, showing that BMPs are also required for the normal formation of skeletal tissue [58].

BMP1 is a protease which takes part in the maturation of fibrillar collagens whereas the other BMPs are secreted molecules of TGF-beta family [59, 60].

BMPs influence the normal development and repair of the synovial joint; therefore alterations in the activity of these molecules could affect the arthritic phenotype [61, 62]. Using genetic association analysis, we have tested BMP5 as the chromosome 6 OA susceptibility gene. Zuzarte-Luis *et al.* [63] provided evidence for a role of this BMP member in the development of limb autopodium through the activation of Smad proteins and MAPK p38. Previous studies demonstrate that secreted signaling molecules in the bone morphogenetic protein (BMP) family play a vital role in both formation and repair of skeletal structures. These molecules are expressed both in early skeletal precursors and in the surface perichondrium and periosteum layers that surround growing cartilage and bone [64].

Wilkins *et al.* [65] identified an SNP and a functional microsatellite associated with OA. It has been exhibited that various alleles of microsatellite are behind the modified transcriptional activity of BMP5 promoter suggestive of *cis*-regulation of *BMP5,* is involved in OA susceptibility.

5. Chemokine (C-C motif) ligand 2 (CCL2) gene

Chemokines are small, secreted proteins that stimulate the directional migration of leukocytes and mediate inflammation (Baggiolini et al, 1997). These are a family of heparin-binding cytokines known for this chemotactic activity. Four subfamilies of chemokines have been identified based on the juxtaposition of cysteine residues in the protein's N-terminus. These families have been named C, C-C, C-X-C, and C-X3-C [66].

Sozzani *et al.* [67] said that the C-C and C-X-C chemokines represent two significant subgroups. The C-X-C chemokines include IL-8 and growth-related oncogene a. The C-C chemokine family comprises families of monocyte chemoattractant proteins (e.g. MCP-1, MCP-2, MCP-3), macrophage inflammatory proteins (e.g., MIP-1a and MIP-1b), and a chemokine designated RANTES (regulated upon activation, normal T cell expressed and secreted). As their names suggest, these chemokines act predominantly on mononuclear cells (e.g., T cells, monocytes, and macrophages).

CCL2; previously known as monocyte chemoattractant protein-1, MCP-1 is a chemoattractant that belongs to the CC chemokine subfamily. Mehrabian *et al.* [68] localized the gene for monocyte chemotactic protein-1 (CCL2) to chromosome 17. Corrigall *et al.* 2001 proposed that CCL2 in the synovial membrane serves to recruit macrophages and perpetuate inflammation in the joints of patients with rheumatoid arthritis. CCL2 plays a crucial role in host defense by recruiting monocytes and macrophages at the site of inflammation [69].

CCL2 production is inducible in various types of cells, including synoviocytes [70]. In several studies, increased expression of CCL2 was observed in patients with inflammatory diseases, including neuropsychiatric syndromes of systemic lupus erythematosus, rheumatoid arthritis, OA and degenerative and inflammatory arthropathies, including gout [71–73].

Reports have demonstrated the involvement of chemokines in cartilage abnormalities in OA [74–76]. Reported studies suggests that CCL2 is involved in inflammatory diseases such as RA and OA, However, in previous studies, the CCL2 gene polymorphism (22510A/G) did not show any association with Spanish and Korean RA [77, 78]. However, Park *et al.* [79] investigate that significant association between polymorphisms of the CCL2 gene and primary knee OA patients in a Korean population.

6. Collagen, type II, alpha 1 (COL2A1) gene

Solomon *et al.* [80] suggest that the human type II collagen gene, COL2A1, has been assigned to chromosome 12.

Law et al. [81, 82] said a cosmid clone containing the entire human type II alpha one collagen gene (COL2A1) was used as a probe in the Southern analysis of DNA from a panel of human/hamster somatic cell hybrids containing different portions of human chromosome 12. Two of the hybrids exhibited a similar terminal deletion q14.3----qt, but one was positive for the gene while the other was negative. Therefore, the gene must reside in the region q14.3.

Holderbaum *et al.* [83, 84] referred a single base change resulting in the substitution of Cys for Arg at position 519 of the type II collagen triple helix is a predisposing factor in the pathogenesis of a precocious-onset form of familial Osteoarthritis associated with a mild chondrodysplasia. Cartilage obtained at the time of total knee replacement in a patient with the Arg-Cys519 mutation was used to investigate the expression of Col2A1 alleles. Using PCR assisted amplification of mRNA with specific amplification of a region of Col2A1 message encompassing exons 31-34, followed by single-strand conformation polymorphism and sequence analyses, we have found transcription products of both mutant and normal type II collagen alleles. Further analysis of the sequence of these exons provides evidence that the Arg-Cys519 mutation arose independently in at least two of the three known affected families. The presence of both mutant and normal alleles of Col2A1 in cDNA derived from cartilage obtained from this patient suggests that Cys519-containing type II collagen may continue to be produced even in advanced stages of Osteoarthritis.

On the other hand, mutations in Type II procollagen (COL2A1) can cause a hereditary form of the joint disorder with a broad spectrum of phenotypes ranging from primary OA with mild chondrodysplasia, mild spondyloepiphyseal dysplasia and osteonecrosis to severe generalized OA, including achondrogenesis and hypochondrogenesis [85].

The COL2A1 gene appears to play a crucial role in OA pathogenesis because the protein encoded is the most abundant in articular cartilage. However, results of the studies searching for a relationship between COL2A1 and OA are controversial, due to that an association with specific genotypes has been reported in some studies, whereas others have denied this. We found no association between OA of the knee and COL2A1 gene polymorphism in the overall sample. Nevertheless, when the association was analyzed according to radiologic grade, a significant relationship was denoted in OA grade 4 with allele p (Pp/pp) [OR (95% CI) 4.1 (1.2¡14.6)] independently of gender, age, and BMI; this indicates that in Mexican Mestizo population, a COL2A1 gene polymorphism is associated with advanced stages of OA of the knee [86–89].

The study of Mu *et al.* [90] does not fully support that COL2A1 could be implicated in primary OA of other non-Asian ethnic groups since ethnic variability in gene susceptibility is very well documented. The clinical features of early-onset OA, mild clinical phenotypes and one patient with osteonecrosis of the femoral head strongly suggested that COL2A1 may also be the underlying cause of OA in our family. Linkage analysis and direct sequencing of COL2A1, however, clearly rule out this possibility. Kannu *et al.* [91] have been described COL2A1 mutations in association with bilateral hip disease and Osteoarthritis in the second decade of life but without ocular abnormalities or short stature. Xu *et al.* [92] investigate the relationships between two COL2A1 single nucleotide polymorphisms (SNPs; T2088C and G4006A) and Osteoarthritis (OA) in Han Chinese women.

7. Interleukin 1 beta (IL1B) gene

Loughlin *et al.* [93] said that IL-1 is the primary catabolic cytokine of the OA joint and can stimulate the synthesis of several proteinases, which can result in the breakdown of cartilage extracellular matrix proteins. The 2 IL-1 genes (*IL1A* and *IL1B*) and the gene encoding IL-1Ra (*IL1RN*) are located on chromosome 2q13 within a 430-kb genomic fragment [94]. Loughlin *et al.* 2002 reported that IL-1R antagonist (IL-1Ra) competes with IL-1 for binding to the IL-1 receptors and can act as an inhibitor of cartilage loss. When the catabolic and anabolic activities of the cytokines are balanced, cartilage integrity is maintained. If there is an imbalance favoring catabolism, however, cartilage destruction can proceed, resulting in OA. It is, therefore, reasonable to propose that a proportion of the genetic susceptibility to OA may be encoded for by variation in the activity of interleukins and that for chromosome 2q this susceptibility could reside within the IL-1 gene clusters. Stern *et al.* [95] reported that IL1B 5810 G > A SNP genotypes marker were not in Hardy-Weinberg equilibrium ($p < 0.05$ in both non-erosive and erosive hand OA subgroups). Statistically significant association with the IL1B 5810 AA genotype was found in the erosive hand OA subgroup (relative risk 3.8, $p = 0.007$). This IL1B 5810 AA genotype association was also significant between erosive and non-erosive hand OA subjects (relative risk 4.01, $p = 0.008$). As expected, significant linkage disequilibrium was present between IL1B 5810 SNP and IL1A (−)889 SNP, other IL1B SNPs, and the nearest IL1RN SNP examined. The IL1B 5810A allele occurs most frequently on haplotypes with the SNP alleles IL1B 1423C, IL1B 1903 T, IL1B 5887C, and IL1A (−)889C. Genotypes at null loci failed to show evidence suggesting population stratification that might account for the spurious association. Sezgin *et al.* [95] said that some researchers had suggested an association between the IL-1 gene cluster and the occurrence of OA.

Previously, Moos *et al.* [96] investigated the distribution of polymorphic alleles of four different genes encoding TNF-alpha, IL1RN, IL1B and IL-6 in the knee or hip OA patients with controls. The analysis of genotype frequencies for the IL1B gene, more OA patients than controls was homozygous for allele 2, although any significant differences for the TNF-alpha, IL1RN and IL-6 polymorphisms were found.

8. SMAD3 (SMAD family member 3) gene

Smad3 was found on human chromosome 15q22.33.The classic TGF-β mediated signaling pathway involves Smad activation. Smads are a family of intracellular proteins that comprise three classes of signaling molecules: - receptor-associated Smads (2 and 3 for TGF- β, 1, 5, and 8 for BMP signaling), the co-factor Smad4, and the inhibitory Smads (6and7) [97]. The receptor-activated SMADs include SMADs 1, 2, 3, 5, and 8. SMADs 2 and 3 respond to TGF- β and activins [97, 98], whereas SMAD1, 5, and 8 function in BMP signaling pathways [99–101]. The receptor-associated Smads bind to the type I receptor, and on ligand binding and activation, are phosphorylated and released into the cytoplasm. The activated receptor-associated Smads form a trimeric heterodimer with the co-factor Smad4, translocate to the nucleus, and influence gene transcription [102].

Yao *et al.* [103] reported that Smad3 gene mutation is a possible predisposing factor for human OA and found gene mutation in OA, providing insight into the function of SMAD3 mediated TGF-b signals in the development of OA and also suggested that Smad3 gene mutation may be a risk factor for genetic susceptibilities to OA.

Ferguson *et al.* [104] has established that the TGF- β Smads inhibit chondrocyte maturation, whereas the BMP-related Smads accelerate maturation. Many studies have shown that transforming growth factor- β (TGF- β) signals function as crucial regulators in bone formation, remodeling and maintenance. [105]. Micheal *et al.* [106] have shown that loss of Smad3 results in impaired immune responses, accelerated wound healing decreased bone density, OA and access to colon cancer.

Yang *et al.* [107] showed that Smad3-mediated TGF- β signals are essential for maintaining articular cartilage in the quiescent state by repressing chondrocyte differentiation and controlling matrix molecule synthesis. Consequently, impairment of TGF-b signals due to Smad3 disruption results in phenotypes resembling human Osteoarthritis.

Wu *et al.* [108] suggested that Loss of Smad3 appears to enhance bone morphogenetic protein signaling in the articular chondrocytes, leading to hypertrophy and OA-like changes. The observation further supports the crucial role of Smad3 that Smurf2 overexpression leads to dephosphorylation of Smad3 and is associated with a spontaneous OA phenotype in transgenic mice.

Also, Cherlet *et al.* [109] reported that Smad3 levels are lower in women than in men, which is consistent with other data showing that estrogens inhibit *SMAD3* transcriptional activity Nevertheless, Valdes *et al.* found that the genetic association of the *SMAD3* intronic SNP with OA was significant in both men and women and that effect sizes were remarkably similar between sexes, confirming the robustness of the result. Valdes *et al.* [109] reported that four SNPs (rs266335, rs12901499, rs6494629, and rs2289263) were found to be nominally significantly associated with knee OA ($P < 0.05$), but only 1 of them, rs12901499, was nominally associated also with hip OA ($P < 0.021$) and also observed that the major allele G was found at a higher frequency among OA patients than among controls [109–112].

9. Growth differentiation factor-5 (*GDF-5*)

Hotten *et al.* [9] determined that the *GDF5* gene contains two exons. Miyamoto *et al.* [113] reported that the gene encoding growth differentiation factor-5 (*GDF-5*) is associated with Osteoarthritis in Asian populations. An SNP in the 5′ untranslated region (UTR) of *GDF-5* showed significant association with hip osteoarthritis in 2 independent Japanese populations. This association was replicated for knee osteoarthritis in Japanese and Han Chinese populations. This SNP, located in the *GDF-5* core promoter, exert allelic differences on transcriptional activity in chondrogenic cells, with the susceptibility allele showing reduced activity. The findings implicated *GDF-5* as a susceptibility gene for Osteoarthritis and suggested that decreased *GDF-5* expression is involved in the pathogenesis of Osteoarthritis. [2] also isolated and characterized human *GDF5*, which they designated *CDMP1*, as well as human *GDF6 (CDMP2)*. *GDF6* is predominantly expressed at sites of skeletal morphogenesis. Al-Yahyaee *et al.* [114–117] identified two mutations in the *GDF5* gene: a silent 1137A-G transition encoding lysine and a 1-bp deletion, 1144delG, predicting a frame shift resulting in loss of the biologically active C terminus of the protein. Thomas *et al.* [118] found that heterozygotes for the C400Y mutation had phenotypes resembling brachydactyly types A1, A4, or C.

10. Estrogen receptor alpha (*ESR- α*) gene

The estrogen receptor (*ESR-α*) is a ligand-activated transcription factor composed of several domains essential for hormone binding, DNA binding, and activation of transcription. Alternative splicing results in several *ESR-α* mRNA transcripts, which differ primarily in their 5′ untranslated regions. The translated receptors show less variability [4, 119]. Ponglikitmongkol *et al.* [120, 121] showed that the human *ESR-α* gene is more than 140 kb long. It contains eight exons, and the position of its introns has been highly conserved, being, for example, remarkably similar to those of one of the chicken thyroid hormone receptor genes. Sputnik *et al.* [122] reported that *ESRα* isoform is a ligand-activated transcription factor composed of several essential domains for hormone binding and activation of transcription. *ESRα* is an essential mediator in the signal transduction pathway. Jin *et al.* [123] reported that Estrogen receptors (*ESR-α*) are known to play an essential role in the pathophysiology of Osteoarthritis. To investigate *ESRα* gene polymorphisms for its association with primary knee osteoarthritis, they conducted a case-control association study in patients with primary knee osteoarthritis and healthy individual in the Korean population. Jin *et al.* (2004) investigated the association between haplotypes of three polymorphism in *PVU II* in intron 1(IVS1-397 T/C), Xba I in intron 1(IVS1-351A/G) and Big I in exon 8 (exon8 229G/A) of *ESR-α* gene and primary knee OA in the Korean population, first two SNP in intron one also investigated by Bergink et al. [124] in Rotterdam population. Ushiyama *et al.* [125] found on *ESR-α* gene association between a genotype of *PVU II* and Xba I polymorphisms in Intron 1 and generalized Osteoarthritis with a severe radiographic change in the Japanese population.

11. Calmodulin 1 (*CALM1*) gene

Rhyner et al. [126] found that the *CALM1* gene contains six exons spread over about 10 kb of genomic DNA. The exon-intron structure was identical to that of *CALM3*. A cluster of transcription-start sites was identified 200 bp upstream of the ATG translation-start codon, and several putative regulatory elements were found in the 5′ flanking region, as well as in intron 1. A short CAG trinucleotide repeat region was identified in the 5-prime untranslated region of the gene. Motoani *et al.* [127] identified susceptibility genes for Osteoarthritis in a large-scale case-control association study using gene-based single-nucleotide polymorphism (SNPs) in a Japanese population. In two independent case-control populations, they found a significant association ($p = 9.8 \times 10^{-7}$) between hip osteoarthritis and an SNP (IVS 3–293 C>T) located in intron 3 of the calmodulin (Cam) 1 gene (*CALM-1). CALM 1* was expressed in cultured chondrocytes, and articular cartilage and its expression were increased in Osteoarthritis. Subsequent linkage – disequilibrium mapping identified five SNPs showing significant equivalent to IVS 3–293 C>T. One of these (-16 C>T) is located in the core promoter region of *CALM 1*. Functional analysis indicated that the susceptibility – 16 T allele decreases *CALM 1* transcription in-vitro and in-vivo. Inhibition of CAM in chondrogenic cells reduced the expression of the major cartilage matrix genes *col 2a1* and *Agc 1*. These results suggested that the transcriptional level of *CALM-1* was associated with susceptibility for hip osteoarthritis through modulation of chondrogenic activity. Their findings revealed that the *CALM-1* mediated signaling pathway is chondrocytes as a novel potential target for the treatment of Osteoarthritis. Loughlin *et al.* [128] studied in a Caucasian population using a cohort of 1672 individuals and concluded that *CALM-1* core promoter polymorphism is not a risk factor for Osteoarthritis.

12. Cartilage oligomeric matrix protein *(COMP)* gene

Briggs *et al.* [129] demonstrated that the *COMP* gene contains 19 exons. Exons 4-19, which encode the EGF-like (type II) repeats, calmodulin-like (type III) repeats (CLRs), and the C-terminal domain, correspond in sequence and intron location to the thrombospondin genes, whereas exons 1-3 are unique to *COMP.* Mabuchi *et al.* [130] reported that hereditary osteochondral dysplasia produces severe early-onset OA Among them are Pseudochondroplasia (PSACH) & multiple epiphyseal dysplasias (MED) both of which are caused by a mutation in the *COMP* gene. Therefore *COMP* may be a susceptibility gene for OA. For these reasons, Mabuchi *et al.* [131] hypothesized that Osteoarthritis is a common disorder may be at the mild end of the phenotypic gradation produced by *COMP* mutations. They ascertained the sequences of the exons and exon-intron boundaries and identified 16 polymorphisms in the *COMP* gene. Using five polymorphisms spanning the entire *COMP* gene (-1417 C/G in promoter region with $P = 0.29$, c.279C/A in exon 4 with $P = 0.19$, IVS5 + 76 T/C with $P = 0.74$, IVS16-45C/T with $P = 0.19$ and IVS18-40 T/C with $P = 0.93$), Mabuchi et al. [132], examined the association of this gene in Japanese patients with Osteoarthritis of the knee and hip joints. Genotype and allele frequencies of the polymorphisms were significantly different between osteoarthritis and control groups, and with the help of this study, they hypothesize that comp gene is a candidature gene for OA Song *et al.* [133] identified mutations in the *COMP* gene in 9 of 9 Korean patients with PSACH and 3 of 5 Korean patients with MED. Three of the eight mutations identified were novel. Deere *et al.* identified 12 mutations in the *COMP* gene, including ten novel mutations in 12 patients with PSACH. The site of the mutations emphasized the importance of the calcium-binding domains and the globular domain to the function of COMP. Kennedy *et al.* reported three SNPs in *COMP*, first in exon-11, nucleotide change A>G at c1156 with allelic frequency 0.03, second in exon -16, nucleotide change G>A at c17.55 with allelic frequency 0.05, and last third one is in 3'UTR, Nucleotide change A>G at c2289 with allelic frequency 0.005 which are associated with OA [133].

CALM1 gene showed a significant association of SNP with the disease. CALM1 gene intronic SNP (rs3213718) was present in our population, and its occurrence was significantly affecting the disease.

A case-control study of 600 subjects showed a significant association of the+104 T/C GDF5 polymorphism with KOA and with individual clinical symptoms of the disease. A study done by **Srivastava et al.** reported that genetic polymorphisms affecting KOA vary between genders and indicate a role for BMP5, COL2A1, CCL2, and IL1B in North Indian Population. Another case-control study of 499 KOA cases and 458 controls exhibited an association between rs1470527, rs9382564 polymorphisms of BMP5 gene with KOA. This association was validated by haplotype analysis. Further, the association between KOA and rs1470527 polymorphism was more robust in both the genders and age groups. However, association with rs9382564 was stronger only in female patients and either gender aged >55 years. Moreover, our data showed a significant association of both SNPs with VAS and WOMAC clinical scores.

Furthermore, a genetic study conducted in our laboratory on SNPs rs921126 (BMP5) and rs12901499 (SMAD3) showed a significant association between these SNPs and risk of KOA. It was found that the risk increased with age (>55) in both the genders. We also conducted a pilot study on VDR gene polymorphism and its association with KOA. It was found that Taq1 polymorphism influenced the clinico-radiological response to vitamin D supplementation in KOA subjects with insufficient 25(OH) vitamin D levels. Validation of the results is in process on large population with insufficient 25(OH) vitamin D levels.

An extensive literature search, we could find only one more study conducted in Indian population other than ours, by Subramanyam et al. They studied the association of rs73297147 and rs73771337 polymorphisms in COL2A1 and CRTL1 genes with primary KOA in South Indian population, and a significant association was observed.

Author details

Rajeshwar Nath Srivastava[1*], Amar Chandra Sharma[1], Sudeepti Ratan Srivastava[1], Saloni Raj[2] and Lavini Raj[3]

1 Department of Orthopaedic Surgery, King George's Medical University, Lucknow, UP, India

2 Westminster College, Utah, USA

3 Hult International Business School, San Francisco, USA

*Address all correspondence to: drrnsrivastava@yahoo.com

References

[1] Jordan JM, Kington RS, Lane NE et al.: Systemic risk factors for Osteoarthritis; In Felson DT (conference chair): Osteoarthritis: new insights. Part 1: The disease and its risk factors. Ann Intern Med 2000; 133: 637-639.

[2] Reginster JY: The prevalence and burden of arthritis. Rheumatology 2002; 41(suppl1): 3-6.

[3] Zeng Q, Huang S, Xiao Z, et al. Osteoarthritis: Clinical and epidemiological investigation. Chinese Journal of Internal Medicine. 1995;**34**:88-90

[4] Yao JY, Wang Y, An J, Mao CM, Hou N, et al. Mutation analysis of the Smad3 gene in human Osteoarthritis. European Journal of Human Genetics. 2003;11:714-717

[5] McAlindon T, Dieppe P. Osteoarthritis: definitions and criteria. Ann Rheum Dis 1989;48:531-532.

[6] Altman R, Asch E, Bloch D, Bole G, Borenstein D, Brandt K, *et al*. Development of criteria for the classification and reporting of Osteoarthritis. Classification of Osteoarthritis of the knee. Diagnostic and Therapeutic Criteria Committee of the American Rheumatism Association. Arthritis Rheum.1986; 29(8):1039-49.

[7] Green GA. Understanding NSAIDs: from aspirin to COX-2. Clin Cornerstone. 2001;3(5):50-60.

[8] Brooks SA, Dwek MV, Schumacher U. Functional and molecular Glycobiology. U. Bios scientific publishers. 2002.

[9] Oegema TR, Lewis JL, Thompson RC. Role of acute trauma in development of Osteoarthritis. BONE AND CARTILAGE. 1993; 40(3-4): 220-3.

[10] Chou P, Soong LN, Lin HY. Community-based epidemiological study on hyperuricemia in Pu-Li, Taiwan. J Formos Med Assoc. 1993; 92(7):597-602.

[11] Herbert RD, Gabriel M. Effects of stretching before and after exercising on muscle soreness and risk of injury: a systematic review. BMJ 2002;325(7362):468.

[12] Creamer P, Lethbridge-Cejku M, Hochberg MC. Factors associated with functional impairment in symptomatic knee osteoarthritis. Rheumatology (Oxford).2000;39(5):4906

[13] Hunter DJ, Felson DT. Osteoarthritis. BMJ. 2006; 332 (7542): 639-42.

[14] Dicesare PE et al. 2005

[15] Dieppe, PA; Lohmander, L.S. Pathogenesis and management of pain in Osteoarthritis. Lancet, 2005, 365, 965-973.

[16] Lawrence, R. C. *et al.* Estimates of the prevalence of arthritis and other rheumatic conditions in the United States. Part II. Arthritis Rheum.2008. 58, 26-35

[17] Felson DT, Anderson JJ, Naimark A, Kannel W, Meenan RF.The prevalence of chondrocalcinosis in the elderly and its association with knee osteoarthritis: the Framingham Study. J Rheumatol 1989; 16:1241-5

[18] Parazzini F; Progretto Menopausa Italia Study Group. Menopausal status, hormone replacement therapy use and risk of self-reported physician-diagnosed Osteoarthritis in women attending menopause clinics in Italy. Maturitas.2003 20;46:207-12

[19] Felson DT, Zhang Y. An update on the epidemiology of knee and hip

osteoarthritis with a view to prevention. Arthritis Rheum 1998;41:1343-55

[20] Guccione AA et al. 1994

[21] Felson DT, Lawrence RC et al. 2009

[22] Mishra A, Sanghi D, Sharma AC, Raj S, Maurya SS, Avasthi S, Singh A, Parmar D, Srivastava RN. "Association of polymorphism in growth and differentiation factor 5 gene with osteoarthritis knee". American Journal of Biochemistry and Biotechnology. 2013; 9(1):1-7

[23] Singh AK, Kalaivani M, Krishnan A, Aggarwal PK et al. Prevalence of Osteoarthritis of Knee Among Elderly Persons in Urban Slums Using American College of Rheumatology (ACR) criteria. J Clin Diagn Res. 2014 Sep; 8(9): JC09–JC11.

[24] Martin JA et al. 2002

[25] Awasthi S, Mishra A, Sanghi D, Singh A, Parihar R, Pankaj R, Ganesh S, Srivastava RN. "Association of Intronic Single nucleotide polymorphism SNP of CALM 1 gene with Osteoarthritis of the Knee in the Indian population, a case-control study". Internet Journal of Medical Update, 2012;7(1)

[26] Sanghi D, Mishra A, Sharma AC, Natu SM, Srivastava RN, Singh A, Agarwal S. "Does Vitamin D Improve Osteoarthritis of the Knee, A Randomized Controlled Pilot Trial".Clin Orthop Relat Res. 2013; 471(8). DOI 10.1007/s11999-013-3201-6

[27] Amar Chandra Sharma, RajeshwarNath Srivastava, SudeeptiRatan Srivastava, DevendraParmar, Ajai Singh, Saloni Raj. Association between Single Nucleotide Polymorphisms of SMAD3 and BMP5 with the Risk of Knee Osteoarthritis. JCDR/2017/22371.10073

[28] Subramanyam K, Poornima S, Khan I A and Hasan Q. Exploration of Genetic Association Studies with Collagen Variants in Clinically Diagnosed Primary Knee Osteoarthritis in South Indian Population: A Non-replication Study. Journal of Arthritis, 2016, 5:5.

[29] Gonzalez A. Osteoarthritis year 2013 in review: genetics and genomics. Asian Biomedicine Vol. 5 No. 1 February 2011; 23-36

[30] Yau Michelle S et al. Genome-Wide Association Study of Radiographic Knee Osteoarthritis in North American Caucasians. Arthritis & Rheumatology. February 2017, Vol. 69, No. 2.

[31] Syddall, CM; Reynard, L.N.; Young, D.A.; Loughlin, J. The identification of trans-acting factors that regulate the expression of GDF5 via the osteoarthritis susceptibility SNP rs143383. PLoS Genet. 2013, 9.

[32] Zhang R, Yao J, Xu P, Ji B, Luck JV, Chin B, et al. A comprehensive meta-analysis of the association between genetic variants of *GDF5* and Osteoarthritis of the knee, hip and hand. Inflammation Research 2015;64(6):405-414

[33] Miyamoto, Y.; Mabuchi, A.; Shi, D. et. al. A functional polymorphism in the 51 UTR of GDF5 is associated with susceptibility to Osteoarthritis. Nat. Genet. 2007, 39, 529-533.

[34] Valdes, A.M.; Spector, T.D.; Tamm, A. et al. Genetic variation in the SMAD3 gene is associated with hip and knee osteoarthritis. Arthritis Rheum. 2010, 62, 2347-2352.

[35] Chapman, K.; Takahashi, A.; Meulenbelt, I. et al. A meta-analysis of European and Asian cohorts reveals a global role of a functional SNP in the 51 UTR of GDF5 with osteoarthritis susceptibility. Hum. Mol. Genet. 2008, 17, 1497-1504.

[36] Kizawa, H.; Kou, I.; Iida, A. et al. An aspartic acid repeat polymorphism

in asporin inhibits chondrogenesis and increases susceptibility to Osteoarthritis. Nat. Genet. 2005, 37, 138-144.

[37] Valdes, A.M.; Evangelou, E.; Kerkhof, H.J.M. et al. The GDF5 rs143383 polymorphism is associated with Osteoarthritis of the knee with genome-wide statistical significance. Ann. Rheum. Dis. 2011, 70, 873-875.

[38] Attur M, Belitskaya-Levy I, Cheongeun, Krasnokutsky S, Greenberg J, Samuels J, et al. Increased Interleukin-1 Gene Expression in Peripheral Blood Leukocytes Is Associated With Increased Pain and Predicts Risk for Progression of Symptomatic Knee Osteoarthritis. Arthritis & Rheumatism 2011; 63 (7):1908-1917. DOI 10.1002/art.30360

[39] Yerges-Armstrong LM, Yau MS, Liu Y, Krishnan S, Renner JB, Eaton CB, et al. Association Analysis of BMD associated SNPs with Knee Osteoarthritis. Journal of Bone and Mineral Research 2014;29(6):1373-1379. DOI: 10.1002/jbmr.2160

[40] Luo SX, Zhang XH, Zhang JJ, Long GH, Dong GF, Su W, et al. Genetic Polymorphisms of Interleukin-16 and Risk of Knee Osteoarthritis. PLOS ONE | .2015. DOI:10.1371/journal.pone.0123442

[41] Torres F, et al. Polymorphic variation of hypoxia-inducible factor-1 A (HIF1A) gene might contribute to the development of knee osteoarthritis: a pilot study. BMC Musculoskeletal Disorders 2015; 16:218 DOI 10.1186/s12891-015-0678-z

[42] Ho KW, Wallace M and Fillingim R. Single-nucleotide polymorphism in COL11A2 associated with thermal hyperalgesia in knee osteoarthritis (P1.210).Neurology 2017;88(16):210

[43] Li H, Zhang X, Cao Y, Hu S, Peng F, Zhou J1 and Li J. Association between EN1 rs4144782 and susceptibility of knee osteoarthritis: A case-control study. Oncotarget, 2017;8(22): 36650-36657

[44] Baggiolini M, Dewald B, Moser B. Human chemokines: an update. Annu Rev Immunol 1997. 15:675-705.

[45] Berckmans RJ, Nieuwland R, Kraan MC, Schaap MC, Pots D, Smeets TJ, Sturk A, Tak PP: Synovial microparticles from arthritic patients modulate chemokine and cytokine release by synoviocytes. *Arthritis Res Ther* 2005, 7:R536-R544.

[46] Bleasel JF, Holderbaum D, Brancolini V, Moskowitz RW, Considine EL, Prockop DJ, Devoto M, Williams CJ. Five families with arginine 519-cysteine mutation in COL2A1: evidence for three distinct founders. Hum Mutat. 1998; 12(3):172-6.

[47] Chou P, Soong LN, Lin HY. Community-based epidemiological study on hyperuricemia in Pu-Li, Taiwan. J Formos Med Assoc. 1993 Jul; 92(7):597-602.

[48] Corrigall, V. M., Arastu, M., Khan, S., Shah, C., Fife, M., Smeets, T., Tak, P.-P., Panayi, G. S. Functional IL-2 receptor beta (CD122) and gamma (CD132) chains are expressed by fibroblast-like synoviocytes: activation by IL-2 stimulates monocyte chemoattractant protein-1 production. J. Immun. 2001;166: 4141-4147.

[49] Creamer P, Lethbridge-Cejku M, Hochberg MC. Factors associated with functional impairment in symptomatic knee osteoarthritis. Rheumatology (Oxford). 2000 May;39(5):490-6.

[50] Ducy P, Karsenty G. The family of bone morphogenetic proteins. Kidney Int 2000; 57:2207-14.

[51] Edwards CJ, Francis-West PH: Bone morphogenetic proteins in the

development and healing of synovial joints. *Semin Arthritis Rheum* 2001, 31:33-42.

[52] Galvez-Rosas A, González-Huerta C, Borgonio-Cuadra VM, Duarte-Salazár C, Lara-Alvarado L, de los Angeles Soria-Bastida M, Cortés-González S, Ramón-Gallegos E, Miranda-Duarte A. A COL2A1 gene polymorphism is related with advanced stages of osteoarthritis of the knee in Mexican Mestizo population. Rheumatol Int. 2010 Jun; 30(8):1035-9. Epub 2009 Sep 16.

[53] Hahn, G. V., Cohen, R. B., Wozney, J. M., Levitz, C. L., Shore, E. M., Zasloff, M. A., Kaplan, F. S. A bone morphogenetic protein subfamily: chromosomal localization of human genes for BMP5, BMP6, and BMP7. Genomics; 1992, 14: 759-762.

[54] Hatakeyama Y, Tuan RS, Shum L: Distinct functions of BMP4 and GDF5 in the regulation of chondrogenesis. *J Cell Biochem* 2004, 91:1204-1217.

[55] Herbert RD, Gabriel M. Effects of stretching before and after exercising on muscle soreness and risk of injury: a systematic review. BMJ. 2002 Aug 31; 325(7362):468.

[56] Hogan B. L.; Bone morphogenetic proteins: multifunctional regulators of vertebrate development. *Genes Dev.* 1996 10: 1580-1594

[57] Holderbaum D, Malemud CJ, Moskowitz RW, Haqqi TM. Human cartilage from late-stage familial Osteoarthritis transcribes type II collagen mRNA encoding a cysteine in position 519. Biochem Biophys Res Commun. 1993 May 14; 192(3):1169-74.

[58] Hunter DJ, Felson DT. Osteoarthritis. BMJ. 2006 Mar 18; 332(7542):639-42.

[59] Hwang SY, Cho ML, Park B, Kim JY, Kim YH, Min DJ, et al. Allelic frequency of MCP-1 promoter 22518 polymorphism in the Korean population and Korean patients with rheumatoid arthritis, systemic lupus erythematosus and adult-onset Still's disease. Eur J Immunogenet 2002; 29:413-16.

[60] Iikuni N, Okamoto H, Yoshio T, Sato E, Kamitsuji S, Iwamoto T, et al. Raised monocyte chemotactic protein-1 (MCP-1)/CCL2 in cerebrospinal fluid of patients with neuropsychiatric lupus. Ann Rheum Dis 2006; 65:253-6.

[61] James M Wilkins*1, Lorraine Southam1, Zehra Mustafa1, Kay Chapman1 and John Loughlin2 Association of a functional microsatellite within intron 1 of the

[62] gene with susceptibility to Osteoarthritis. *BMC Medical Genetics* 2009, 10:141

[63] Jones C. M., Lyons K. M., and Hogan B. L., Involvement of Bone Morphogenetic Protein-4 (BMP-4) and *Vgr-1* in morphogenesis and neurogenesis in the mouse. Development 1991, **111**:531-542

[64] Kannu P, Bateman JF, Randle S, Cowie S, du Sart D, McGrath S, Edwards M, Savarirayan R. Premature arthritis is a distinct type II collagen phenotype. Arthritis Rheum. 2010 May; 62(5):1421-30.

[65] Kellgren JH, Lawrence JS. Radiological assessment of osteo-arthrosis. Ann Rheum Dis. 1957 Dec;16(4):494 502

[66] King, J. A., P. C. Marker, K. J. Seung, and Kingsley D. M., BMP5 and the molecular, skeletal, and soft-tissue alterations in *short ear* mice. Dev. Biol. 1994 **166**:112-122

[67] King J. A., Storm E. E., Marker P. C, Dileone R. J., and. Kingsley D. M, The role of BMPs and GDFs in development of region-specific skeletal structures. Ann. NY Acad. Sci. 1996; **785**:70-79

[68] Kingsley DM. What do BMPs do in mammals? Clues from the mouse short-ear mutation. Trends Genet 1994; 10: 16-21.

[69] Kingsley D. M., The TGF-ß superfamily: new members, new receptors, and new genetic tests of function in different organisms. Genes Dev. 1994a; **8**:133-146

[70] Knight JC: Regulatory polymorphisms underlying complex disease traits. *J Mol Med* 2005, 83:97-109.

[71] Knight PG, Glister C: (2006) TGF-beta superfamily members and ovarian follicle development. *Reproduction* 2006, 132:191-206.

[72] Knowlton RG, Katzenstein PL, Moskowitz RW, Weaver EJ, Malemud CJ, Pathria MN, Jimenez SA, Prockop DJ. Genetic linkage of a polymorphism in the type II procollagen gene (COL2A1) to primary Osteoarthritis associated with mild chondrodysplasia. N Engl J Med. 1990 Feb 22;322(8):526-30

[73] Koch AE, Kunkel SL, Harlow LA, Johnson B, Evanoff HL, Haines GK, et al. Enhanced production of monocyte chemoattractant protein-1 in rheumatoid arthritis. J Clin Invest 1992; 90:772-9.

[74] Koch AE, Kunkel SL, Shah MR, Fu R, Mazarakis DD, Haines GK, et al. Macrophage inflammatory protein-1b: a C-C chemo-CHEMOKINES AND CARTILAGE DETERIORATION 1069 kine in Osteoarthritis. Clin Immunol Immunopathol 1995; 77:307-14.

[75] Largo R, Diez-Ortego I, Sanchez-Pernaute O, Lopez-Armada MJ, Alvarez-Soria MA, Egido J, et al. EP2/EP4 signalling inhibits monocyte chemoattractant protein-1 production induced by interleukin 1b in synovial fibroblasts. Ann Rheum Dis 2004; 63:1197-204.

[76] Law ML, Tung L, Morse HG, Berger R, Jones C, Cheah KS, Solomon E. The human type II collagen gene (COL2A1) assigned to 12q14.3. Ann Hum Genet. 1986 May; 50(Pt 2):131-7.

[77] Leonard EJ, Yoshimura T. Human monocyte chemoattractant protein-1 (MCP-1). Immunol Today 1990; 11:97-101.

[78] Lee YH, Kim HJ, Rho YH, Choi SJ, Ji JD, Song GG. Functional polymorphisms in matrix metalloproteinase-1 and monocyte chemoattractant protein-1 and rheumatoid arthritis. Scand J Rheumatol 2003; 32:235-9.

[79] Lisignoli G, Toneguzzi S, Pozzi C, Piacentini A, Grassi F, Ferruzzi A, et al. Chemokine expression by subchondral bone marrow stromal cells isolated from Osteoarthritis (OA) and rheumatoid arthritis (RA) patients. Clin Exp Immunol 1999; 116:371-8.

[80] Loughlin J, Dowling B, Mustafa Z, et al. Association of the interleukin-1 gene cluster on chromosome 2q13 with knee osteoarthritis. Arthritis Rheum 2002;46:1519-27

[81] Luo G., Hofmann C., Bronckers A. L., Sohocki M., and Bradley A. *et al.*, BMP-7 is an inducer of nephrogenesis and is also required for eye development and skeletal patterning. Genes Dev 1995. **9**:2808-2820

[82] Lyons K. M., Pelton R. W., and Hogan B. L., Patterns of expression of murine Vgr-1 and BMP-2a RNA suggest that transforming growth factor-ß-like genes coordinately regulate aspects of embryonic development. Genes Dev1989. **3**:1657-1668

[83] Mailhot G, Yang M, Mason-Savas A, Mackay CA, Leav I, Odgren PR: BMP-5 expression increases during chondrocyte differentiation in vivo and in vitro and promotes proliferation and

cartilage matrix synthesis in primary chondrocyte cultures. *J Cell Physiol* 2008, 214:56-64.

[84] Matsukawa A, Miyazaki S, Maeda T, Tanase S, Feng L, Ohkawara S, Yoshinaga M, Yoshimura T: Production and regulation of monocyte chemoattractant protein-1 in lipopolysaccharide- or monosodium urate crystal-induced arthritis in rabbits: roles of tumour necrosis factor-alpha, interleukin-1, and interleukin-8. *Lab Invest* 1998, 78:973-985.Scand J Rheumatol 2007;36:299-306

[85] McAlindon T, Dieppe P. Osteoarthritis: definitions and criteria. Ann Rheum Dis. 1989 Jul; 48(7):531-2.

[86] Mehrabian, M., Sparkes, R. S., Mohandas, T., Fogelman, A. M., Lusis, A. J. Localization of monocyte chemotactic protein-1 gene (SCYA2) to human chromosome 17q11.2-q21.1. Genomics 1991; 9: 200-203.

[87] Meulenbelt I, Seymour AB, Nieuwland M, et al. Association of the interleukin-1 gene cluster with radiographic signs of Osteoarthritis of the hip. Arthritis Rheum 2004;50:1179-86

[88] Meulenbelt I, Williams CJ, Te Koppele JM, Van de Giessen GC, Slagboom PE. Population haplotype analysis and evolutionary relations of the COL2A1 gene. Ann Hum Genet. 1996 May;60(Pt 3):189-99.

[89] Moos V, Rudwaleit M, Herzog V, et al. Association of genotypes affecting the expression of interleukin- 1beta or interleukin-1 receptor antagonist with Osteoarthritis.Arthritis Rheum 2000;43: 2417-22

[90] Nelson F, Dahlberg L, Laverty S, Reiner A, Pidoux I, Ionescu M, Fraser GL, Brooks E, Tanzer M, Rosenberg LC, Dieppe P, Robin Poole A. Evidence for the altered synthesis of type II collagen in patients with Osteoarthritis. J Clin Invest. 1998 Dec 15;102(12):2115-25.

[91] Nothwang HG, Strahm B, Denich D, Kübler M, Schwabe J, Gingrich JC, et al. Molecular cloning of the interleukin-1 gene cluster: construction of an integrated YAC/PAC contig and a partial transcriptional map in the region of chromosome 2q13.Genomics, 1997; 41:370-8.

[92] Oegema TR Jr, Lewis JL, Thompson RC Jr. Role of acute trauma in development of Osteoarthritis. Agents Actions. 1993 Nov;40(3-4):220-3.

[93] Park HJ, Yoon SH, Zheng LT, Lee KH, Kim JW, Chung JH, Lee YA, Hong SJ: Association of the 22510A/G chemokine (C–C motif) ligand two polymorphism with knee osteoarthritis in a Korean population. Scand J Rheumatol 2007;36:299-306

[94] Pulsatelli L, Dolzani P, Piacentini A, Silvestri T, Ruggeri R, Gualtieri G, et al. Chemokine production by human chondrocytes. J Rheumatol 1999;26:1991-2001.

[95] Reddi AH. The interplay between bone morphogenetic proteins and cognate binding proteins in bone and cartilage development: noggin, chordin and DAN. Arthritis Res 2001; 3:1-5.

[96] Reddi A. H. and Huggins C., Biochemical sequences in the transformation of normal fibroblasts in adolescent rats. Proc. Natl. Acad. Sci. USA 1972 69:1601-1605

[97] Rollins BJ. Chemokines. Blood 1997;90:909-28

[98] Seitz M, Loetscher P, Dewald B, Towbin H, Ceska M, Baggiolini M. Production of interleukin-1 receptor antagonist, inflammatory chemotactic proteins and prostaglandin E by rheumatoid and osteoarthritic synoviocytes: regulation by IFN-gamma and IL-4. J Immunol 1994;152:2060-5.

[99] Sezgin M, Erdal E M, Altintas M Z, Ankarali C H, Barlas O I, Turkmen E, Sahin G: Lack of association polymorphisms of the IL1RN, IL1A, and IL1B genes with knee osteoarthritis in Turkish patients. *Clin Invest Med* 2007; 30 (2): E86-E92

[100] Smith AJ, Keen LJ, Billingham MJ, et al. Extended haplotypes and linkage disequilibrium in the IL1R1- IL1A-IL1B-IL1RN gene cluster: association with knee osteoarthritis. Genes Immun 2004;5:451-60

[101] Solomon E, Hiorns LR, Spurr N, Kurkinen M, Barlow D, Hogan BL, Dalgleish R. Chromosomal assignments of the genes coding for human types II, III, and IV collagen: a dispersed gene family.Proc Natl Acad Sci USA.1985 May; 82(10):33304.

[102] Solloway MJ, Dudley AT, Bikoff EK, Lyons KM, Hogan BL, et al., Mice lacking Bmp6 function. Dev Genet 1998; 22: 321-339.

[103] Somi S, Buffing AA, Moorman AF, Van Den Hoff MJ: Dynamic patterns of expression of BMP isoforms 2, 4, 5, 6, and 7 during chicken heart development. *Anat Rec A Discov Mol Cell Evol Biol* 2004, 279:636-651.

[104] Southam L, Chapman K, Loughlin J: Genetic association analysis of *BMP5* as a potential osteoarthritis susceptibility gene. *Rheumatology,* 2003, 42:911-912.

[105] Sozzani S, Locati M, Allavena P, van Damme J, Mantovani A. Chemokines: a superfamily of chemotactic cytokines. Int J Clin Lab Res 1996; 26:69-82.

[106] Stern AG, de Carvalho MR, Buck GA, Adler RA, Rao TP, Disler D, Moxley G; I-NODAL Network. Association of erosive hand osteoarthritis with a single nucleotide polymorphism on the gene encoding interleukin-1 beta. Osteoarthritis Cartilage. 2003 Jun;11(6):394-402.

[107] Thomas J. T., Lin K., Nandedkar M., Camargo M., and Cervenka J. *et al.*, A human chondrodysplasia due to a mutation in a TGF-ß superfamily member. Nat. Genet. 1996;12:315-317

[108] Urist M. R., Bone: formation by autoinduction. Science 1965: 150:893-899

[109] Volin MV, Shan MR, Tokuhira M, Haines GK, Woods JM, Koch AE. RANTES expression and contribution to monocyte chemotaxis in arthritis. Clin Immunol Immunopathol 1998;89:44-53.

[110] Xu P, Yao J, Hou W. Relationships between COL2A1 gene polymorphisms and knee osteoarthritis in Han Chinese women. Mol Biol Rep. 2011 Apr; 38(4):2377-81. Epub 2010 Nov 19.

[111] Zuzarte-Luis V, Montero JA, Rodriguez-Leon J, Merino R, Rodriguez-Rey JC, Hurle JM: A new role for BMP5 during limb development acting through the synergic activation of Smad and MAPK pathways. *Dev Biol* 2004, 272:39-52

[112] Nakayama, T., M.A. Snyder, S.S. Grewal, K. Tsuneizumi, T. Tabata, and J.L. Christian. *Xenopus* Smad8 acts downstream of BMP-4 to modulate its activity during vertebrate embryonic patterning. *Development.*1998. 125:857-867.

[113] Parfitt AM. Quantum concept of bone remodelling and turnover: implications for the pathogenesis of osteoporosis. Calcif Tissue Int 1979; 28:1-5.

[114] Sainz J, Van Tornout JM, Loro ML, Sayre J, Roe TF, Gilsanz V. Vitamin D-receptor gene polymorphisms and bone density in prepubertal American girls of Mexican descent. N Engl J Med 1997;337; 77-82.

[115] Suzuki, A., C. Chang, J.M. Yingling, X.F. Wang, and A. Hemmati-Brivanlou.

1997. Smad5 induces ventral fates in *Xenopus* embryo. *Dev. Biol*. 184:402-405.

[116] Valdes A M, Spector T D *et. al.* Genetic Variation in the *SMAD3* Gene Is Associated With Hip and Knee Osteoarthritis. Arthritis & rheumatism. 2010. Vol. 62, No. 8

[117] White –O Connor B, Sobal J. Nutrient intake and obesity in a multidisciplinary assessment of Osteoarthritis. Clin Thr 1986:9: S30-S40.

[118] Wu Q, Kim KO, Sampson ER, Chen D, Awad H, O'Brien T, et al. Induction of an osteoarthritis-like phenotype and degradation of phosphorylated Smad3 by Smurf2 in transgenic mice. Arthritis Rheum 2008;58:3132-44.

[119] Yang X, Chen L, Xu X, Li C, Huang C, and Deng C. TGF-β/Smad3 Signals Repress Chondrocyte Hypertrophic Differentiation and Are Required for Maintaining Articular Cartilage. The Journal of Cell Biology, 2001.Volume 153, Number 1

[120] Zeng Q, Huang S, Xiao Z et al.: Osteoarthritis: clinical and epidemiological investigation. Chin J Internl Med 1995; 34: 88-90.

[121] Nevitt MC, Lane NE, Scott JC, Hochberg MC, Pressman AR, Genant AK, et al., and the Study of Osteoporotic Fractures Research Group. Radiographic Osteoarthritis of the hip and bone mineral density. Arthritis Rheum 1995; 38:907-16.

[122] Ralston SH. Genetic determinants of osteoporosis. Curr Opin Rheumatol. 2005; 17:475-9.

[123] Reneland RH, Mah S, Kammerer S, Hoyal CR, Marnellos G, Wilson SG, Sambrook PN, Spector TD, Nelson MR, Braun A. Association between a variation in the phosphodiesterase 4D gene and bone mineral density. BMC Med Genet. 2005; 6:9.

[124] Riham G.Mahfouz, Azza M. Abdu Allah, Seham A. Khodeer, Waleed F. Abd Elazeem, Mostafa Al Nagar and Walid A. Shehab-Eldin3 Frequency of Distribution of Interleukin 6 Gene 174G/C Polymorphism in obese Egyptian Cohort Journal of American Science, 2011;7(8).

[125] Simonet WS, Lacey DL, Dunstan CR, Kelley M, Chang MS, Lüthy R, Nguyen HQ, Wooden S, Bennett L, Boone T, Shimamoto G, DeRose M, Elliott R, Colombero A, Tan HL, Trail G, Sullivan J, Davy E, Bucay N, Renshaw-Gegg L, Hughes TM, Hill D, Pattison W, Campbell P, Sander S, Van G, Tarpley J, Derby P, Lee R, Boyle WJ (April 1997). "Osteoprotegerin: a novel secreted protein involved in the regulation of bone density". Cell 89 (2): 309-19.

[126] Styrkarsdottir U, Cazier JB, Kong A, Rolfsson O, Larsen H, Bjarnadottir E, Johannsdottir VD, SigurdardottirMS,BaggerY,ChristiansenC, Reynisdottir I, Grant SF, Jonasson K, Frigge ML,Gulcher JR, Sigurdsson G, Stefansson K 2003 Linkage of osteoporosis to chromosome 20p12 and association to BMP2. PLoS Biol1:E69.

[127] Styrkarsdottir U, et al. Linkage of Osteoporosis to Chromosome 20p12 and Association to BMP2. PLoS Biol. 2003; 1:E69.

[128] Sakao K, Takahashi KA, Arai Y, Saito M, Honjo K, Hiraoka N, Asada H, Shin-Ya M, Imanishi J, Mazda O, Kubo T: Osteoblasts derived from osteophytes produce interleukin-6, interleukin-8, and matrix metalloproteinase-13 in Osteoarthritis. J Bone Miner Metab 2009, 27:412-423.

[129] Sowers M, Lachance L, Jamadar D, Hochberg MC, Hollis B, Crutchfield M,

et al. The associations of bone mineral density and bone turnover markers with Osteoarthritis of the hand and knee in pre- and perimenopausal women. Arthritis Rheum 1999;42:483-9.

[130] Scheidt Nave C, Bismar H, Leidig-Bruckner G, Woitge H, Seibel MJ, Ziegler R, Pfeilschifter J 2001 Serum interleukin 6 is a major predictor of bone loss in women specific to the first decade past menopause. J Clin Endocrinol Metab 86:2032-2042.

[131] Xiong D, Shen H, Zhao L, Xiao1 P, Yang T, Guo Y, Wang W, Guo Y, Liu Y, Recker R, Deng H. A Robust and Comprehensive Analysis of 20 Osteoporosis Candidate Genes by Very High-Density Single-Nucleotide Polymorphism Screen among 405 Caucasian Nuclear Families Identified Significant Association and Gene-Gene Interaction. J Bone Miner Res. 2006.

[132] Z.S. Özkan, D. Deveci, E. Önalan Etem and H. Yüce, Lack of effect of bone morphogenetic protein 2 and 4 gene polymorphisms on bone density in postmenopausal Turkish women Genet. Mol. Res. 9 (4): 2311-2316 (2010)

[133] Zhou Houde, BU Yanhong, Tang Aiguo, Xie Hui, Liao, Eryuan Time course of the osteoprotegerin gene expression in human primary osteoblasts as well as MG-63 cell lines and the effect of 17β-estradiol on it journal of Chinese clinical medicine volume 1 November 2 July 2006.

Chapter 14

Serum Hepcidin Hormone Level and Its Genes Polymorphism

Safa A. Faraj and Naeem M. Al-Abedy

Abstract

This chapter sheds light on hepcidin, historical view of hepcidin, and the time of its discovery in the first section. Then this chapter gives information about the genetic aspect and the importance of gene knowledge of hepcidin in explaining many disorders in human beings, supported by illustration figures. The regulation of iron in the human body as an essential function of hepcidin is discussed in this chapter. Examples of the genes of hepcidin (HAMP and HFE) are highlighted in detail as they are essential in regulating iron as well as discussing the genetic mutations that occur in these genes and their medical and clinical impacts for many diseases such as thalassemia. Finally, the inherited disorders related to hepcidin that lead to genetic diseases are discussed.

Keywords: iron, gene mutation, HAMP gene, HFE gene, hemochromatosis

1. Introduction

Hepcidin is presently regarded as the key to the iron balance regulator. The balance of intracellular iron is preserved by proteins that regulate iron. Hepcidin, encoded by the HAMP gene is a 25 amino acid peptide that has been lately found [1]. Several mutations in the HAMP gene have been reported. The G71D mutation is probable to be linked to reduced hepcidin activity [2]. Mutations in iron-regulating proteins cause the disorder (HFE, TfR2, and HAMP) genes. Fekri et al. showed that H63D mutation of the HFE gene could play some role in disease evolution. In iron homeostasis, the HFE gene plays a very significant role by regulating iron absorption [3]. HFE mutations are currently referred to as the reason for decreased absorption of iron, iron overload, and hereditary hemochromatosis [4]. Many types of research have shown that patients with HFE mutations in beta-thalassemia are likely to create hemochromatosis that will require early chelation of iron even in heterozygous conditions [5]. Among the mutations discovered most frequently are the three missense mutations (SNPs), which are found in the HFE gene. The most prevalent mutation within the HFE gene exon 4, leading in a shift of cysteine-to-tyrosine amino acid at position 282 (C282Y), 60% of hereditary hemochromatosis instances in Mediterranean populations accounted for this mutation. H63D is also a mutation leading to the replacement of histidine with aspartic acid during a C-G shift at nucleotide 187 of exon 2 of the HFE gene. In combination with the C282Y allele (C282Y/H63D), hemochromatosis is most pronounced. The HFE gene's third mutation is a substitute for 193AT in exon a pair of with an ensuing serine to cysteine replacement in amino acid position 65 (S65C) [6]. The interaction concerning the mutations over genes influencing blood homeostasis including thalassemia may want to hold a synergistic result, increasing the iron storage [7].

2. Hepcidin

2.1 Discovery of hepcidin

The hepcidin molecule ("hep" hepatic origin, "cidin" antimicrobial activity) was described in the year 2000; it is an antimicrobial peptide that acts in parts in innate immunity and iron metabolism [8]. It is a peptide hormone the liver produces, and it works as a regulator of iron [9]. Hepcidin is a regulator of iron homeostasis. Its production is increased by iron excess and inflammation and decreased by hypoxia and anemia. Hepcidin inhibits the flow of iron into the plasma from duodenal enterocytes that absorb dietary iron, macrophages that recycle iron from senescent erythrocytes, and iron-storing hepatocytes. Iron-loading anemias are diseases in which hepcidin is controlled by ineffective erythropoiesis and concurrent iron overload impacts [10]. Hepcidin was isolated from the human urine and blood, especially from plasma after filtration [8]. Hepcidin was produced by macrophages, adipocytes, neutrophils, lymphocytes, renal cells, and β-cells [11].

Hepcidin is produced by macrophages, adipocytes, neutrophils, lymphocytes, renal cells, and β-cells [11]. In the studies of experiment on mice used for the determination of hepcidin regulation, the expression, function, and structure showed that severe iron overload is occurring due to the gene responsible for hepcidin production, and the gene has the role of iron regulation. Hepcidin has several functions such as inflammation, hypoxia, hypoxia, and iron stores [12]. Hepcidin reacts with ferroportin, and the ferroportin is found in spleen, duodenum, and placenta. If the ferroportin decreases, it results in reduced iron intake and macrophage release of iron and using of the iron, which is stored in the liver [13].

2.2 The hepcidin antimicrobial peptide (HAMP) gene and the structure of hepcidin

Hepcidin was initially identified as a liver-expressed antimicrobial peptide (LEAP1) with direct antimicrobial activity against a number of bacterial and fungal species [14]. It was soon discovered that hepcidin plays a major role in the regulation of iron homeostasis, being overexpressed in the liver with an induced (dietary or parenteral) iron overload [15]. Hepcidin acts as a negative regulator of iron stores; in response to increased iron levels, the liver increases hepcidin synthesis which then acts on the sites of absorption (enterocytes of the duodenum), storage (primarily hepatocytes of the liver), or recycling (macrophages of the reticuloendothelial system) leading to a decrease in the release of iron from these tissues. Hepcidin exerts its influence by binding to and inducing the internalization and degradation of ferroportin (FPN), the only known exporter of iron [16].

Gene of human hepcidin is carried out by chromosome 19q13.1. It consists of a (2637) nucleated base [17]. HAMP gene was founded in the liver cells in the brain, trachea, heart, tonsils, and lung [18]. HAMP gene encodes preprohepcidin, which consists of (84) amino acids; hepcidin has three forms: 25 aa, 22 aa, and 20 aa peptides. All the types are founded in urine, while 25 and 20 are founded in human serum [19]. **Figure 1** shows the structure of hepcidin-25, which consists of (8) cysteine linked by disulfide link [21]. Nuclear magnetic resonance spectroscopy (NMR) is used for the analysis of the structure of hepcidin; it has four disulfide bonds [22].

DOI: http://dx.doi.org/10.5772/intechopen.93622

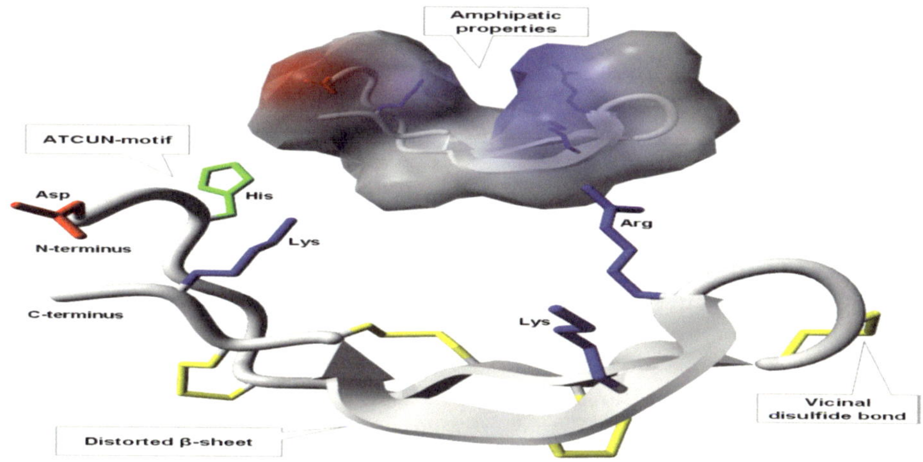

Figure 1.
Molecule structure of human synthetic hepcidin-25. Front: an overview of the structure of hepcidin-25. Distorted β-sheets are shown as gray arrows, and the peptide backbone is colored gray. The disulfide bonds are colored yellow, highlighting the position of an unusual vicinal bond between adjacent cysteines at the hairpin turn. Positive residues of arginine (Arg) and lysine (Lys) are pictured in blue, the negative residue of aspartic acid (asp) in red, and the histidine containing amino-terminal Cu2 ± Ni2 + (ATCUN)-binding motif in the N terminal region is colored green. Background: Hepcidin-25 molecule displayed with a solvent-accessible surface that illustrates the amphipathic structure of the molecule [20].

2.3 Iron regulation by hepcidin

The iron content in biological fluids is strictly regulated in all organisms to provide iron as required and to prevent toxicity, as excess iron can contribute to reactive oxygen species production. Iron homeostasis in mammals is controlled at the level of intestinal absorption, as iron is not excreted. Hepcidin, a circulating peptide hormone, is the master systemic iron homeostasis regulator that combines iron use and processing with the iron acquisition [22].

This hormone is released primarily by hepatocytes and is a negative regulator of plasma iron entry (**Figure 2**).

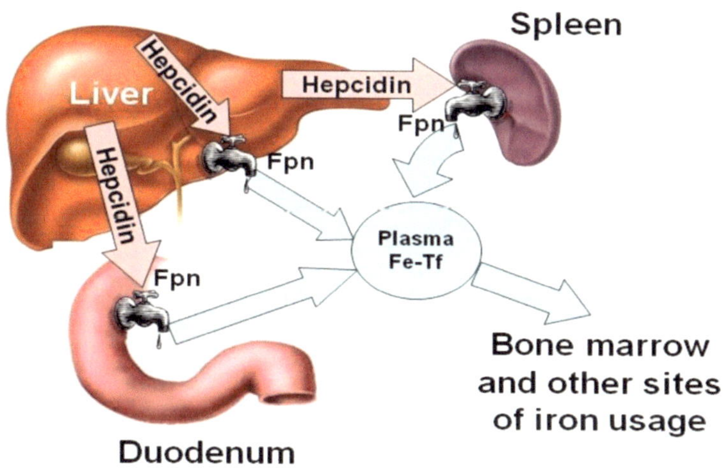

Figure 2.
Hepcidin internalization and degradation [22].

Hepcidin works by binding to ferroportin, an iron carrier present in intestinal duodenum cells, macrophages, and placenta cells. Hepcidin binding leads to the internalization and degradation of the ferroportin [16]. Cell surface ferroportin deficiency prevents iron from entering the plasma. Reduced iron entry into plasma results in low transferrin saturation, and less iron is delivered to the production of erythroblast [23]. In addition, reduced hepcidin expression results in increased ferroportin cell surface and increased iron absorption. Hepcidin dysregulation leads to iron disorders [24].

Overexpression of hepcidin results in chronic disease anemia, whereas low production of hepcidin results in hereditary hemochromatosis with consequent accumulation of iron in vital organs [25] (**Figure 2**). Most inherited iron disorders arise from inadequate development of hepcidin relative to the degree of accumulation of tissue iron. Impaired hepcidin expression was found to result from mutations in any of four genes: transferrin receptor 2 (TFR2), hemochromatosis (HFE), type 2 hemochromatosis (HFE2), and antimicrobial hepcidin peptide (HAMP) [26]. Mutations in HAMP, the gene encoding hepcidin, lead to iron overload disease because the absence of hepcidin allows for constitutively high absorption of iron.

Hepcidin-mediated iron homeostasis regulation. As in condition with high levels of hepcidin in the bloodstream result in the iron exporter ferroportin which is being internalized and degraded. Loss of ferroportin cell surface results in the loading of macrophage iron, low levels of plasma iron, and decreased erythropoiesis due to decreased iron-bound transferrin. Decreased erythropoiesis causes chronic disease anemia and regulates the level of iron imports into plasma, normal transferrin saturation, and normal levels of erythropoiesis in response to iron demand (**Figure 3**). Hemochromatosis, or iron overload, results from the insufficient levels of hepcidin, resulting in increased plasma iron imports, high transferrin saturation, and excess liver iron deposition [28].

2.4 The HAMP gene mutation

Mutation in the HAMP gene will produce a change in the hepcidin function. HAMP gene consists of exon 3, the last exon is encoded proteins, and it is considered the most important and largest area in the gene and is contained on many polymorphisms [29]. Polymorphisms are in the HFE gene more than in the HAMP gene. There are about 16 types of single nucleotide polymorphism founded in different studies [30]. Many reports detected (8) mutations in the gene. The persons who

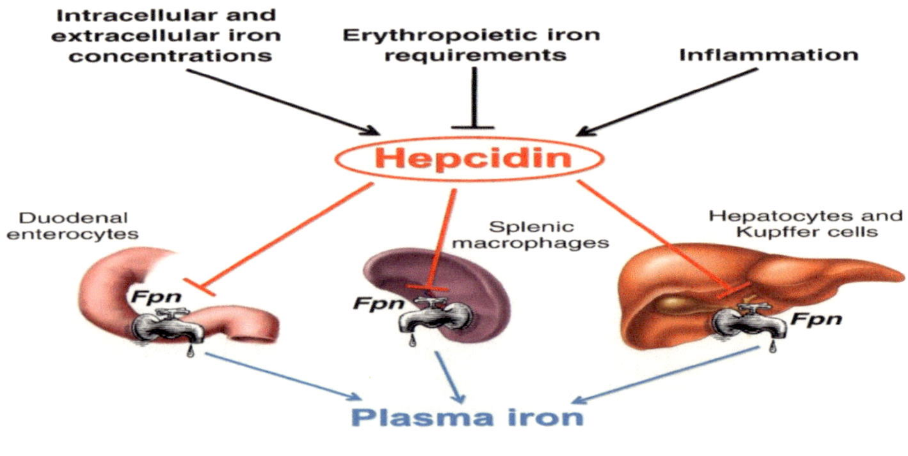

Figure 3.
Regulation of iron balance [27].

carry mutations in the HAMP gene show juvenile hemochromatosis, which occurs at 10–30 years [31]. The microsatellite marker probes are used as the first genetic alteration in the HAMP gene [32]. In some time, the mutation occurs in c.233G > A after exchanging of some amino acids' inactive peptide or substitution of C78 by a tyrosine, C78T. The mutation makes possible the bisulfite bonds binding of ferroportin to hepcidin, which results in the accumulation absorbed of iron [33].

Mutation of C70R causes malformation in the bisulfite bond of the cysteines. The arginine becomes an alternative for the cysteine that its exchanging does not allow the production of bisulfite bridge between (3) and (6) in hepcidin peptide [34]. Alteration of C to T was occurring at position (166) (166C-T) of the HAMP gene and alteration of arginine at position (56) for a stopping codon (R56X), 193A to (T). As well as the ferroportin also does not bind to hepcidin, producing more iron. While defect in guanine exon 2 at position 93 leads to a mutation in RNA [35], defect in Met50del from exon 2 causes a disorder in the expression of the active peptide and causes changes in reading frames. Another mutation, G71D, alters to amino acid 71, which is between (3–4) cysteine, which prevents linking with ferroportin [36].

HFE-H63D variant is associated with the HAMP-G71D variant in sickle cell disease patients and increases iron overload [37]. The polymorphism (G to A) at the +14 position of the 5′-UTR regions produce new initiation codon, inducing yielding of a new abnormal protein and change in the reading frame. Unstable protein will be generated, which is analyzed after the translation of the mRNA [38]. The haplotype of the HAMP gene is caused by the linked of polymorphisms NC-582A > G and NC-1010C > T, with ferritin concentration more than 300 μg/L [39].

The association of HFE gene polymorphisms and HAMP is common. In some clinical cases, there are several mixed clinical signs, with iron overload. The variants C-582A > G and C-153C > T decrease hepcidin expression, but the mechanism of action of peptide stays the same without transferrin saturation and increasing ferritin [40]. The body organs contain iron at large amounts such as the heart and liver, and it will be affected and damaged [41]. Any change in the HAMP gene may cause a defective hepcidin protein, and it became no action. Accumulation of iron and ferritin in the organs helps to develop diseases in different organs such as coronary diseases, diabetes mellitus, HIV, HBV, and HCV [42]. It was noticed that some neurodegenerative diseases are associated with a high level of hepcidin in plasma such as Alzheimer, Parkinson, and sclerosis [43].

2.5 Homeostatic iron regulation gene (HFE) human

Human homeostatic iron regulator protein, also known as the HFE protein (HighmFE2+), is a protein that in humans is encoded by the HFE gene. The HFE gene is located on the short arm of chromosome 6 at location 6p21.3 [44]. Simon and colleagues, in the 1970s, noted that hemochromatosis is relatively common, associated with markers of human leukocyte antigen (HLA), and transmitted as an autosomal recessive trait [45]. In 1996, Feder and colleagues used positional cloning to classify HFE, the gene of hemochromatosis, associated with the main chromosome 6p histocompatibility complex (MHC) [46]. The HFE membrane protein is similar to the proteins of the MHC class I and binds beta-2 microglobulin (β2M) [46]. HFE binds the extracellular α1-α2 domain to the transferrin receptor (TFRC) [47]. HFE is needed for normal hepatic synthesis regulation of hepcidin, the principal iron metabolism controller [25].

Common HFE mutations represent approximately 90% of phenotypes of hemochromatosis in Western European descent whites. Feder and colleagues named the gene HLA-H [46] although the name had been published earlier to designate a presumed pseudogene in the HLA class I region [48]. Marsh demanded a more suitable designation [49]. The HFE symbol (H = high; FE = iron) was accepted by both the WHO Nomenclature Committee for HLA System Factors and the HUGO Genome Nomenclature Committee.

2.6 Structure and function of HFE gene human

HFE has seven exons of 12 kb [39]. HFE includes 9.6 kb of DNA on chromosome 6p in the extended region of HLA class I. Histone genes on both sides of HFE are present [50]. Exon 1 corresponds to the peptide of the signal and exons 2–4, respectively, to the domains α1, α2, and α3. For the transmembrane domain, exon 5 accounts. The cytoplasmic tail is encoded by exon 6's 5′ portion, which includes a native stop codon. The full-length transcript, therefore, represents six exons [51]. HFE is a protein containing 343 amino acids, including a signal peptide, an extracellular transferrin-binding region (α1 and α2), an immunoglobulin-like domain (α3), a transmembrane region, and a short cytoplasmic tail. HFE binds β2 M to form a cell surface heterodimer [46].

HFE is glycosylated at asparagine residues 110, 130, and 234 during transport to the cell membrane. For normal intracellular trafficking and function, glycosylation is important. HFE interacts with TFRC [51, 52]. HFE's structure revealed that its TFRC ligand binds in a molar ratio of 2:1 TFRC: HFE [47]. There is a peptide-binding groove in most class I MHC molecules. Since the α1 and α2 helices are closer to HFE, the HFE analog site is too narrow to bind peptides [47]. TFRC and HFE bind strongly to the essential pH of cell surfaces but not to the intracellular vesicle acid pH [47]. The structure of a complex between the extracellular portions of HFE and TFRC shows that binding affects both HFE and its ligand configurations. In their domain arrangements and dimer interfaces, the structures of TFRC alone and TFRC complexed with HFE differ [53].

Studies of 293 cells cultivated to express wild-type HFE proteins showed that with TFR, HFE forms stable complexes. The association of HFE protein with TFR was significantly reduced in 293 cells over the expression of HFE C282Y, as shown in **Figure 4** [54]. Through inhibiting TFRC: TF-Fe interaction in an experiment using purified proteins and a biosensor chip [47], normal HFE protein decreased the affinity of TFRC to TF. HFE changes the conformation of the Tf-Fe binding site as observed by biosensor assays when HFE binds to TFRC in vitro, decreasing the entry of iron into Chinese ovarian hamster cells [47]. No evidence of binding of HFE and TFR2 was detected in coimmunoprecipitation or surface resonance-based testing experiments using soluble HFE and TFR2 [55].

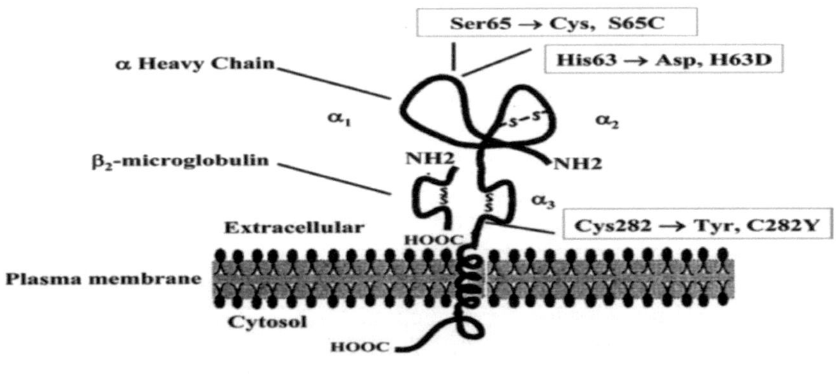

Figure 4.
The HFE gene diagram. The image was changed after getting permission from the author. Cys282 -> Tyr282 exchanging mutation of C282Y and His63 -> Asp63 exchanging mutation of H63D [54].

The relevance, if any, of these in vitro results to in vivo iron homeostasis is unclear. Many factors, including HFE, act as hepcidin transcription upstream regulators [18]. HAMP expression in untreated patients with hemochromatosis, homozygosity of C282Y, and iron overload was significantly lower than controls [56].

2.7 Mutation of the HFE gene

The three most common mutations of HFE in the coding region are: C282Y (exon 4; c.845GA; rs1800562), H63D (exon 2; c.187CG; rs1799945), and S65C (exon 2; c.193AT; rs1800730) [57]. The C282Y mutation interrupts a critical disulfide bond in HFE's α3 domain, abrogating its binding to β2 M and limiting its location to the cytoplasm [58]. H63D and S65C affect the α1 binding groove but do not prevent HFE on cell surfaces. HFE C282 is conserved because cysteine 282 is essential to β2M binding and extracellular presentation of HFE. H63 is also conserved. Histidine 63 forms a salt bridge in the α2 domain that binds TFRC, suggesting that the salt bridge is important for HFE function [59].

Histidine 116 and 145 and tyrosine 140 are widely conserved. A cluster of four histidine residues (H109, H111, H116, and H145) is associated with Y140 in the α1 domain. This configuration resembles functional sites in other iron-binding proteins [40]. Hereditary hemochromatosis type-1 is caused by HFE gene mutation. Allelic is most prevalent among the individuals of Europe [60]. Single point simple change in exon 4, 845G to 845A in the HFE gene results in an exchange of cysteine by tyrosine. Also, another gene mutation included a change of the allele 187C to 187G, wherever histidine is exchanged. The substitution of serine for cysteine is considered the third mutation of the HFE gene S65C [61]. Also, H63D mutation is present in HFE protein, causing a decrease in the transferrin receptor [62]. When hepcidin protein cannot bind with the transferrin receptor, another factor has a great role in hepcidin protein, therefore, any changes in hepcidin protein help to aggregation iron by inhibiting the transcription of hepcidin and do not allow iron absorption from the intestine [63]. Also, the modifying of the H63D genotype associated between the metabolism of iron and lead, wherever there is increased iron in the body, is associated with a high level in the same body [64]. Iron overload has been identified in individuals with the digenic inheritance of one or more HFE mutations and a non-HFE gene mutation that is also involved in iron metabolism [44]. One example is the development of hemochromatosis in individuals who are double heterozygous for one or more HFE mutations and a hepcidin gene mutation (HAMP) [65].

Iron loading occurred in people with HFE mutation digenic inheritance and either a hemojuvelin gene (HJV) mutation [58] or a TFR2 gene (TFR2) mutation [66].

2.8 Iron overload and genetic alterations

Activation of the HFE gene works with hemochromatosis that includes iron accumulation that occurs with heart failure, cirrhosis, diabetes, and hepatocarcinoma. Interaction between β2-microglobulin and HFE protein is the main hypothesis that explains the development of hemochromatosis in the body. β2-microglobulin and HFE protein react together and form the transferrin-1 receptor, which helps to absorbed iron from the diet [67]. Many reports founded new genes responsible for the iron metabolism [hereditary hemochromatosis (HH)]. It included two types, the first type is HFE hemochromatosis that occurs due to mutation of the HFE gene, and its spread commonly in the Caucasian people. The second type is non-HFE hemochromatosis. Non-HFE hemochromatosis, including HAMP, ferroportin, TFR, and HJV gene, is detected among the diseases. Genetic alterations in ceruloplasmin and ferritin encourage iron accumulation with the occurrence of hyperferritinemia [44, 68]. The accumulation of iron in the tissues interferes with tissues' function, such as hepatic failure, cardiac problems, cirrhosis, and diabetes. Furthermore, there are pituitary gland diseases and disorder in the function of joints, gonads, abdominal pain, and hyperpigmentation of the skin [69].

Hemochromatosis diagnosis is made by deepening on clinical signs, biochemical markers, genetic examination, and Liver biopsy [70]. Some indicators, such as ferritin, transferrin iron elevated, and liver enzymes, are elevated in hemochromatosis patients. The level of ferritin in serum is greater 300 μg/L in men and 200 μg/L in women, also the same speech on the level of transferrin. If the level of ferritin and transferring are elevated, it should make a genetic analysis of some genes such as HAMP, HJV, HFE, transferrin receptor genes, and ferroportin [71].

3. Conclusions

The knowledge about hepcidin and its genetic structure, as well as a common mutation that occurs in it, is vital to understand the iron metabolism and iron disorders. This chapter helps the reader to get ideas about that.

Acknowledgements

The authors of this chapter wish to thank the workers in the molecular biology department at the Al-Karama Teaching Hospital in Wasit, Iraq, for the credibility and keenness they have achieved in carrying out the work assigned to them.

Conflict of interest

The authors have no conflict of interest.

Author details

Safa A. Faraj[1,2]* and Naeem M. Al-Abedy[3]

1 Department of Pediatrics, College of Medicine, Wasit University, Kut, Iraq

2 Children Welfare Teaching Hospital, Baghdad Medical City, Iraq

3 Al-Karama Teaching Hospital, Wasit, Iraq

*Address all correspondence to: safaafaraj@uowasit.edu.iq

References

[1] Ganz T. Hepcidin and iron regulation, 10 years later. Blood. 2011;**117**(17):4425-4433

[2] Merryweather-Clarke AT, Cadet E, Bomford A, Capron D, Viprakasit V, Miller A, et al. Digenic inheritance of mutations in HAMP and HFE results in different types of haemochromatosis. Human Molecular Genetics. 2003;**12**(17): 2241-2247

[3] Fekri K, Rasouli NA, Zavareh SA, Jalil M, Moradi F, Hosseinpour M, et al. Hepcidin and HFE polymorphisms and ferritin level in β-thalassemia major. International Journal of Hematology-Oncology and Stem Cell Research. 2019;**13**(1):42

[4] Dhillon BK, Prakash S, Chandak GR, Chawla YK, Das R. H63D mutation in HFE gene is common in Indians and is associated with the European haplotype. Journal of Genetics. 2012;**91**(2):229-232

[5] Sharma V, Panigrahi I, Dutta P, Tyagi S, Choudhry VP, Saxena R. HFE mutation H63D predicts risk of iron over load in thalassemia intermedia irrespective of blood transfusions. Indian Journal of Pathology & Microbiology. 2007;**50**(1):82-85

[6] do Rio Preto J, Oliveira TM, Souza FP, Jardim AC, Cordeiro JA, Pinho JR, et al. HFE gene mutations in Brazilian thalassemic patients. Brazilian Journal of Medical and Biological Research. 2006;**39**:12

[7] Jazayeri M, Bakayev V, Adibi P, Rad FH, Zakeri H, Kalantar E, et al. Frequency of HFE gene mutations in Iranian beta-thalassaemia minor patients. European Journal of Haematology. 2003;**71**(6):408-411

[8] Abbas I. Development of LC-MS/MS methods for the quantitative determination of hepcidin-25, a key regulator of iron metabolism. Humboldt-Universität zu Berlin, Mathematisch-Naturwissenschaftliche Fakultät; 2018

[9] Petrak J, Havlenova T, Krijt M, Behounek M, Franekova J, Cervenka L, et al. Myocardial iron homeostasis and hepcidin expression in a rat model of heart failure at different levels of dietary iron intake. Biochimica et Biophysica Acta (BBA) - General Subjects. 2019; **1863**(4):703-713

[10] Leitch HA, Thachil J. Iron physiology, iron overload, and the porphyrias. Am Soc Hematol Self-Assessment Program, Sixth Ed David P Steensma, Md, Adam Cuker, Md, Ms, Christine L Kempton, Md, Msc, Grzegorz S Nowakowski, MD Download Cit file Ris Ref Manag EasyBib Bookends Mendeley Pap EndNote Ref Works BibTex Close Search. 2016

[11] Skalnaya MG, Skalny AV. Essential Trace Elements in Human Health: A physician's View. Tomsk: Publishing House of Tomsk State University; 2018. p. 224

[12] Dlouhy AC, Bailey DK, Steimle BL, Parker HV, Kosman DJ. Fluorescence resonance energy transfer links membrane ferroportin, hephaestin but not ferroportin, amyloid precursor protein complex with iron efflux. Journal of Biological Chemistry. 2019;**294**(11):4202-4214

[13] Collins JF, Wessling-Resnick M, Knutson MD. Hepcidin regulation of iron transport. The Journal of Nutrition. 2008;**138**(11):2284-2288

[14] Park CH, Valore EV, Waring AJ, Ganz T. Hepcidin, a urinary antimicrobial peptide synthesized in the liver. Journal of Biological Chemistry. 2001;**276**(11):7806-7810

[15] Gao G, Li J, Zhang Y, Chang YZ. Cellular iron metabolism and regulation. In: Brain Iron Metabolism and CNS Diseases. Singapore: Springer; 2019. pp. 21-32

[16] Nemeth E, Ganz T. The role of hepcidin in iron metabolism. Acta Haematologica. 2009;**122**(2-3):78-86

[17] Kwapisz J, Slomka A, Zekanowska E. Hepcidin and its role in iron homeostasis. eJIFCC. 2009;**20**(2):124

[18] Elnabaheen EM. Hepcidin Status among Iron Deficient Anemic Pregnant Women in Gaza strip: A Case Control Study. Hepcidin Status among Iron Deficient Anemic Pregnant Women in Gaza strip: A Case Control Study. 2017

[19] Stasiak I, Lillie B, Acvp D, Crawshaw G, Aczm D, Ganz T, et al. The Role of Hepcidin in Regulation of Iron Balance in Bats. AAZV Annual Conference. 2012;2012

[20] Kemna EH, Kartikasari AE, van Tits LJ, Pickkers P, Tjalsma H, Swinkels DW. Regulation of hepcidin: Insights from biochemical analyses on human serum samples. Blood Cells, Molecules, and Diseases. 2008;**40**(3):339-346

[21] Huang Y, Gu W, Wang B, Zhang Y, Cui L, Yao Z, et al. Identification and expression of the hepcidin gene from brown trout (Salmo trutta) and functional analysis of its synthetic peptide. Fish & Shellfish Immunology. April 1 2019;**87**:243-253

[22] Ganz T. Erythropoietic regulators of iron metabolism. Free Radical Biology and Medicine. 2019;**133**:69-74

[23] Koury MJ. Erythroferrone: A missing link in iron regulation. The Hematologist. 2015;**12**(1):10

[24] Kleven MD, Jue S, Enns CA. Transferrin receptors TfR1 and TfR2 bind transferrin through differing mechanisms. Biochemistry. 2018;**57**(9):1552-1559

[25] Nemeth E. Hepcidin and β-thalassemia major. Blood - The Journal of the American Society of Hematology. 2013;**122**(1):3-4

[26] Fillebeen C, Charlebois E, Wagner J, Katsarou A, Mui J, Vali H, et al. Transferrin receptor 1 controls systemic iron homeostasis by fine-tuning hepcidin expression to hepatocellular iron load. Blood - The Journal of the American Society of Hematology. 2019;**133**(4):344-355

[27] Ramos E, Ruchala P, Goodnough JB, Kautz L, Preza GC, Nemeth E, et al. Minihepcidins prevent iron overload in a hepcidin-deficient mouse model of severe hemochromatosis. Blood - The Journal of the American Society of Hematology. 2012;**120**(18):3829-3836

[28] De Domenico I, Ward DM, Kaplan J. Hepcidin regulation: Ironing out the details. The Journal of Clinical Investigation. 2007;**117**(7):1755-1758

[29] Pandey S, Pandey SK, Shah V. Role of HAMP genetic variants on pathophysiology of iron deficiency anemia. Indian Journal of Clinical Biochemistry. 2018;**33**(4):479-482

[30] Arts HH, Eng B, Waye JS. Multiplex allele-specific PCR for simultaneous detection of H63D and C282Y HFE mutations in hereditary hemochromatosis. Journal of Applied Laboratory Medicine. 2018;**3**(1):10-17

[31] Fitzsimons EJ, Cullis JO, Thomas DW, Tsochatzis E, Griffiths WJ. Diagnosis and therapy of genetic haemochromatosis (review and 2017 update). British Journal of Haematology. 2018;**181**(3):293-303

[32] Brissot P, Loréal O, Jouanolle AM. Molecular testing in hemochromatosis.

In: Diagnostic Molecular Pathology. Elsevier; 2017. p. 245-253

[33] Delatycki MB, Allen KJ, Gow P, MacFarlane J, Radomski C, Thompson J, et al. A homozygous HAMP mutation in a multiply consanguineous family with pseudo-dominant juvenile hemochromatosis. Clinical Genetics. 2004;**65**(5):378-383

[34] Roetto A, Daraio F, Porporato P, Caruso R, Cox TM, Cazzola M, et al. Screening hepcidin for mutations in juvenile hemochromatosis: Identification of a new mutation (C70R). Blood. 2004;**103**(6):2407-2409

[35] Ekanayake D, Roddick C, Khanbhai M, Powell LW. Homozygosity for the C282Y Substitution in the Hfe Gene: the Incomplete Penetrance and Variable Expressivity. European Medical Journal Hepatology. 2015;**3**(January):79-85

[36] Reichert CO, da Cunha J, Levy D, Maselli LM, Bydlowski SP, Spada C. Hepcidin: SNP-like polymorphisms present in Iron metabolism and clinical complications of Iron accumulation and deficiency. Genetic Polymorphisms. 6 September 2017:221-224

[37] Rahman HA, Abou-Elew HH, El-Shorbagy RM, Fawzy R, Youssry I. Influence of iron regulating genes mutations on iron status in Egyptian patients with sickle cell disease. Annals of Clinical and Laboratory Science. 2014;**44**(3):304-309

[38] Matthes T, Aguilar-Martinez P, Pizzi-Bosman L, Darbellay R, Rubbia-Brandt L, Giostra E, et al. Severe hemochromatosis in a portuguese family associated with a new mutation in the 5′-UTR of the HAMP gene. Blood. 2004;**104**(7):2181-2183

[39] Silva B, Pita L, Gomes S, Gonçalves J, Faustino P. The hepcidin gene promoter nc.-1010C> T;- 582A> G haplotype modulates serum ferritin in individuals carrying the common H63D mutation in HFE gene. Annals of Hematology. 2014;**93**(12):2063-2066

[40] Krause A, Neitz S, Mägert HJ, Schulz A, Forssmann WG, Schulz-Knappe P, et al. LEAP-1, a novel highly disulfide-bonded human peptide, exhibits antimicrobial activity. FEBS Letters. 2000;**480**(2-3):147-150

[41] Gehrke SG, Herrmann T, Kulaksiz H, Merle U, Bents K, Kaiser I, et al. Iron stores modulate hepatic hepcidin expression by an HFE-independent pathway. Digestion. 2005;**72**(1):25-32

[42] Pechlaner R, Weiss G, Bansal S, Mayr M, Santer P, Pallhuber B, et al. Inadequate hepcidin serum concentrations predict incident type 2 diabetes mellitus. Diabetes/Metabolism Research and Reviews. 2016;**32**(2):187-192

[43] Crichton RR, Dexter DT, Ward RJ. Brain iron metabolism and its perturbation in neurological diseases. Monatshefte für Chemie - Chemical Monthly. 2011;**142**(4):341-355

[44] Pietrangelo A, Torbenson M. Disorders of Iron overload. In: MacSween's Pathology of the Liver E-Book. Vol. 275. Elsevier Health Sciences; 2017

[45] Simon M, Le Mignon L, Fauchet R, Yaouanq J, David V, Edan G, et al. A study of 609 HLA haplotypes marking for the hemochromatosis gene:(1) mapping of the gene near the HLA-A locus and characters required to define a heterozygous population and (2) hypothesis concerning the underlying cause of hemochromatosis-HLA association. American Journal of Human Genetics. 1987;**41**(2):89

[46] Feder JN, Gnirke A, Thomas W, Tsuchihashi Z, Ruddy DA, Basava A,

et al. A novel MHC class I–like gene is mutated in patients with hereditary haemochromatosis. Nature Genetics. 1996;**13**(4):399-408

[47] Lebrón JA, Bennett MJ, Vaughn DE, Chirino AJ, Snow PM, Mintier GA, et al. Crystal structure of the hemochromatosis protein HFE and characterization of its interaction with transferrin receptor. Cell. 1998;**93**(1):111-123

[48] Venditti CP, Harris JM, Geraghty DE, Chorney MJ. Mapping and characterization of non-HLA multigene assemblages in the human MHC class I region. Genomics. 1994;**22**(2):257-266

[49] Marsh SG. Nomenclature for factors of the HLA system, update July 1996. Tissue Antigens. 1996;**48**(5):607

[50] Dorak MT. HFE (hemochromatosis). Atlas Genet Cytogenet Oncol Haematol. 2011;**13**(p. 75-77):11-14

[51] Bhatt L, Murphy C, O'Driscoll LS, Carmo-Fonseca M, McCaffrey MW, Fleming JV. N-glycosylation is important for the correct intracellular localization of HFE and its ability to decrease cell surface transferrin binding. The FEBS Journal. 2010;**277**(15):3219-3234

[52] Salter-Cid L, Brunmark A, Li Y, Leturcq D, Peterson PA, Jackson MR, et al. Transferrin receptor is negatively modulated by the hemochromatosis protein HFE: Implications for cellular iron homeostasis. Proceedings of the National Academy of Sciences. 1999;**96**(10):5434-5439

[53] Bennett MJ, Lebrón JA, Bjorkman PJ. Crystal structure of the hereditary haemochromatosis protein HFE complexed with transferrin receptor. Nature. 2000;**403**(6765):46-53

[54] Feder JN, Penny DM, Irrinki A, Lee VK, Lebrón JA, Watson N, et al. The hemochromatosis gene product complexes with the transferrin receptor and lowers its affinity for ligand binding. Proceedings of the National Academy of Sciences. 1998;**95**(4):1472-1477

[55] West AP, Bennett MJ, Sellers VM, Andrews NC, Enns CA, Bjorkman PJ. Comparison of the interactions of transferrin receptor and transferrin receptor 2 with transferrin and the hereditary hemochromatosis protein HFE. Journal of Biological Chemistry. 2000;**275**(49):38135-38138

[56] Bridle KR, Frazer DM, Wilkins SJ, Dixon JL, Purdie DM, Crawford DH, et al. Disrupted hepcidin regulation in HFE-associated haemochromatosis and the liver as a regulator of body iron homoeostasis. The Lancet. 2003;**361**(9358):669-673

[57] Barton JC, Acton RT. Hepcidin, iron, and bacterial infection. In: Vitamins and Hormones. Elsevier; 2019. p. 223-242

[58] Waheed A, Parkkila S, Zhou XY, Tomatsu S, Tsuchihashi Z, Feder JN, et al. Hereditary hemochromatosis: Effects of C282Y and H63D mutations on association with β2-microglobulin, intracellular processing, and cell surface expression of the HFE protein in COS-7 cells. Proceedings of the National Academy of Sciences. 1997;**94**(23):12384-12389

[59] Fleming RE, Sly WS. Mechanisms of iron accumulation in hereditary hemochromatosis. Annual Review of Physiology. 2002;**64**(1):663-680

[60] European Association for the Study of the Liver. EASL clinical practice guidelines for HFE hemochromatosis. Journal of Hepatology. 2010;**53**(1):3-22

[61] McDonald CJ, Wallace DF, Crawford DH, Subramaniam VN. Iron storage disease in Asia-Pacific

populations: The importance of non-HFE mutations. Journal of Gastroenterology and Hepatology. 2013;**28**(7):1087-1094

[62] Cézard C, Singh AR, Le Gac G, Gourlaouen I, Ferec C, Rochette J. Phenotypic expression of a novel C282Y/R226G compound heterozygous state in HFE hemochromatosis: Molecular dynamics and biochemical studies. Blood Cells, Molecules, and Diseases. 2014;**52**(1):27-34

[63] Traeger L, Enns CA, Krijt J, Steinbicker AU. The hemochromatosis protein HFE signals predominantly via the BMP type I receptor ALK3 in vivo. Communications Biology. 2018;**1**(1):1-7

[64] Fan G, Du G, Li H, Lin F, Sun Z, Yang W, et al. The effect of the hemochromatosis (HFE) genotype on lead load and iron metabolism among lead smelter workers. PLOS One. 2014;**9**(7):e101537

[65] Barton JC, LaFreniere SA, Leiendecker-Foster C, Li H, Acton RT, Press RD, et al. HFE, SLC40A1, HAMP, HJV, TFR2, and FTL mutations detected by denaturing high-performance liquid chromatography after iron phenotyping and HFE C282Y and H63D genotyping in 785 HEIRS study participants. American Journal of Hematology. 2009;**84**(11):710-714

[66] Biasiotto G, Belloli S, Ruggeri G, Zanella I, Gerardi G, Corrado M, et al. Identification of new mutations of the HFE, hepcidin, and transferrin receptor 2 genes by denaturing HPLC analysis of individuals with biochemical indications of iron overload. Clinical Chemistry. 2003;**49**(12):1981-1988

[67] Salgia RJ, Brown K. Diagnosis and management of hereditary hemochromatosis. Clinics in Liver Disease. 2015;**19**(1):187-198

[68] Pericleous M, Kelly C, Ala A, Schilsky ML. The epidemiology of rare hereditary metabolic liver diseases. In: Clinical Epidemiology of Chronic Liver Diseases. Cham: Springer; 2019. pp. 307-330

[69] Quigley P. Hereditary hemochromatosis: Dealing with iron overload. Nursing. 2016;**46**(5):36-43

[70] Castiella A, Zapata E, Alústiza JM. Non-invasive methods for liver fibrosis prediction in hemochromatosis: One step beyond. World Journal of Hepatology. 2010;**2**(7):251

[71] Porto G, Brissot P, Swinkels DW, Zoller H, Kamarainen O, Patton S, et al. EMQN best practice guidelines for the molecular genetic diagnosis of hereditary hemochromatosis (HH). European Journal of Human Genetics. 2016;**24**(4):479-495

Chapter 15

Palindromic Rheumatism: Biology and Treatment Options

Ayesha Noor, Ali Raza Ishaq, Fatima Noor, Tahira Younis, Afshan Syed Abbas, Fatima Jalal, Rahmawaty Samad, Sara Mumtaz and Faiza Jabeen

Abstract

Palindromic rheumatism is a syndrome characterized by recurrent, self-resolving, and inflammatory attacks in and around the joints that have long recognized association with rheumatoid arthritis. PR attacks mostly start in small joints i.e. knees, shoulder, and small joints of the hand. Whether PR should be considered as a single disease or prodrome of RA remains a thought-provoking question. Multiple genetic and environmental factors contribute to the development of PR. Many studies have explained the relationship between a high concentration of Anti-CCP antibodies and PR. Potential benefits of Gold therapy have been recognized in literature but still, there are some questions about toxicity and efficacy that need further considerations. In addition to that anti-malarial drugs, Abatacept, Tofacitinib, and Rituximab showed the variable result in different patients and needed further study to validate their medical use. Moreover, yarrow, oat, colchicum, dill, fennel, wild rue, bitter melon, willow, garlic, and burdock seem suitable candidates to treat rheumatoid although their use in PR is still not reported. Additional experimental researches on these drugs lead to an increase in our knowledge to fight against PR in the future using novel therapeutic approaches. We have attempted to cover this topic in a chapter form to provide a comprehensive view and hope that it will serve as a reference for clinicians who treat patients with PR.

Keywords: palindromic rheumatism, rheumatoid arthritis, environmental risk factors, genetic risk factors, therapies

1. Introduction

Palindromic rheumatism constitutes episodic and recurrent attack of articular inflammation that lasts from a few hours to several days which conclude without residual joint damage [1]. PR tends to affect small joints mostly, so the Knees, Shoulder, and small joints of hands are more prone to attack. Characteristic symptoms of PR include pain, swelling, redness, and disability of joints. This idiopathic condition was firstly described by Hench and Rosenberg in 1944 [2–6].

Distinctive features of PR include reoccurrence of attack at regular intervals and symptoms-free periods between attacks. Several studies have shown that about half of patients with PR develop Rheumatoid Arthritis (RA) and other joint diseases in later life [7]. PR is a single disease or spectrum of RE a leading question that is unanswered for 70 years. However, the target tissues are mostly the same in PR and RA [6, 8].

A high concentration of Rheumatoid factor and Anti -CCP antibodies in both PR and RA strengthen the correlation between the two diseases. However, despite these similarities, PR is different from RA in that joints are free of symptoms between attacks. According to research attacks of PR usually affect one joint but other structures can be affected in 30% cases and Rheumatoid nodules also appear in one-third cases of PR. Time of attack is not definite however according to Research in London 50% of patients develop attack of PR in the late afternoon and some others at night time [7–9].

2. Epidemiology

Indeed mounting studies highlight the frequency of PR is significantly lower than RA [9]. Epidemiological data from Canadian research suggest that females are most likely to develop PR than men in both conditions of Arthritis. However, it has been estimated that the average age was 56 years in RA and 49 in PR [10].

3. Etiology

Etiological factors for PR are under investigation and uncertain however intrinsic (gene mutations), extrinsic (external factor lifestyle and smoking) and idiopathic factors seem to be important in PR. Initially, it was believed that allergic agents and infectious agents may provoke the symptoms, but recent studies have shown that even the injection of histamine did not cause PR. It is thought that trauma, stress, anxiety, and cold can stimulate the flares of PR however recent data support the thought. Consumption of nitrate-containing food triggers PR [11].

Several etiological studies suggested that the mutated MEFV Gene seems to be an aggregating factor for the severity of PR [12]. According to Iranian research during the attack of PR level of C-reactive protein was increased in about 50%of cases. Erythrocyte sedimentation rate was also elevated during the attack of PR [13]. Anti-CCP antibodies level was also found high in PR patients [3, 14]. Another study showed that autoantibodies RF and anti-CCP concentration appear to be elevated in PR patients and is thought to be responsible for developing RA and other connective tissue diseases [15]. Further research conducted in PR patients also showed uplift RF concentration in 33.3% of patients and a high concentration of anti-CCP in 38.9% of patients. Another research showed the follow-up of 43 patients to other connective tissue diseases and of 28 patients to RA out of a total of 160 patients of PR [16, 17]. The recently high concentration of anti-CCP antibodies and anti-keratin was found in the patients of PR [11]. Studies have documented that ultrasonography of synovitis of PR patients showed a high concentration of ACPA antibodies in PR patients. These studies suggest the strong relationship between Anti-Cyclic Citrullinted Protein antibodies and Rheumatoid factor in PR and RA [17–19].

Another report has elucidated the role of the HLA Gene on Chromosome 6 which is thought to be responsible for about 30% of all cases of RA [20–22]. It is also reported that the HLA-DRB1 alleles encode for a shared epitope [10] which may be a risk factor for PR and RA [23]. Recent studies also showed a strong prevalence of HLA-DR shared epitope alleles in PR. The homozygosity of SE alleles in PR patients is responsible for the progress of half of the PR cases to RA. A Korean study showed a great prevalence of HLA-DRB1*0803 and HLA-DRB1*1302 in PR patients and these alleles are distinct in PR [24].

Many investigations have concluded that gene involvement in gene–environment interaction is not only one factor for a mutation in HLA-DRB1 but factors affecting gene linkage equilibrium may also cause variation in the gene [25].

Other researchers have demonstrated the role of PADI4 (Protein-arginine deiminase type-4) which is a gene that encodes for enzymes that are responsible for the formation of Citrulline from arginine. It has recently been found that any effect in the stability of PADI4 results in a high level of Anti-CCP antibodies [26]. According to a Chinese study, periodic and episodic attacks of PR show a strong link with PADI4 [18, 22, 26].

TNFα (Tumor Necrosis factor) which are short-lived pro-inflammatory cytokines showed a strong relationship with PR. Another case study investigated TNFRSF1A and TNFRSF1B mutations in PR patients in the Chinese population [27, 28]. Another novel study has indicated that the concentration of cytokines like IL-6 and TNFα was elevated in synovium and serum of patients [29, 30]. It has been reported that TNFα microsatellite polymorphism indicates a close connection with the disease by its association with HLA-DRB1 SE. Autoinflammatory diseases like PR are caused by deregulation in inflammasome components [31]. According to Novel research, the PYCARD\ASC Splice variant has been found in PR patients. This inflammasome-associated mutation may be a risk factor for PR patients. According to this research exon2, PYCARD\ASC is more expressed in patients with PR. PYCARD\ASC mature IL-IB for innate immunity. The inflexibility of PYCARD\ASC leads to more secretions of IL-IB so the exon2 splice variant may be a risk factor for disease in PR patients [32]. The presence of conserved exon2 in DNA of all patients of PR, high amount of NLRP3 (Nucleotide-binding oligomerization domain, Leucine-rich Repeat, and Pyrin domain), and high concentration of IL-1B and IL-18 show the strong association between PR and this gene [33]. Comprehensive pathophysiology of Palindromic rheumatism has shown in **Figure 1**.

Polymorphism in the promoter sequence of Stromelysin 1(MMB- gene) and HLA gene have been studied in recent years. Recent studies indicated a close

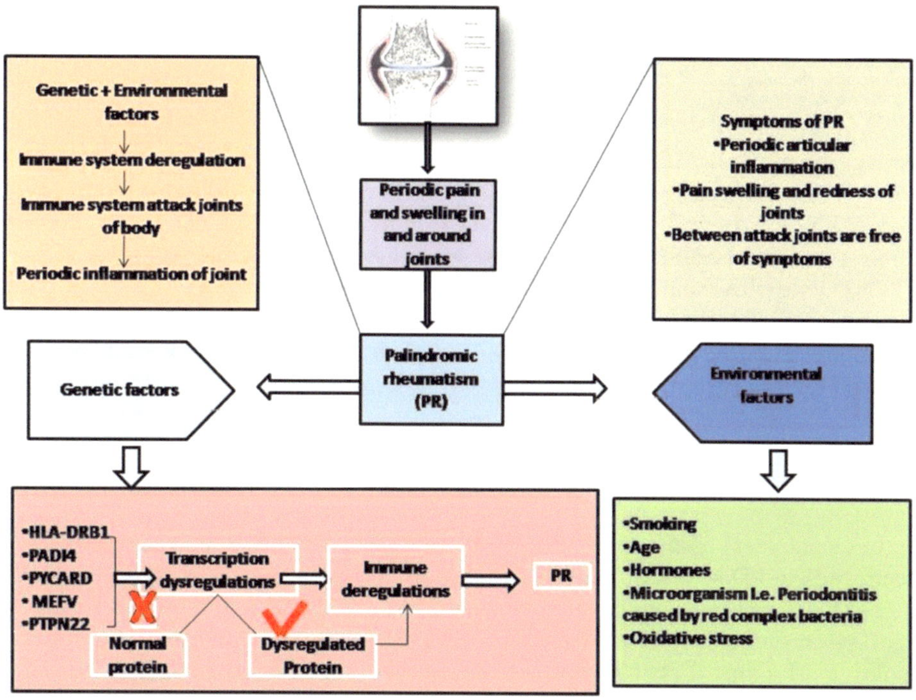

Figure 1.
Overviews of Palindromic rheumatism.

Disease	Clinical characteristics	Gene mutation	References
Gout (Urate crystals deposing disease)	Presence of needle-like crystals in synovial fluid of joints	URAT1, GLUT-9 (Transporters) involvement	[36]
Reactive Arthritis	Infections history in genitourinary and Gastrointestinal tract (GI)	HLA-B27(In 80% of population)	[37]
Arthritis associated with Bowel disease	Gut inflammation,10%arthritis precedes enteritis	HLA-B27,NOD2, *ATG16L1*	[37]
Whipple's disease	Weight loss, diarrhea, and fever caused by *Tropheryma whipplei*	HLA-B27(50–75% of population)	[37]
Behcet disease	T cells abnormality and neutrophils hyperfunction, inflammatory lesion	HLA-B51(10–80%)	[37]
Sarcoidosis	Granuloma formation and the possibility of erosive bone lesion	HLA-DRB1 Involvement	[37]
Celiac disease	Nonerosive arthritis	Transglutaminase antibodies and malabsorption parameters	[37]
Familial Mediterranean fever	Familial history, MEFV mutation, Chronic arthritis (5%)	MEFV mutation	[37]
TRAPS	Autosomal dominant	TNF alpha	[37]
Hyperlipidemia	Xanthomas	High cholesterol and triglycerides	[37]
Intermittent Hydrarthrosis	No inflammatory sign	MEFV involvement	[37]
Relapsing polychondritis	Cartilaginous structure involves only	Nonspecific	[37]

Table 1.
Palindromic rheumatism and other relapsing diseases.

association between MEFV and PR. Another research investigated that smoking and PTPN22 Showed an association with an increase in ACPA. It is also reported that PTPN22 (Protein Tyrosine Phosphatase Non-receptor Type 22) which encodes for protein tyrosine phosphate clearly shows an association of microsatellite STP PTPN22gene with rheumatism [34]. According to a Chinese report Anti-MCV (Anti Mutated Citrullinated Vimentin) antibodies have also been reported as a biomarker in patients with Rheumatoid Arthritis although their role in PR is not studied [35] (**Table 1**).

4. Environmental factors

In recent year's association of periodontitis (PD) and PR association has been studied. PD is inflammation of periodontal tissues caused by red-complex bacteria i.e.*P.gingivalis* which affects the process of Citrullination by expressing peptidyl- arginine deiminase enzyme (PAD) [38]. According to research in Israel avoiding offending diet and intake of proper diet may affect the flares of PR. Smoking can also trigger the process of Citrullination by affecting the immune reactions of the HLA Gene [39]. Most of the case study reports that the onset of

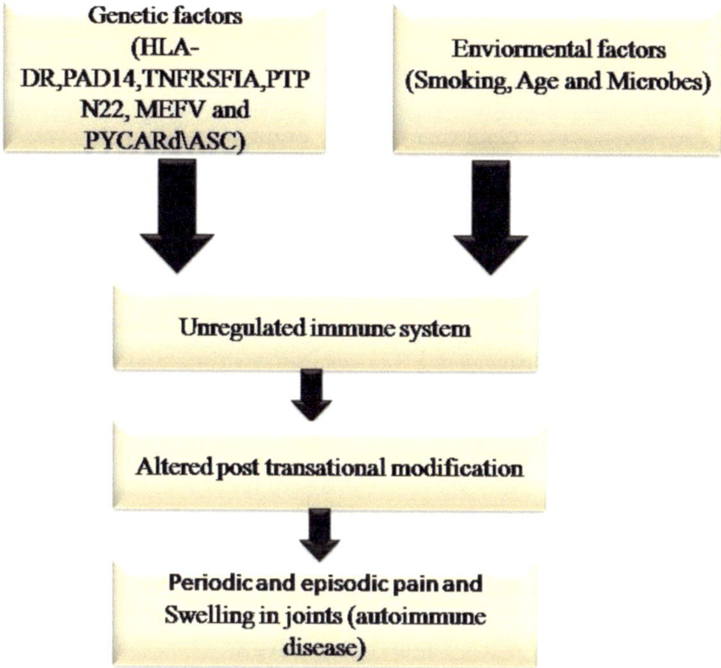

Figure 2.
Schematic representation of factors for development of Palindromic Rheumatism.

PR frequently begins in the late afternoon and early morning [40, 41]. Several factors responsible for the progression of PR as shown in **Figure 2**.

5. Palindromic rheumatism and diet

It has been observed that certain types of food can trigger a periodic attack of Palindromic rheumatism and elimination of that type of food from the diet has resulted in a reduction of attacks. According to a clinical trial conducted for evaluation of the role that certain type of food can play in PR, patients show that patients who were offended to eat eggs, cheese, fish, and canned vegetables resulted in the complete cessation of attack, and those patients who were presented with these food show more reoccurrence of attack. Therefore, offending food should be avoided to reduce the occurrence of PR although this needs more research that which type of specific food should be avoided [4, 42].

6. PR progression to RA

According to research in Japan, Anti-CPA in PR patients who developed RA was higher than those who did not develop RA in a future life [15]. According to a British case study of 39 patients of PR 19 showed progression to RA. Another study also indicated that most cases of PR progressed to other chronic arthritis and 35\60 progressed to RA [43]. The reason for this progression is multifactorial and one of the factors for this progression is a misdiagnosis of PR as there is no specific test and diagnosis is made mostly on physician's judgment and others may include progression duration which may vary from months to20 years [14, 44, 45].

7. Treatment options for PR

There is no specific treatment for PR for several reasons described earlier. Research has reported Non- Steroidal anti-inflammatory drugs appeared to delay flares of PR. In recent years Gold therapy is emerging as a promising treatment option for relieving joint pain and swelling. According to a study, about 60% of patients showed an improved result of gold salt. However, other studies have reported a high mucocutaneous side effect of gold therapy [8, 46]. The use of sulfasalazine also proves to be good for treating episodic flares of PR. In another study use of Chloroquine marked good results in the severity of an attack of PR where 41 patients out of 51 showed improvement [47]. Another study also focused on the delay effect of antimalarial drugs in PR flares [48]. Antimalarial drugs prove to be good in treating PR by their inhibitory effect on TNFα and IL-1. The case of the application of antimalarial in 71 patients mostly showed good results by decreasing flares of PR [47].

According to recent research PR patients who did not respond well to drugs mostly use in PR showed a very good response to Rituximab [49]. Biological DMARDs (Abatacept) affect the immune system by inhibiting T-cells stimulation and prove to be good for patients who poorly respond to methotrexate. Tofacitinib is also used in patients of PR who poorly or intolerantly respond to DMARDS [50]. The possible treatment options are shown in **Figure 3**.

According to research in Japan successful use of Kampo therapy (a Chinese herbal medicine) in three patients with Rheumatoid reveal its pharmaceutical potential in treating rheumatism although it needs a deep study of these findings to uncover its biological potential [51]. According to research in Iran yarrow, oat, colchicum, dill, fennel, wild rue, bitter melon, willow, garlic, and burdock help treat rheumatoid although their use in PR is still not reported [52]. Heat therapy is a medication-free way to relieve muscle pain and stiffness and is also recommended to treat PR [41].

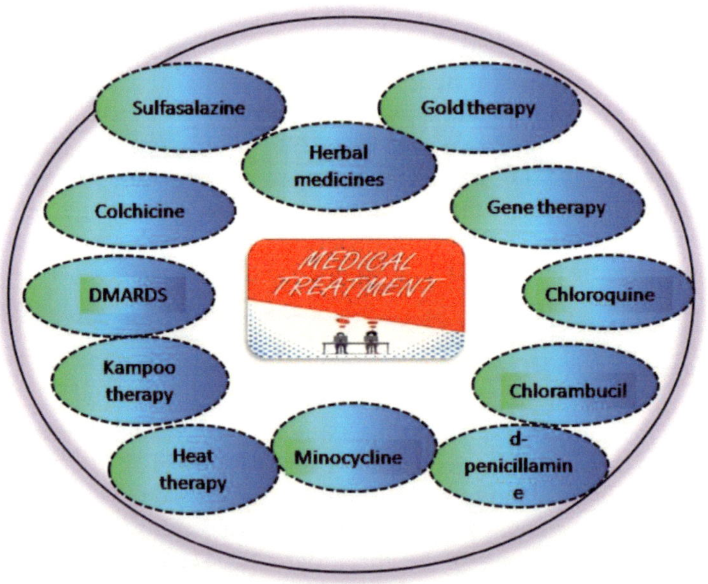

Figure 3.
Possible treatment options for PR.

8. Conclusion

Several risk factors such as genetic and environmental factors favor PR development. Most exposable genes are HLA-DRBI, PTPN22, TNFα, and PYCARD which mutate because of unbalancing in environmental factors. This study updates the information that Non- Steroidal anti-inflammatory drugs, Heat therapy, Chloroquine, and sulfasalazine show good results in the treatment of PR. Although many studies have validated these emerging therapies, still there is a need for further research to figure out their efficacy and precision. Side effects of these drugs and therapies must be considered before clinical applications for achieving stunning gains in the future. Additionally, these summarized genes might be capable of improving the therapeutic inventions for PR hence will serve as a significant pioneer for researchers who wants to identify the associative pathways involve in the pathogenesis of PR.

Authors' contributions

All Authors contributed equally.

Funding

No funding.

Competing interests

The authors declare that they have no competing interests.

Consent for publication

Not applicable.

Ethics approval and consent to participate

Not applicable.

Availability of data and material

Not applicable.

Acronyms and abbreviations

PR	Palindromic Rheumatism
RA	Rheumatoid Arthritis
Anti-CCP	Anti-cyclic citrullinated peptide
ACPA	Anti-citrullinated protein antibodies
HLA-DRB1	Human Leukocyte Antigen-DRB1

PAD	Peptidyl- arginine deiminase enzyme
PADI	Protein-arginine deiminase type-4
TNFα	Tumor Necrosis factor
URAT1	Urate transporter1
GLUT9	Glucose transporter of member 9
HLA	Human leukocyte antigen Complex gene
MEFV	Mediterranean fever
Anti-MCV	Anti-Mutated Citrullinated Vimentin
PTPN22	Protein Tyrosine Phosphatase Non-receptor Type 22
NLRP3	Nucleotide-binding oligomerization domain, Leucine-rich Repeat, and Pyrin domain-3
TNFRSF1A/1B	Tumor necrosis factor receptor 1 A/B
PYCARD	PYD And CARD Domain Containing
ASC	Apoptosis-associated speck-like protein containing a CARD
IL-6	Inter-leukine-6
DMARDS	Disease-modifying antirheumatic drugs

Author details

Ayesha Noor[1], Ali Raza Ishaq[2], Fatima Noor[3], Tahira Younis[1], Afshan Syed Abbas[4], Fatima Jalal[1], Rahmawaty Samad[5], Sara Mumtaz[6] and Faiza Jabeen[1*]

1 Department of Zoology, Government College University Faisalabad, Pakistan

2 State Key Laboratory of Biocatalysis and Enzyme Engineering, Environmental Microbial Technology Center of Hubei Province, College of Life Science, Hubei University, Wuhan, China

3 Department of Bioinformatics and Biotechnology, Government College University Faisalabad, Pakistan

4 Department of Zoology, University of Education, Lahore, Pakistan

5 Faculty of Medicine, Universitas Negeri Gorontalo, Indonesia

6 National University of Medical Sciences, Rawalpindi, Pakistan

*Address all correspondence to: acancerian@hotmail.com

References

[1] Eliakim A, Neumann L, Horowitz J, Buskila D, Kleiner-Baumgarten A, Sukenik S. Palindromic rheumatism in Israel—a disease entity? A survey of 34 patients. Clinical rheumatology. 1989;8(4):507-511.

[2] Kaushik P. Palindromic rheumatism: a descriptive report of seven cases from North Dakota and a short review of literature. Clinical rheumatology. 2010;29(1):83.

[3] Hannonen P, Müttönen T, Oka M. Palindromic rheumatism: a clinical survey of sixty patients. Scandinavian journal of rheumatology. 1987;16(6):413-420.

[4] Gislason S. Managing Inflammatory Arthritis: Environmed Research Inc; 2018.

[5] Gran J, Husby G, Thorsby E. HLA antigens in palindromic rheumatism, nonerosive rheumatoid arthritis and classical rheumatoid arthritis. The Journal of rheumatology. 1984;11(2):136.

[6] Vaughan WT. Palindromic rheumatism among allergic persons. Journal of Allergy. 1943;14(3):256-264.

[7] Mankia K, Emery P. Palindromic rheumatism as part of the rheumatoid arthritis continuum. Nature Reviews Rheumatology. 2019;15(11):687-695.

[8] Wajed M, Brown D, Currey H. Palindromic rheumatism. Clinical and serum complement study. Annals of the rheumatic diseases. 1977;36(1):56-61.

[9] Nielen MM, van Schaardenburg D, Reesink HW, Twisk JW, van de Stadt RJ, van der Horst-Bruinsma IE, et al. Simultaneous development of acute phase response and autoantibodies in preclinical rheumatoid arthritis. Annals of the rheumatic diseases. 2006;65(4):535-537.

[10] Powell A, Davis P, Jones N, Russell AS. Palindromic rheumatism is a common disease: comparison of new-onset palindromic rheumatism compared to new-onset rheumatoid arthritis in a 2-year cohort of patients. The Journal of rheumatology. 2008;35(6):992-994.

[11] Salvador G, Gomez A, Vinas O, Ercilla G, Canete J, Munoz-Gomez J, et al. Prevalence and clinical significance of anti-cyclic citrullinated peptide and antikeratin antibodies in palindromic rheumatism. An abortive form of rheumatoid arthritis? Rheumatology. 2003;42(8):972-975.

[12] Barton A, Bowes J, Eyre S, Spreckley K, Hinks A, John S, et al. A functional haplotype of the PADI4 gene associated with rheumatoid arthritis in a Japanese population is not associated in a United Kingdom population. Arthritis & Rheumatism: Official Journal of the American College of Rheumatology. 2004;50(4):1117-1121.

[13] Gonzalez-Lopez L, Gamez-Nava JI, Jhangri GS, Ramos-Remus C, Russell AS, Suarez-Almazor M. Prognostic factors for the development of rheumatoid arthritis and other connective tissue diseases in patients with palindromic rheumatism. Journal of rheumatology. 1999;26(3):540-545.

[14] Russell AS, Devani A, Maksymowych WP. The role of anti-cyclic citrullinated peptide antibodies in predicting progression of palindromic rheumatism to rheumatoid arthritis. The Journal of rheumatology. 2006;33(7):1240-1242.

[15] Maksymowych WP, Suarez-Almazor ME, Buenviaje H, Cooper B-L, Degeus C, Thompson M, et al. HLA and cytokine gene polymorphisms in relation to occurrence of palindromic

rheumatism and its progression to rheumatoid arthritis. The Journal of rheumatology. 2002;29(11):2319-2326.

[16] Guerne P-A, Weisman MH. Palindromic rheumatism: part of or apart from the spectrum of rheumatoid arthritis. The American journal of medicine. 1992;93(4):451-460.

[17] Chen H-H, Lan J-L, Hung G-D, Chen Y-M, Lan HH-C, Chen D-Y. Association of Ultrasonographic Findings of Synovitis With Anti–Cyclic Citrullinated Peptide Antibodies and Rheumatoid Factor in Patients With Palindromic Rheumatism During Active Episodes. Journal of Ultrasound in Medicine. 2009;28(9):1193-1199.

[18] Tamai M, Kawakami A, Iwamoto N, Arima K, Aoyagi K, Eguchi K. Contribution of anti-CCP antibodies, proximal interphalangeal joint involvement, HLA-DRB1 shared epitope, and PADI4 as risk factors for the development of rheumatoid arthritis in palindromic rheumatism. Scandinavian journal of rheumatology. 2010;39(4):287-291.

[19] Thompson B, Mohammed I, Holborow E, Currey H. Palindromic rheumatism. II. Failure to detect circulating immune complexes during acute episodes. Annals of the rheumatic diseases. 1979;38(4):329-331.

[20] Gabriel SE. The epidemiology of rheumatoid arthritis. Rheumatic Disease Clinics of North America. 2001;27(2):269-281.

[21] Pines A, Kaplinsky N, Orgad S, Gazit E, Frankl O. Familial palindromic rheumatism: a possible association with HLA. Annals of the rheumatic diseases. 1983;42(6):631-633.

[22] Fisher L, Kirk A, Awad J, Festenstein H, Alonso A, Perry J, et al. HLA antigens in palindromic rheumatism and palindromic onset rheumatoid arthritis. Rheumatology. 1986;25(4):345-348.

[23] MacGregor A, Ollier W, Thomson W, Jawaheer D, Silman A. HLA-DRB1* 0401/0404 genotype and rheumatoid arthritis: increased association in men, young age at onset, and disease severity. The Journal of Rheumatology. 1995;22(6):1032-1036.

[24] Kim S, Lee H-S, Lee K, Bae SC, Jun J-B. Palindromic rheumatism: different genetic background implies a distinct disease entity. Annals of the rheumatic diseases. 2006;65(11):1539-1540.

[25] Gregersen PK, Silver J, Winchester RJ. The shared epitope hypothesis. An approach to understanding the molecular genetics of susceptibility to rheumatoid arthritis. Arthritis & Rheumatism: Official Journal of the American College of Rheumatology. 1987;30(11):1205-1213.

[26] Kang CP, Lee HS, Ju H, Cho H, Kang C, Bae SC. A functional haplotype of the PADI4 gene associated with increased rheumatoid arthritis susceptibility in Koreans. Arthritis & Rheumatism: Official Journal of the American College of Rheumatology. 2006;54(1):90-96.

[27] Hall SK, Perregaux DG, Gabel CA, Woodworth T, Durham LK, Huizinga T, et al. Correlation of polymorphic variation in the promoter region of the interleukin-1β gene with secretion of interleukin-1β protein. Arthritis & Rheumatism: Official Journal of the American College of Rheumatology. 2004;50(6):1976-1983.

[28] Tolusso B, Sacco S, Gremese E, La Torre G, Tomietto P, Ferraccioli G. Relationship between the tumor necrosis factor receptor II (TNF-RII) gene polymorphism and sTNF-RII plasma levels in healthy controls and in rheumatoid arthritis. Human immunology. 2004;65(12):1420-1426.

[29] Arend WP, Dayer JM. Cytokines and cytokine inhibitors or antagonists in rheumatoid arthritis. Arthritis and rheumatism. 1990;33(3):305.

[30] Feldmann M, Brennan F, Chantry D, Haworth C, Turner M, Abney E, et al. Cytokine production in the rheumatoid joint: implications for treatment. Annals of the rheumatic diseases. 1990;49:480.

[31] Suganuma Y, Tanaka H, Kawase A, Kishida A, Yamaguchi M, Yabuuchi A, et al. Expression of a PYCARD/ASC variant lacking exon 2 in Japanese patients with palindromic rheumatism increases interleukin-1β secretion. Asian Pacific Journal of Allergy and Immunology. 2019.

[32] Vannini A, Cheung K, Fusconi M, Stammen-Vogelzangs J, Drenth J, Dall'Aglio A, et al. Anti-cyclic citrullinated peptide positivity in non-rheumatoid arthritis disease samples: citrulline-dependent or not? Annals of the rheumatic diseases. 2007;66(4):511-516.

[33] Kokkonen H, Mullazehi M, Berglin E, Hallmans G, Wadell G, Rönnelid J, et al. Antibodies of IgG, IgA and IgM isotypes against cyclic citrullinated peptide precede the development of rheumatoid arthritis. Arthritis research & therapy. 2011;13(1):R13.

[34] Taylor LH, Twigg S, Worthington J, Emery P, Morgan AW, Wilson AG, et al. Metaanalysis of the association of smoking and PTPN22 R620W genotype on autoantibody status and radiological erosions in rheumatoid arthritis. The Journal of Rheumatology. 2013;40(7):1048-1053.

[35] Reyes-Pérez IV, Sánchez-Hernández PE, Muñoz-Valle JF, Martínez-Bonilla GE, García-Iglesias T, González-Díaz V, et al. Cytokines (IL-15, IL-21, and IFN-γ) in rheumatoid arthritis: Association with positivity to autoantibodies (RF, anti-CCP, anti-MCV, and anti-PADI4) and clinical activity. Clinical Rheumatology. 2019;38(11):3061-3071.

[36] Robinson PC. Gout–An update of aetiology, genetics, co-morbidities and management. Maturitas. 2018;118:67-73.

[37] Cabrera-Villalba S, Sanmartí R. Palindromic rheumatism: a reappraisal. International Journal of Clinical Rheumatology. 2013;8(5):569.

[38] Klareskog L, Stolt P, Lundberg K, Källberg H, Bengtsson C, Grunewald J, et al. A new model for an etiology of rheumatoid arthritis: smoking may trigger HLA–DR (shared epitope)–restricted immune reactions to autoantigens modified by citrullination. Arthritis & Rheumatism: Official Journal of the American College of Rheumatology. 2006;54(1):38-46.

[39] Roh S. Smoking as a Preventable Risk Factor for Rheumatoid Arthritis: Rationale for Smoking Cessation Treatment in Patients with Rheumatoid Arthritis. Journal of Rheumatic Diseases. 2019;26(1):12-19.

[40] Mattingly S, Jones D, Robinson W. Palindromic rheumatism. Journal of the Royal College of Physicians of London. 1981;15(2):119.

[41] Takata S, Harada S, Mitsunobu F, Mifune T, Hosaki Y, Asida K, et al. A Patient with severe palindromic rheumatism and frequent episodes of pain. Internal medicine. 2001;40(2):140-143.

[42] Nesher G, Mates M. Palindromic rheumatism: Effect of dietary manipulation. Arthritis & Rheumatism. 1997;40(9).

[43] Koskinen E, Hannonen P, Sokka T. Palindromic rheumatism: longterm

outcomes of 60 patients diagnosed in 1967-84. The Journal of rheumatology. 2009;36(9):1873-1875.

[44] Emad Y, Anbar A, Abo-Elyoun I, El-Shaarawy N, Al-Hanafi H, Darwish H, et al. In palindromic rheumatism, hand joint involvement and positive anti-CCP antibodies predict RA development after 1 year of follow-up. Clinical rheumatology. 2014;33(6):791-797.

[45] Sanmartí R, Cabrera-Villalba S, Gómez-Puerta JA, Ruiz-Esquide V, Hernández MV, Salvador G, et al. Palindromic rheumatism with positive anticitrullinated peptide/protein antibodies is not synonymous with rheumatoid arthritis. A longterm followup study. The Journal of rheumatology. 2012;39(10):1929-1933.

[46] Rakieh C, Nam J, Hunt L, Hensor E, Das S, Bissell L, et al. Predicting the development of clinical arthritis in anti-CCP positive individuals with non-specific musculoskeletal symptoms: a prospective observational cohort study. Annals of the rheumatic diseases. 2015;74(9):1659-1666.

[47] Youssef W, Yan A, Russell AS. Palindromic rheumatism: a response to chloroquine. The Journal of rheumatology. 1991;18(1):35-37.

[48] Gonzalez-Lopez L, Gamez-Nava JI, Jhangri G, Russell AS, Suarez-Almazor ME. Decreased progression to rheumatoid arthritis or other connective tissue diseases in patients with palindromic rheumatism treated with antimalarials. Journal of rheumatology. 2000;27(1):41-46.

[49] Edwards JC, Szczepański L, Szechiński J, Filipowicz-Sosnowska A, Emery P, Close DR, et al. Efficacy of B-cell–targeted therapy with rituximab in patients with rheumatoid arthritis. New England Journal of Medicine. 2004;350(25):2572-2581.

[50] Van Vollenhoven RF, Fleischmann R, Cohen S, Lee EB, García Meijide JA, Wagner S, et al. Tofacitinib or adalimumab versus placebo in rheumatoid arthritis. New England Journal of Medicine. 2012;367(6):508-519.

[51] Nighojkar PA, Momin M, Phadatare SP. Rheumatoid arthritis–Pharmacology and its management with traditional herbs. 2017.

[52] Zarei L, Naji-Haddadi S, Pourjabali M, Naghdi N, Tasbih-Forosh M, Shahsavari S. Systematic review of anti-rheumatic medicinal plants: An overview of the effectiveness of articular tissues and joint pain associated with rheumatoid arthritis. Journal of Pharmaceutical Sciences and Research. 2017;9(5):547.

Chapter 16

Metagenomics and Pandemic Viruses

Paulo Vitor Marques Simas and Clarice Weis Arns

Abstract

Humanity's history contains many pandemic reports and now the scientific community has the possibility to identify the pathogens before the disease emergency. In this perspective, it is essential to carry out large-scale epidemiological studies in key animals that are in constant contact with humans. For this, the next generation sequencing (NGS) by the metagenomic approach—genetic material recovered directly from samples without previous amplification—is able to reveal the hidden microbial diversity. Metagenomes' work aims to contribute to the facilitation of epidemiological studies through the adoption of simple effective strategies for the pathogens' identification, understanding the evolutionary dynamic of them before the pandemic time. Here, we have presented some examples related to the successful metagenomic approaches and the continuous advice of the researchers to identify viruses and other possible pandemic pathogens.

Keywords: environmental genetic material, phylogenetic network, biogeography, one health

1. Introduction

Emerging pathogens (those that have recently been introduced, discovered, or recognized; that have recently evolved; or that have increased in incidence through geographic expansion or adaptation to a greater diversity of hosts—spillover; or that have shown changes in their pathogenic properties), especially viral agents, present a unique challenge for science and medicine because little is known about them before they cause epidemics from zoonotic sources. Zoonotic transmission can occur through an overflow event from one animal species to another, eventually causing infections in humans as well. For most of these viruses, therapies and/or vaccination strategies have not been developed, and therefore, clinical treatment options for infected patients are limited to nonspecific supportive therapy (adapted from [1]).

In general, the epidemiological studies are conducted passively, after the establishing horizontal transmission of each viral infectious disease or after spillover events. Such a conduct will certainly not be sufficient to meet the new demands for infectious diseases. Therefore, one health, a science that promotes ecological health through the interaction between human, animal, and environmental health, arises with the proposal that all health should be thought together so that it has a sustainable condition of existence.

In this perspective, it is essential to carry out large-scale epidemiological studies in key animals such as bats and rodents in general. For this, the next generation sequencing (NGS) by the metagenomic approach—genetic material recovered directly from samples without previous amplification—is able to reveal the hidden microbial diversity. Metagenomes' work aims to contribute to the facilitation of epidemiological studies through the adoption of simple effective strategies for the pathogens' identification, understanding the evolutionary dynamic of them before the pandemic time.

2. Viral metagenomics: understanding the evolutionary history of viruses

There are many purposes that can be achieved through the use of metagenomics using NGS. Thomas et al. [2] described that this methodology can be a source of genetic information about the possible new biocatalysts or enzymes in cases of bioprospecting; can realize the genomic connections between the function and phylogeny of "non-cultivable" organisms; can identify the evolutionary profiles and associate the function and structure in microbial communities; and also can inference about the new hypotheses of microbial functions.

The Baltimore classification considers the virus taxonomy from seven groups: I (double-stranded DNA viruses), II (single-stranded DNA viruses), III (double-stranded RNA viruses, dsRNA), IV [positive-sense single-stranded RNA viruses, (+) ssRNA], V [negative-sense single-stranded RNA viruses, (−) ssRNA], VI (single-stranded RNA viruses with a DNA intermediate in their life cycle, ssRNA-RT), and VII (double-stranded DNA viruses with an RNA intermediate in their life cycle, dsDNA-RT). However, metagenomics approaches have contributed to understand the real universal viral taxonomy.

Since the RNA viruses are the most abundant to infect eukaryotes in general and all these viruses contain a specific gene, the RNA-dependent RNA polymerase (RdRp), responsible to produce the protein for their replication, it was possible to reconstruct their evolutionary history. Using 4617 different RdRps, they found five key subdivisions, two inclosing dsRNA and (−) ssRNA viruses, the other one containing + ssRNA and dsRNA, and two including only (+) ssRNA viruses. In these analyzes, it was showed that the (+) ssRNA viruses possibly have had a single ancestor and that the dsRNA viruses have emerged from (+) ssRNA in two independent events and the (−) ssRNA and could have emerged later from dsRNA. In addition, phylogenetic analysis including other genes showed the broad horizontal transference between the hosts remotely related, including between animals and plants. All these findings allow the continuous update of the viruses' arrangement by International Committee on Taxonomy of Viruses (ICTV) [3]. Considering this continuous host-change, also analyzed from diversity of data previously unavailable and now discovered by metagenomics assays, possibly, the evolution of RNA viruses is associated with the evolutionary history of the host, which dates from hundreds of millions of years ago, from the ocean to the terrestrial environment. In this sense, it should be imprudent to associate the origin of some RNA viruses' families of vertebrates from mosquitoes and ticks [4].

In general, when unrecognized zoonotic viral pathogens emerge from wildlife, it can induce a strong impact on the public health and generate a pandemic risk. In these cases, there are no molecular or serological diagnostic assays specific available for detecting them and the metagenomic appears like the strategy unique to start the developing of new diagnostic test. Temmam et al. [5], in a review paper, already suggested that to predict emerging zoonotic infections always have been an important challenge for public health. Furthermore, the viral metagenomics was pointed by them as the promising alternative to surveillance, especially to identify viromas in selected animals and arthropods closely in contact with humans.

For the diagnosis of viral diseases, the gold standard test is the viral isolation. However, it has the disadvantage of being laborious, consuming many days, in addition to the biosecurity issues of handling in the laboratory. Since the 1980s, with the advent of molecular techniques of polymerase chain reaction (PCR) and sequencing, the scientific community has gradually adopted these tests for the investigation and pathogens diagnosis, mainly the emerging ones. The international scientific community's decision to sequence the human genome provided the rapid evolution of all molecular techniques and allowed the emergence of techniques such as real-time PCR and high-performance sequencing NGS.

Before the appreciation of metagenomics into the scientific community, the molecular diagnostic methods most common on the routine were practically performed using the PCR system and its variations and Sanger sequencing. Therefore, several commercial molecular diagnostic tests to detect viral, bacterial, and fungal pathogens were made available using real-time PCR, loop-mediated isothermal amplification (LAMP), and conventional PCR, among others.

These approaches are increasingly used to detect new viral pathogens as well as to generate complete genomes of non-cultivable viruses. The *in silico* identification of complete viral genomes from sequence data would allow a rapid phylogenetic characterization of these new viruses [6]. Thus, the methodology emerges as an alternative for the discovery of new viral agents in animals, allowing the expansion of knowledge of viral diversity as well as of potentially emerging zoonosis. The identified material can provide insights into the evolution and viruses phylogenies. These data, associated with studies on the biology of the hosts, can contribute to the understanding of the eco-epidemiology of several viruses that, through complementary analyzes, help in the identification of potentially pathogenic agents, a perspective action from One Health.

Important advances in the knowledge of microbial diversity through metagenomic tools in several ecosystems have allowed to associate the sequence data with the complex biological characteristics of an ecosystem, be it the human intestine, for example, whether aquatic or terrestrial environmental. This can be an essential source to understanding the microbial ecology and biogeography. Different works of metagenomics have shown that the biogeography of viruses, in general, is associated with the characteristics of habitats. As such, similar viral biogeography can be found in similar habitats that can be geographically far. On the other hand, also was identified substantial sharing of the human viroma between unrelated population and widespread viral quasi-species distinctly dissimilar habitats [7]. So, is really important to implement the metagenomic assay in several places around the world in order to deepen the understanding of this complex ecological dynamic. There are many antibiotics and antiviral molecules produced in these ecosystems. To identify the biosynthetic origin of these drugs can improve treatment alternatives for existing diseases or even for emerging infectious diseases.

Viromas of different animal species as well as different human organs have been obtained through this methodological approach. Brotman et al. [8] described the temporal relationship between vaginal microbiota and detection of human papilloma virus (HPV). A small DNA virus, the Torque Teno Virus (TTV), was unexpectedly found in samples from patients with ophthalmitis (negative bacterial and fungal cultures) [9]. Enteric metaviroma of shrews (genus *Crocidura*, one of nine genera of the shrew subfamily *Crocidurinae*), small insectivorous mammals similar to rodents, were determined and identified with new insect and virus viruses, including cyclovirus, picornavirus, and picorna-like virus. In addition, several cycloviruses, including human variants, were detected in wild shrews with a high prevalence rate. Complete or almost complete genomic sequences of these new viruses were determined and subjected to genetic characterization [10, 11].

Metagenomic analyzes have shown that bats can be reservoirs to several species of RNA viruses. Many of these viruses are highly pathogenic and exhibit broad cell tropism, being able to infect a wide variety of cells and hosts (Hendra virus, HEV; Nipah virus, NIV; Ebola virus, EBOV; Meddle East Respiratory Syndrome virus, MERS; Severe Acute Respiratory Syndrome Coronavirus type 1, SARS-CoV; Severe Acute Respiratory Syndrome Coronavirus type 2, SARS-CoV-2). Viruses such as HEV, rabies virus (RABV), and NIV show high genomic conservation within their bat hosts, which suggests that they are under strong selective restrictions [7].

Considering that bats can be host a great viruses diversity and that little is known about their viroma, Dacheux et al. [12] determined the viral diversity of five different French insectivorous bats species (nine specimens). They detected viruses from many viral families that infect bacteria, plants, fungi, insects, and vertebrate animals and mammals (Retroviridae, Herpesviridae, Bunyaviridae, Poxviridae, Flaviviridae, Reoviridae, Bornaviridae, and Picobirnaviridae). They described new mammalian viruses, including rotavirus, gammaretrovirus, bornavirus, and bunyavirus, as well as the first nairovirus identification in bats.

Bat metaviroma from Myanmar (China) revealed the presence of new mammalian viruses. The analysis was conducted using organs of 853 bats of six species, identifying known sequences belonging to 24 viral families. Of the viral contigs (2% of the total sequences), 45% were related to vertebrate viruses, 28% to insect viruses, 27% to phages, and 95 contigs to plant viruses. The validation performed by PCR followed by phylogenetic analyzes led to the discovery of some new bat viruses of the genera *Mamastrovirus*, *Bocavirus*, *Circovirus*, *Iflavirus*, and *Orthohepadnavirus* [13].

In African fruit bats populations (*Eidolon helvum*), it was identified by metagenomics a great abundance and diversity of new herpes and papillomavirus. The authors also described a new adenovirus and detected, for the first time in Chiroptera, sequences of a poxvirus closely related to contagious mollusk [14].

Herman Tse et al. [15], carrying out studies of 156 apparently healthy rectal swab samples from bats also using a metagenomic approach, discovered a new Papillomavirus strain, *Miniopterus schreibersii Papillomavirus type 1* (MscPV1), with a 7.5 kb long genome. In addition to the new agent characterization, the researchers also carried out several phylogenetic studies that allowed us to infer that MscPV1 and *Erethizon dorsatum* papillomavirus (EdPV1) are more closely related, with an approximate divergence of 60.2–91.9 million years.

He and collaborators (2014) identified hundreds of sequences related to alpha and Betacoronavirus sequencing 268 rectal swabs from 68 bats from four counties in Yunnan province. They also reported the complete genome of a new SARS-CoV (LYRa11) containing 29,805 nucleotides in length, 13 ORFs, 91% nucleotide identity with human and civet SARS CoVs, 89% similarity to another bat SARS-CoV-like. One of the most interesting reports was obtained through recombination analyzes. Such analyzes indicated that LYRa11 is a probable recombinant descendant of parental strains evolved from SARS-CoVs-like bats.

An outbreak of respiratory infection of unknown origin began to manifest in many people in Wuhan-Hubei-China in late 2019. Difficulties in controlling the disease by conventional methods of treatment suggested a new infectious disease with viral characteristics and effective transmission of person to person. A short time later and with the support of the international scientific community, it was confirmed that the new disease called Coronavirus Disease 2019 (COVID-19) was caused by a new coronavirus initially called 2019 Novel Coronavirus (2019-nCoV). It is not the first time that a human coronavirus has caused a major disease with risk of global spread. Severe Acute Respiratory Syndrome Coronavirus type 1 (SARS-CoV-1) and, in 2013, Middle East Respiratory Syndrome (MERS-CoV) emerged, both in two different places, of zoonotic origin and evolved from bats having intermediary animals as hosts, the civets and camelids, respectively. Using metagenomics approaches, in a few days, the scientific community has managed to obtain the complete genome of this virus. Based on genetic, evolutionary and molecular studies, the 2019-nCoV virus was named Severe Acute Respiratory Syndrome Coronavirus type 2 (SARS-CoV-2), a Sarbecovirus, Betacoronavirus, brother of SARS-CoV-1.

However, it is an important question without conclusive answer: what is the real reservoir for SARS-CoV-2? Until now, no virus isolation was successful from the none animal source but Lam et al. [16], by metagenomics approaches, identified several pangolin coronavirus lineages suggesting that these animals could be considered as possible hosts in the emergence of new coronaviruses and must be removed from wet markets to prevent zoonotic transmission.

In the COVID-19 pandemic, until August 2020, more than 20,000,000 people were infected and almost 800,000 death in more than 190 countries that detected the virus [17]. Peddu et al. [18] described the metagenomic analysis as a good alternative to investigate and to response to this and future viral pandemics; they evaluated, by metagenomic sequencing, positive and negative samples from Seattle, WA. Part of these samples showed superinfection or colonization with human parainfluenza virus 3 or *Moraxella* species, emphasizing to be essential to conduct molecular testes using a viral respiratory panel. In addition, negative samples for SARS-CoV-2 by RT-PCR were positive for Rhinovirus A and C, showing that the metagenomic analysis of these SARS-CoV-2 negative samples was able to identify candidate etiological agents for the respiratory signs in those patients.

Did we learn from the past epidemics? Are we prepared for the worst? Gonzalez et al. [19] stablished these key questions related our learning from the humanity history. According to them, the ultimate goal should be develop a resilient global health infrastructure; like the bio-surveillance using geographic information systems (GIS) and metagenomics to trace the molecular changes in pathogens during their emergence, and mathematical models to assess risk should be "critical point" for preventing a pandemic.

3. Epidemiological surveillance in bats by metagenomic approaches: a powerful tool for conducting large-scale studies

In recent years, emerging and serious infectious diseases have caused worldwide fear. It is also known that many of these diseases are caused by viruses from bats, such as Ebola, Marburg, SARS coronavirus (SARS-CoV), MERS coronavirus (MERS-CoV), Nipah (NIV) and Hendra (HEV) [20], and nowadays, SARS-CoV-2. The growing recognition of the bats importance as reservoirs for new diseases is due to the fact that they constitute 20% of known mammal species, have unique and diverse lifestyles, including the ability to fly, often presenting gregarious social structures achieving incredible abundance and densities (some cave bats reaching up to 500 individuals per square meter) and long life [21].

As more information has been obtained regarding the factors or causes of emergence, there has been an expectation that it is possible to predict the emergence of new pathogens. These and other factors have significantly increased the demand for new viral pathogens, especially at the human-animal interface in species of wild and domestic animals [22].

With the exception of studies focusing on lyssavirus, most viruses' investigations in bats have been limited to one particular zoonotic agent involved in an outbreak of geographically localized disease [23]. COVID-19 showed the need to form an international front for active surveillance of bats different populations to detect potential zoonotic agents as well as low pathogenicity unknown viruses that can recombine/mutate and become pathogenic.

The emergence of highly pathogenic viruses such as SARS and MERS-CoV has identified coronaviruses as agents of high interest in epidemiological surveillance. In addition to concluding that SARS-CoV may have originated from bats, it is suggested that several other new viruses exist in animals and some of them pose a risk to public health [24–27].

Although great advances have been made in the knowledge of these viruses, there is much to learn about the evolution of highly pathogenic agents in reservoir animals such as bats [1]. Several studies have pointed out a great diversity of coronaviruses belonging to the genus α- and β-*coronavirus* of the subfamily Orthocoronavirinae that occur widely in bat species around the world, including Africa, Europe, the Americas, and Asia. Interestingly, an analysis of viruses isolated from bats in Mexico showed that host species were driving forces in the evolution of coronaviruses, and that a single bat species can contain several coronaviruses. In addition, the phylogenetic association of CoVs with the species/genus was particularly evident in allopatric populations separated by significant geographical distances [22].

Simas and Arns [28] described a metagenomic methodology using bat common specie from urban areas in the Americas, the *Tadarida brasiliensis,* in order to establish a rapid methodology for active epidemiological surveillance in bats as the best reservoir animal model. The assay aimed to identify viral agents in oral and rectal swabs collected from asymptomatic *T. brasiliensis* bats from a colony in the Campinas-SP, Brazil. From this, these researchers described the diversity and abundance of the identified viral agents and could relate phylogenetically the identified. The workflow is described in **Figure 1**.

The most important steps are the pretreatment and the validation because these can remove the host genetic material and confirm the sequences of dataset identified. For the pretreatment, the researchers used filtration and treatment with

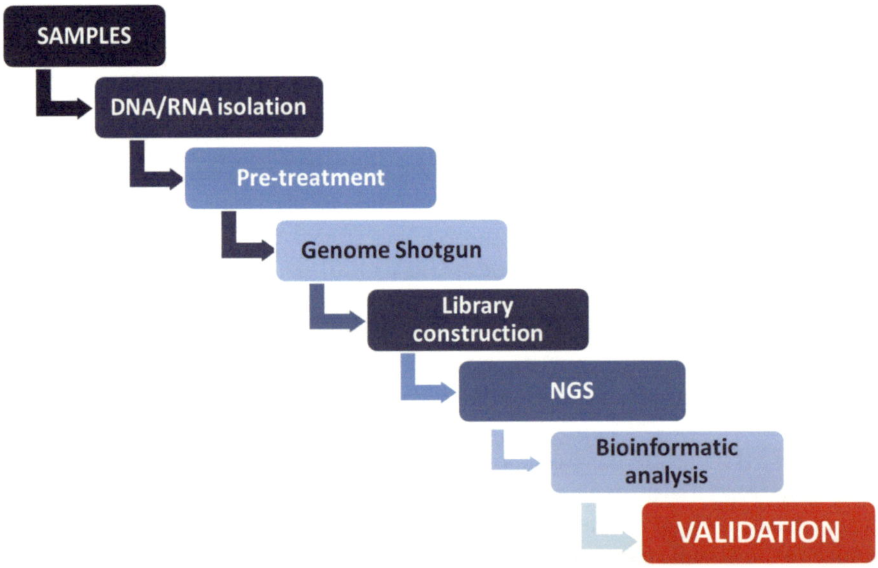

Figure 1.
Workflow described by Simas and Arns [28] to conduct active epidemiological surveillance from model reservoir animals like bats using metagenomic approach.

DNAse and proteinase K enzymes. These procedures help to eliminate the genetic material "contaminants" and to assure the most viral genetic material into the sequence dataset.

Using these assays, a large number of excellent quality paired-end sequences were obtained in the HiSeq 2500 Illumina platform (345,409,110 reads paired-ends—76.47% $Q \geq 30$). In the reading assembly procedures with the MetaVelvet and Metavir 2 genome assemblies' platforms, many viral genetic materials from several pathogenic viruses were identified. It can be noted that the different platforms used provided complementary data, indicating the need to carry out similar procedures in studies that use the same metagenomic methodology.

Although the search for similarity carried out by MetaVelvet in different databases provided a small number of viral matches (97; 2 for coronavirus), these results were validated and allowed the identification of a coronavirus with a strong phylogenetic relationship with Porcine Epidemic Diarrhea virus (PEDV), a high pathogenicity swine virus, and human coronavirus, HCoV-NL63. PEDV has been reported in many other countries, including Germany, France, Switzerland, Hungary, Italy, China, South Korea, Thailand and Vietnam and was first identified in the United States in May 2013. The US outbreak occurred in 23 states, with 2692 confirmed cases leading to major economic losses. Studies have shown that all American PEDV strains are closely related to a strain from China, AH2012 [29]. However, the identification of PEDV in wild animals common in the Americas, such as bats *Tadarida brasiliensis*, can help to understand the evolution of these agents in animal reservoirs and to understand the eco-epidemiology from the genetic diversity studies like this.

Metavir 2 identified sequences of viruses associated with various pathogens in humans. Many sequences have been classified as belonging to the Herpesviridae family. Several viral agents in this family are known to cause a wide variety of human diseases, including various types of cancer. In addition, since they have a great capacity to infect many types of cells or tissues [30], bats may be serving as a reservoir for recombination and the emergence of new strains capable of infecting other animals even cause human infections.

Several viruses of the Order Caudovirales were also identified, most from the Siphoviridae family. These phages are capable of infecting several species of human pathogenic bacteria (*Enterobacteria*, *Shigella*, *Mycobacterium*, and *Bacillus*), so it is an indirect evidence of the presence of these bacteria also in bats. The concomitant detection of herpes and phages indicates that bats can act as important agents in the evolution of these viral agents, since the existence of recombination between them has already been described [31].

Many betaretroviruses, viruses that cause various types of tumors in primates, sheep, and rats have been detected. Sano et al. [32] also identified several viral agents from the Retroviridae and Herpesviridae families in bats in the Philippines. Dacheux et al. [12] determined the viral diversity of five different species of French insectivorous bats (nine specimens). All of these results suggest that retroviruses and herpesviruses are widely distributed in bat populations.

The detection of several dsRNA virus sequences, a virus group that cause gastroenteritis in children (rotavirus) and others that are pathogenic for cattle and sheep, their identification in bats contributes to the understanding of their circulation in ecosystems. Another Brazilian study also reported the presence of rotavirus in bat feces. Phylogenetic analyzes indicated the formation of a clade with sequences of bovine and human origins, suggesting recombination between the strains in animal hosts, events that precede transmission to humans in zoonotic viral diseases [33].

4. Conclusions

The metagenomic proceeding is fast and highly sensitive to access the genetic diversity on the ecosystem in general. With the use of the metagenomic approach, in a few days, the scientific community has managed to obtain the complete SARS-CoV-2 genome. Because epidemiological studies are still conducted from the onset of diseases, outbreaks are still being worked on, not prevention. From this perspective, this methodology showed to be able to be applied to conduct epidemiological surveys and it should be widely applied to understand, by the genetic diversity, the molecular eco-epidemiology of viral agents before the pandemic time.

Author details

Paulo Vitor Marques Simas[1*] and Clarice Weis Arns[2]

1 The National University of San Marcos, Faculty of Veterinary Medicine, Laboratory of Avian Pathology, Peru

2 University of Campinas, Institute of Biology, Department of Genetics, Evolution, Immunology and Microbiology, Brazil

*Address all correspondence to: simaspvm@gmail.com

References

[1] Coleman CM, Frieman MB. Coronaviruses: Important emerging human pathogens. Journal of Virology. 2014;**88**(10):5209-5212

[2] Thomas T, Gilbert J, Meyer F. Metagenomics—A guide from sampling to data analysis. Microbial Informatics and Experimentation. 2012;**2**(3). DOI: 10.1186/2042-5783-2-3

[3] Wolf YI, Kazlauskas D, Iranzo J, Lucía-Sanz A, Kuhn JH, Krupovic M, et al. Origins and evolution of the global RNA virome. MBio. 2018;**9**(6): e02329-e02318. DOI: 10.1128/mBio. 02329-18. PMID: 30482837; PMCID: PMC6282212

[4] Shi M, Lin XD, Chen X, Tian JH, Chen LJ, Li K, et al. The evolutionary history of vertebrate RNA viruses. Nature. 2018;**556**(7700):197-202. DOI: 10.1038/s41586-018-0012-7. Epub: 04 April 2018. Erratum in: Nature. 2018 Jun 26: PMID: 29618816

[5] Temmam S, Davoust B, Berenger JM, Raoult D, Desnues C. Viral metagenomics on animals as a tool for the detection of zoonoses prior to human infection? International Journal of Molecular Sciences. 2014;**15**(6):10377-10397. DOI: 10.3390/ ijms150610377. PMID: 24918293; PMCID: PMC4100157

[6] Wood-Charlson EM, Weynberg KD, Suttle CA, Roux S, van Oppen MJ. Metagenomic characterization of viral communities in corals: Mining biological signal from methodological noise. Environmental Microbiology. 2015;**17**(10):3440-3449. DOI: 10.1111/1462-2920.12803

[7] Paez-Espino D, Eloe-Fadrosh E, Pavlopoulos G, et al. Uncovering Earth's virome. Nature. 2016;**536**:425-430. DOI: 10.1038/nature19094

[8] Brotman RM, Shardell MD, Gajer P, Tracy JK, Zenilman JM, Ravel J, et al. Interplay between the temporal dynamics of the vaginal microbiota and human papillomavirus detection. The Journal of Infectious Diseases. 2014;**210**(11):1723-1733

[9] Lee AY, Akileswaran L, Tibbetts MD, Garg SJ, Van Gelder RN. Identification of torque teno virus in culture-negative endophthalmitis by representational deep DNA sequencing. Ophthalmology. 2015;**122**(3):524-530

[10] Sasaki M, Orba Y, Ueno K, Ishii A, Moonga L, Hang'ombe BM, et al. Metagenomic analysis of the shrew enteric virome reveals novel viruses related to human stool-associated viruses. The Journal of General Virology. 2015;**96**(Pt 2):440-452

[11] Smith I, Wang LF. Bats and their virome: An important source of emerging viruses capable of infecting humans. Current Opinion in Virology. 2013;**3**(1):84-91

[12] Dacheux L, Cervantes-Gonzalez M, Guigon G, Thiberge JM, Vandenbogaert M, Maufrais C, et al. A preliminary study of viral metagenomics of French bat species in contact with humans: Identification of new mammalian viruses. PLoS One. 2014;**9**(1):e87194

[13] He B, Li Z, Yang F, Zheng J, Feng Y, Guo H, et al. Virome profiling of bats from Myanmar by metagenomic analysis of tissue samples reveals more novel mammalian viruses. PLoS One. 2013;**8**(4):e61950

[14] Baker KS, Todd S, Marsh GA, Crameri G, Barr J, Kamins AO, et al. Potentially zoonotic paramyxoviruses from the African straw-colored fruit bat *Eidolon helvum*. Journal of Virology. 2013;**87**(3):1348-1358

[15] Tse H, Tsang AKL, Tsoi WH, Leung ASP, Ho CC, Lau SKP, et al. Identification of a novel nat papillomavirus by metagenomics. PLoS One. 2012;**7**(8):1-9

[16] Lam TT, Jia N, Zhang YW, Shum MH, Jiang JF, Zhu HC, et al. Identifying SARS-CoV-2-related coronaviruses in Malayan pangolins. Nature. 2020;583(7815):282-285. DOI: 10.1038/s41586-020-2169-0. Epub 2020 Mar 26. PMID: 32218527

[17] World Health Organization (WHO) [Internet]. EMERGENCY—Coronavirus Disease (COVID-19) Pandemic. 2020. Available from: https://www.who.int/emergencies/diseases/novel-coronavirus-2019. [Accessed: 14 August 2020]

[18] Peddu V, Shean RC, Xie H, Shrestha L, Perchetti GA, Minot SS, et al. Metagenomic analysis reveals clinical SARS-CoV-2 infection and bacterial or viral superinfection and colonization. Clinical Chemistry. 2020;**66**(7):966-972. DOI: 10.1093/clinchem/hvaa106. PMID: 32379863; PMCID: PMC7239240

[19] Gonzalez JP, Souris M, Valdivia-Granda W. Global spread of hemorrhagic fever viruses: Predicting pandemics. Methods in Molecular Biology. 2018;**1604**:3-31. DOI: 10.1007/978-1-4939-6981-4_1. PMID: 28986822; PMCID: PMC7120037

[20] Han HJ, Wen HL, Zhou CM, Chen FF, Luo LM, Liu JW, et al. Bats as reservoirs of severe emerging infectious diseases. Virus Research. 2015;**205**(2):1-6

[21] Wood JLN, Leach M, Waldman L, MacGregor H, Fooks AR, Jones KE, et al. A framework for the study of zoonotic disease emergence and its drivers: Spillover of bat pathogens as a case study. Philosophical Transactions of the Royal Society B. 2012;**367**:2881-2892

[22] Mackenzie JS, Jeggo M. Reservoirs and vectors of emerging viruses. Current Opinion in Virology. 2013;**3**(2):170-179

[23] Sonntag M, Mühldorfer K, Speck S, Wibbelt G, Kurth A. New adenovirus in bats, Germany. Emerging Infectious Diseases. 2009;**15**(12):2052-2055

[24] Anderson LJ, Tong S. Update on SARS research and other possibly zoonotic coronaviruses. International Journal of Antimicrobial Agents. 2010;**36**(Suppl 1):S21-S25

[25] Fouchier RA, Hartwig NG, Bestebroer TM, Niemeyer B, de Jong JC, Simon JH. A previously undescribed coronavirus associated with respiratory disease in humans. Proceedings of the National Academy of Sciences of the United States of America. 2004;**101**:6212-6216

[26] Van der Hoek L, Pyr CK, Jebbink MF, Vermeulen-Oost W, Berkhout RJ, Wolthers KC. Identification of a new human coronavirus. Nature Medicine. 2004;**10**:368-373

[27] Woo PC, Lau SK, Chu CM, Chan KH, Tsoi HW, Huang Y. Characterization and complete genome sequence of a novel coronavirus, coronavirus HKU1, from patients with pneumonia. Journal of Virology. 2005;**79**:884-895

[28] Simas PVM. Metaviroma de Morcegos Tadarida brasiliensis = Viral metagenomic of Tadarida brasiliensis bats. 2015. 1 [Internet]. Thesis (PhD) – University of Campinas, Institute of Biology, Campinas, Brazil. Available at: http://www.repositorio.unicamp.br/handle/REPOSIP/316613. [Accessed: 28 August 2020]

[29] Wang L, Byrum B, Zhang Y. New variant of porcine epidemic diarrhea virus, United States, 2014. Emerging Infectious Diseases. 2014;**20**(5):917-919

[30] Chayavichitsilp P, Buckwalter JV, Krakowski AC, Friedlander SF. Herpes simplex. Pediatrics in Review. 2009;**30**(4):119-129

[31] Weller SK, Sawitzke JA. Recombination promoted by DNA viruses: Phage λ to herpes simplex virus. Annual Review of Microbiology. 2014;**68**:237-258

[32] Sano K, Okazaki S, Taniguchi S, Masangkay JS, Puentespina R Jr, Eres E, et al. Detection of a novel herpesvirus from bats in the Philippines. Virus Genes. 2015;**51**(1):136-139

[33] Luchs A, Timenetsky MDCST. G8P[6] rotaviruses isolated from Amerindian children in Mato Grosso do Sul, Brazil, during 2009: Close relationship of the G and P genes with those of bovine and bat strains. The Journal of General Virology. 2014;**95** (Pt 3):627-641